College Preparatory Mathematics 1

(Algebra 1)
Second Edition

Managing Editor:
Leslie Dietiker
    Phillip and Sala Burton High

Contributing Editors:
Susan Baskin
    Oakland Unified
Elizabeth Coyner
    Bear River Elementary
Brian Hoey
    Christian Brothers High
Pat King
    Holmes Junior High
Dina Luetgens
    Andros Karperos Middle
Micheal Marsh
    Culver City High
Bob Petersen
    Sacramento High

Illustrator:
Eric Ettlin
    Menlo-Atherton High

Technical Assistance:
Dean Hickerson
    Consultant
Thu Pham
    The CRESS Center
    University of California, Davis

Program Directors:
Tom Sallee
    Department of Mathematics
    University of California, Davis
Judy Kysh
    Northern California Mathematics Project
    The CRESS Center
    University of California, Davis
Elaine Kasimatis
    Department of Mathematics
    California State University, Sacramento
Brian Hoey
    CPM Educational Program

Credits for the First Edition

Editors
Scott Holm, Cloverdale High
Elaine Kasimatis, UC Davis
Bob Petersen, Sacramento High

Consultant
Joel Teller, College Preparatory School

Technical Assistance
Thu Pham, UC Davis
Crystal Mills, Consultant

First Edition Contributors

Annette Bartos	Churchill Middle School
Duane Blomquist	Roseville High
Beverly Braverman	C. K. McClatchy High
Nancy Clark	Woodland Senior High
James Friedrich	Valley High
Carol Grossnicklaus	Oxnard High
Maria Herndon	Valley High
Ted Herr	Roseville High
Brian Hoey	Christian Brothers High
Scott Holm	Cloverdale High
Gail Holt	El Camino Fundamental High
Sylvia Huffman	Del Campo High
Yury Lokteff	San Juan High
Grant McMicken	San Juan High
Richard Melamed	El Camino Fundamental High
Crystal Mills	Sacramento Waldorf High
Bob Petersen	Sacramento High
Brad Schottle	Elk Grove High
Jeanne Shimizu-Yost	San Juan High
Bonnie Sieber	Valley High
Pat Stowers	O. W. Holmes Junior High
Sharon Swanson	John F. Kennedy High
Clark Swanson	Sacramento High
Linda Tucker	Bear River High
Joe Veiga	Elk Grove High
Michael White	Temecula Valley High
Malcolm Wong	Luther Burbank High

Copyright © 1994, 1998, 2000, 2002 by CPM Educational Program. All rights reserved. No part of this publication may be reproduced or transmitted in any form or by any means, electronic or mechanical, including photocopy, recording, or any information storage and retrieval system, without permission in writing from the publisher. Requests for permission should be made in writing to: CPM Educational Program, 1233 Noonan Drive, Sacramento, CA 95822. E-mail: cpm@cpm.org.

3 4 5 6 7 8 9 10   04 03 02 01

Printed in the United States of America   ISBN 1-885145-69-1

A Note to Parent(s) and Guardian(s)

Hello. My name is Chris Ott. While I am not an author of the College Preparatory Mathematics (CPM) textbook series, I have tutored dozens of CPM students. Based on this experience, I have written a study guide for CPM Algebra 1 that is incorporated into the student text. If you want an extended overview of this course, read all the PZL problems in the first five units (Units 0-4). They are easy to find, since each PZL problem has a puzzle piece (like the one above) around the problem number in the left margin. These problems will explain the goals of CPM Algebra 1, its methods, the structure of the textbook, the role and duties of the teacher, and the responsibilities of the student. It has numerous practical suggestions for students to maximize the prospect for success in this course.

The authors have also written a more extensive unit by unit *Parent's Guide with Review for Math 1 (Algebra 1)*. This document contains annotated solutions for important problems that help you and your child understand the core ideas of each unit. Most of these sections are followed by a few additional sample problems and their solutions and/or answers. The guide also makes suggestions about how to help your child in this course. At the end of each unit there are several dozen additional practice problems to supplement the textbook. The guide concludes with a brief outline of how the five content threads of the course--graphing, writing equations, ratios, solving equations, and manipulating symbols--are developed through the fourteen units in this textbook. Some schools make the guide available through the classroom teacher or school library. You may order the guide online at www.cpm.org, get an order form there, or mail your request for the *Parent's Guide with Review for Math 1 (Algebra 1)*. 2nd. ed., **version 6.1,** with a check payable to "CPM Educational Program" for $20 plus local tax (CA only) and $3 shipping and handling, to 1233 Noonan Drive, Sacramento, CA 95822-2569. Allow ten days for delivery.

College Preparatory Mathematics 1: Second Edition
(Algebra 1)

Table of Contents

Unit 0 **Getting Started: Working in Teams** . . . . GS-1 to GS-27
page 1   Our introductory unit is intended both to show that this course is different from a traditional Algebra 1 course and to develop some important skills. Students work in teams and pairs as they interpret graphs, solve number puzzles, explore perimeter and area, and do a "big problem" that introduces them to basic probability.

Unit 1 **Difference of Squares: Organizing Data** . . . . SQ-1 to SQ-91
page 19   We begin the course by introducing manipulatives to review operations with integers. Algebra tiles assist students with an introduction to combining like terms. Students interpret graphs, and conclude the unit by solving word problems using the Guess and Check method in tandem with Making a Table and Looking for a Pattern.

Unit 2 **The Kitchen Floor: Area and Subproblems** . . . . KF-1 to KF-127
page 51   Area and perimeter problems help introduce the next problem solving strategy: Subproblems. Algebra tiles are used to review Order of Operations and to introduce the Distributive Property. Students investigate some functions and special uses of their scientific calculator.

Unit 3 **The Burning Candle: Patterns and Graphs** . . . . BC-1 to BC-89
page 90   Using scientific calculators and patterns, students explore the relationships among graphs, tables, and rules. The emphasis is on linear and quadratic equations, but other types of curves are explored as well. Students practice writing algebraic expressions and continue to solve word problems by Guess and Check in preparation for writing equations.

Unit 4 **Choosing a Phone Plan: Writing and Solving Equations** . . CP-1 to CP-123
page 123   Building from patterning skills and practice with Guess and Check tables, students begin to write equations for standard word problems. We teach different ways of solving linear equations, beginning with the concrete approach of "cups and tiles."

Unit 5 **Estimating Fish Populations: Numerical, Geometric, and Algebraic**
page 155   **Ratios** . . . . EF-1 to EF-127
We examine numerical and geometric ratios in a variety of activities. We explore the concept of similarity by enlarging and reducing simple figures on dot paper and then compare ratios of corresponding sides, perimeters, and areas of similar figures. Then we examine similar right triangles in connection with ratios and graphs of linear equations, in preparation for an introduction to the concept of slope later in the year. Finally, we write and solve equations for problems involving equivalent ratios.

Unit 6 **World Records: Graphing & Systems of Linear Equations** WR-1 to WR-110
page 191   This unit ties together the first semester by applying graphing, problem solving strategies, and algebra to solving word problems that involve systems of linear equations. Students learn to solve systems by substitution. The Distributive Property is extended to multiplying binomials using tiles and generic rectangles.

Unit 7 **The Big Race: Slopes and Rates of Change** . . . . **BR-1 to BR-104**
page 233   We continue the topics of the last unit, study x- and y-intercepts, and introduce slope and the slope-intercept form of a linear equation. The emphases are on understanding the physical interpretation of a graph and solving systems of linear equations graphically and algebraically.

Unit 8 **The Amusement Park: Factoring Quadratics** . . . . **AP-1 to AP-94**
page 267   Students use Algebra Tiles and generic rectangles to factor trinomials and special products. The Zero Product Property is introduced with graphing quadratics and more work with x-intercepts.

Unit 9 **The Birthday Party Piñata: Writing Equations from Diagrams**
page 301   . . . . **BP-0 to BP-120**
In Unit 4, we learned how to translate verbal descriptions to equations. Here the emphasis becomes translating verbal descriptions to diagrams and then to equations. Quadratic equations and square roots arise naturally from problems written using the Pythagorean Theorem. Students also develop writing equations of lines given two points.

Unit 10 **Yearbook Sales: Exponents and Quadratic Equations** . . **YS-1 to YS-110**
page 333   This unit examines quadratics in more depth: factoring when $a \neq 1$, solving them, culminating with the introduction of the Quadratic Formula. The remainder of the unit develops exponents and works with rational expressions.

Unit 11 **The Cola Machine: Functions and Equality** . . . . **CM-1 to CM-136**
page 361   We complete our study of systems of equations by introducing the elimination method. Students also practice multiplying and dividing rational expressions. Then we begin to tie the year together by formalizing our work with relations and functions. Students study two more families of functions: absolute value equations and equations with square roots. The unit concludes by tying our work with equations to the properties of real numbers.

Unit 12 **The Grazing Goat: Problem Solving and Inequality** . . . **GG-1 to GG-135**
page 399   Students learn to solve and graph linear, quadratic, and absolute value inequalities. We conclude our study of rational expressions by introducing addition and subtraction. The unit includes an overview of the problem solving strategies introduced during the year, then students apply them to various types of word problems. Area and perimeter problems are spiraled throughout the lessons.

Unit 13 **The Rocket Show: More About Quadratic Equations** . . **RS-1 to RS-110**
page 429   Topics in the unit include completing the square, exploring graphs of parabolas, and the derivation of the Quadratic Formula. Students will also study data points, trend lines, and lines and parabolas of best fit.

Skill Builders (Extra Practice) page 457   **Glossary** page 519   **Index** page 529

Unit 0 Objectives
Getting Started: WORKING IN TEAMS

In this unit, you will have the opportunity to:

- develop good work habits with homework.

- become accustomed to working in a team to solve mathematics problems.

- become acquainted with different forms of systematic lists.

- investigate basic probability.

- explore area and perimeter.

- analyze data and represent it with graphs and charts.

- organize your notebook for this course.

```
         Problem Solving  ▐▌
                Graphing  ▐▌
Writing and Solving Equations ▐
                  Ratios  ▐▌
                Geometry  ▐▌
      Symbol Manipulation ▐
```

Unit 0
Getting Started: WORKING IN TEAMS

 WELCOME TO YOUR ALGEBRA 1 COURSE!

Hi! I'm Chris Ott from "The Learning Connection" in Davis, California. As a college student I tutored quite a few students in math and science. I wrote the PZL parts of the course from the point of view of a student, not a teacher. These "puzzle pieces" are an honest attempt to level with you about what you will need to do to be successful in Algebra. Put all the pieces together and you'll increase your chances of success in this course and in many others. Start by reading the authors' description of the course, then complete the "mission" at the end of the reading. I'll go into more detail later.

This course is designed to help you learn the main ideas and procedures of algebra as well as help you understand them. You will also have numerous opportunities to see the math you learn at work in realistic situations.

You live in a technological time when computers are in virtually every business and good calculators cost $15. These tools do much of the repetitive kinds of tasks that used to be done with pencil and paper. To use these tools effectively, people have to do the thinking. That is why more and more people need to understand and be able to use algebra. In many places algebra is now a high school graduation requirement. **This course will teach you the basic content and skills of traditional algebra** and do it in ways that will help you understand what you are doing. It will teach you strategies for doing math that will be useful in later math courses and in life.

What can you expect? To begin with, the course is **built around problems.** You will spend more of your class time doing and discussing problems in study teams and not very much time listening to your teacher tell you rules. Don't misunderstand--rules are important, but they will emerge as you develop an understanding of the mathematics they represent. You will continue to use each rule you learn throughout this course and others.

Your textbook is designed to help you think about and develop algebraic ideas. Your teacher will help you stay on track as you do your daily assignments and will also summarize the points of your lessons. He or she will also highlight the points of your lessons by giving brief lectures, leading class discussions, and reviewing selected problems. However, **a key part of learning algebra is thinking carefully about ideas yourself.** Expect your teacher to ask you leading questions that will help you answer your own questions. By learning, with the guidance of your teacher, to ask and answer your own questions, you will become a better, more confident learner in math as well as in other courses. **In this course you will be learning to think mathematically and becoming a better problem solver.**

In this Algebra 1 course we will:
- deal with the major concepts taught in all Algebra 1 courses;
- develop your problem solving abilities;
- connect algebra to other areas of mathematics and science;
- emphasize <u>understanding</u> mathematics; and,
- include many problems that require a team effort.

While computers and calculators are good at doing repetitive tasks, part of understanding math is learning to do what computers cannot do. Algebraic thinking includes learning to:
- write algebraic equations from words or diagrams;
- interpret the graph of a function;
- set up and use ratios in problem solving; and
- solve equations and relate these solutions to graphing.

Our goal is to **have mathematics make sense.** Making sense of new ideas takes time (sometimes <u>lots</u> of time) and discussion with others. It is not always easy. You need to do lots of problems AND to talk about the algebra problems. No one can learn to play a musical instrument by simply watching and listening to others play; they need to ask questions, talk about music and practice to get good at it. Likewise, the **ONLY** way you can become good at algebra is by doing problems and talking about them. **If you do both, you will learn algebra.**

Do your best to meet these expectations each day. You will find that algebra is not only useful, it is fun. Have a successful year!

MISSION POSSIBLE (something to do)

With your team, find <u>at least</u> three of the main ideas the authors wanted you to know about this course. You can find these ideas in the "Welcome Note" you just read. Make a list of them. Be sure that you put a copy in your algebra notebook.

GS-1. Much of your class time will be spent working with your fellow students in study teams of two or four. Study teams are the first source of support for learning in this course. Read the guidelines for study teams in the box below, as well as the paragraph that follows.

Record the names of the members of your study team on your paper. Then copy these four guidelines into your notebook.

GUIDELINES FOR STUDY TEAMS post in CR

1) Each member of the team is responsible for his or her own behavior.
2) Each member of the team must be willing to help any other team member who asks for help.
3) You should only ask the teacher for help when all team members have the same question.
4) Use your team voice.

Often, a problem will direct you to share the work within your study team. This does not mean one person does all the work and then tells the rest of the team members the answers. Decide how to divide the labor, then share each individual's work to help solve the problem. Your team can verify whether your solutions are reasonable and provide an opportunity for all of you to discuss different ways to solve the same problem.

GS-2. KEEPING A NOTEBOOK

You will need to keep an organized notebook for this course. Here is one method of keeping a notebook. Your teacher may alter these guidelines.

- The notebook should be a sturdy three-ring loose-leaf binder with a hard cover.
- It should have dividers in it to separate it into at least six sections:

    TEXTBOOK   NOTES
    HOMEWORK/CLASSWORK TESTS and QUIZZES
    ➤TOOL KIT   ➤GRAPH PAPER
    — Mastery sheets

    — classwork & notes
    — hw/c
    — quizzes
    — graph paper

Because you do not want to lose your notebook, you should put your name inside the front cover. If you lose it on the bus, you will want it returned to you, so you should also put your phone number and address (or the school's) inside the cover. If you cover your book, you may wish to put your name in large clear letters on the outside cover. Also, write your name and the name of your teacher inside the back cover of your textbook.

STUDY TOOLS FOR SUCCESS

MATERIALS: Your notebook will be your written record of what you have done in the course and will be the chief way to study for tests, so take good care of it. You will also need a scientific calculator and plenty of graph paper throughout the year.

We also recommend that you carry two colored pencils or pens, a #2 pencil and a pen (blue or black ink), and a ruler. All of these tools, along with your calculator, can be kept in a plastic pouch inside your binder.

ATTITUDE and EFFORT: You will increase your chances for success in algebra by using your class time to complete the day's lesson. We expect you will develop your ability to contribute to and benefit from working together in teams. You must be willing to both ask and answer questions within your study team. Furthermore, you will need to work on problems for more than a minute or two--sometimes for several days! In addition, this course will often ask you to explain your thinking--sometimes orally, frequently in writing. In short, YOU must take responsibility for doing your work. Simply stated, you need to do all of your assignments regularly. If you do all of this, you will be much more successful in this course.

GRAPH INTERPRETATION

GS-3.   DIAMOND PROBLEMS

With your study team, see if you can discover a pattern in the three diamonds below. In the fourth diamond, if you know the numbers (#), can you find the unknowns (?) ? Explain how you would do this. Note that "#" is a standard symbol for the word "number".

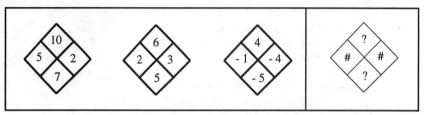

Patterns are an important problem solving skill we use in algebra. The patterns in Diamond Problems will be used later in the course to solve algebraic problems.

Copy the Diamond Problems below and use the pattern you discovered to complete each of them.

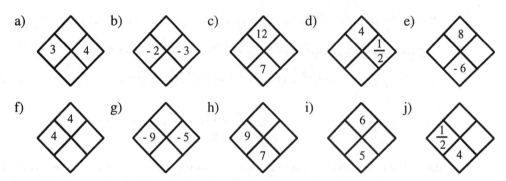

GS-4. PROGRESS REPORTS*

Jack, Donna, Nancy, and Lee are students in Ms. Speedi's math class. Each student's progress report is represented by one of the points on the graph below.

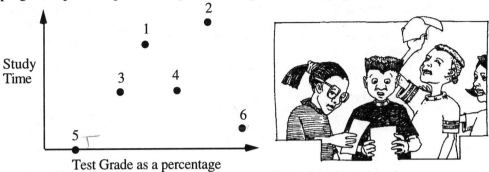

"Jack never studies and has a poor test percentage."

"Nancy is very able, but because of her active social life, she seldom studies outside of class. Her test percentage is OK, but could be better."

"Donna has worked hard both in and out of class. Her test grades are great."

"Lee studies often outside of class and has done reasonably well on the tests."

a) With your team, decide which student's report corresponds with which point.

b) Make up reports for the remaining two points on the graph.

c) Individually, think about your own study skills and predict your math grade at the end of the first quarter. Copy the graph (without the numbers 1 through 6) and place a point on it that reflects your prediction based on your study habits. Explain your choice of location in a sentence or two.

---

* Adapted from *The Language of Functions and Graphs*, Joint Matriculation Board and the Shell Centre for Mathematical Education, University of Nottingham, England.

Getting Started: Working in Teams

THE KEY TO SUCCESS IN CPM ALGEBRA 1

Hello, again. This is Chris. From time to time I'll be talking to you about how to be successful in this course.

How many times have you read a math book? Maybe a better question is, "Do you think it is necessary to read a math book?" You probably think a math book is a collection of examples to look up and lists of problems to do, but certainly not something to read! That's what I thought until I finally figured out that I was missing lots of valuable information. I learned the hard way that I had to be responsible for my own education. I had to use the textbook resources and work with other students in order to learn.

I am not trying to scare you. But since the name of the program you are in is, "College Preparatory Mathematics," I just want you to consider what this course will require of you.

CPM Algebra requires that you read the textbook. If you understand that, you are on your way to doing well in this class. Start learning how to read a text book NOW. Otherwise, one of the hardest transitions to college will be how to read a textbook for important information.

Information is the key to success in this program. In this course, you will collect pieces to a puzzle that, when put together, you call algebra. Remember, this is a college-required math course. It will have different expectations than some of the courses you have taken before.

I want to point something out right away. Look at the first sentence in the seventh paragraph of the welcome note in PZL-1. It says, "Our goal is to have **mathematics make sense.**" Does this sound like such a terrible thing? Of course your first response is to laugh and say, "Yes--nobody wants to learn math!" You know what? I have heard that before, and I don't believe it--the same way you don't believe it when you hear teachers say how wonderful math is. Get beyond the disbelief and really ask yourself, "What's in it for me? Why is the goal of having math make sense an advantage for me?"

While you are considering this question, I am going to share with you one student's experience in Math 1. Shannon Springmeyer of Carnegie Middle School wrote to her teacher in an assignment reflecting on what she had learned in Math 1. She said, "Above all I have learned how to logically think through a problem and use information to produce a solution." She continued with, "I now have many more skills than I did at the beginning of the year and most important, I understand!" Shannon found ways that made math make sense. She was able to make use of the several approaches to learning algebra in this course. **Shannon took responsibility for her education by actively working at learning algebra.** In turn, she was able to understand algebra and get good grades. Shannon will use these skills for the rest of her life. If you **expend reasonable effort**, you will get what you want out of this class.

MISSION POSSIBLE (something to do)

Read the questions below, then collect some clues to help answer them by re-reading the welcome note in PZL-1. Then write your answers as directed by your teacher. They will help you to understand what is in this course for you.

a) If you were to write a math textbook, what would you do to have mathematics make sense?

b) How would you use your book to teach a class, and what would the classroom setting be?

GS-5. MATHOGRAPHY   *Letter to self*

Write a letter about yourself that will help me get to know you as an individual. The letter should address the three general topics below (in bold). The questions provided below are merely suggestions to help you get started.

About You: By what name do you like to be called? With whom do you live? What languages do you speak? What languages are spoken at home? What are your interests, talents, hobbies? What are you proud of? When is your birthday?

You as a Math Student: Describe your math memories from kindergarten until now. What experiences in math have you liked? Why? How do you feel about taking algebra?

You as a Student: What are you like as a student? What are your academic strengths and weaknesses? What are your favorite courses? What have you liked about them? Have you worked with others before? Do you like working as part of a team? Why or why not?

TEAM BUILDING, AREA, AND PERIMETER

GS-6. The cooperative card game was an opportunity for you to begin working together as a study team to solve a problem. Use complete sentences to describe what happened with your team while you were solving the cooperative logic problem(s). What worked well? Do you think that all problems can be solved in study teams? Think back to your previous learning experiences and explain your answer in a short paragraph.

Coop Learing Puzzle

Using whole numbers only, draw every possible rectangle with an area of 16 on graph paper. Write A = 16 inside each figure. Label the dimensions of the length and width for each rectangle.

a) What is area? How is it measured on your graph paper?

b) What is perimeter? Calculate the perimeters and write P = (answer) below each figure. Check with your study team to see if you have all possible rectangles. Add any rectangles you may have missed.

c) Using complete sentences, describe the shape of the rectangle with the largest perimeter. Include a sketch of this rectangle (remember to label the length, width, area, and perimeter for each sketch). Do you think the rectangle with the largest perimeter will always look this way? Why or why not?

d) Describe the shape of the rectangle with the smallest perimeter. Include a sketch of that rectangle (label your sketch completely). Do you think the rectangle with the smallest perimeter will always look this way? Why or why not?

PZL-3 BUILDING A SUCCESSFUL MATH COURSE

Look again at the welcome note. Why is this note important? Right away it tells you what you can expect. Paragraph three is probably the single most important paragraph you will need to read this year. Here it is again.

What can you expect? To begin with, the course is **built around problems.** You will spend more of your class time doing and discussing problems in study teams and not very much time listening to your teacher tell you rules. Don't misunderstand--rules are important, but they will emerge as you develop an understanding of the mathematics they represent. You will continue to use each rule you learn throughout this course and others.

There are no surprises here. The course will require you to come up with strategies for solving problems. **You have to actively think and work on problems.** You cannot just sit back and wait to be told everything. It won't happen. Maybe your last math class was like that, but not this one. Your role as a student is different here.

At first you may say, "The teacher is not teaching me," but this is only if you expect the teacher to be explaining at the board or overhead, then assigning practice problems. In this course the teacher will lecture and solve examples when appropriate, usually <u>after</u> you have had a chance to explore an idea. <u>Then</u> you should have some real questions based on your initial study to ask your teacher.

Part of your role is to do some thinking. Again, don't worry. Your teacher and this text will help guide you through the thinking you need to learn math. Often your teacher will ask you questions when you are stuck to get you working toward a solution yourself. When your teacher <u>is</u> explaining something, it will be very important that you listen carefully and take good notes.

By understanding the development of the rules in algebra, you will be better able to apply them when you need them. How many times have you memorized something, only to forget it when the test came--or right after the test? How many times have you memorized a certain type of problem, but when you saw a slightly different version of it, you could not do it? **Your goal should be to eliminate the problem of forgetting by developing strong problem solving skills**. This way, when you face a slightly different version of a problem, you will be able to figure out the key math idea and use it rather than saying, "I don't understand."

The teacher and the textbook will provide all the pieces to the puzzle you need to make sense of algebra. There are three things you need to do:

- first, you **will have to be actively involved in** <u>doing</u> **math every day;**
- second, you will have to **learn how to read the textbook;** and
- third, you will have to **communicate with the teacher every day.**

We have already begun to collect information from the text by reading the welcome note in PZL-1. Notice that we pulled three major points from this note. They are:

- **The course is built around problems.**
- **You are being asked to learn** <u>more</u> **about the math you study and to use some different ways to learn.**
- **The goal is to have mathematics make sense.**

MISSION POSSIBLE (something to do)

Go back to the welcome note (PZL-1) one more time and, with the three major points in mind:

a) make a list of at least three things that you will need to do in order to succeed in this class.

b) Circle the things on your list that are different from what you are used to doing.

GS-8. On graph paper draw all the rectangles with an area of 20. Use only whole numbers for the dimensions. Label completely. Identify the ones that have the largest and smallest perimeters. Describe these shapes. Remember to use complete sentences.

GS-9. In high school mathematics, you need to be able to work with numbers, words, and geometric representations. Use diagrams to represent percent in the following problems.

a) Draw a figure that represents 100%. Label 100% below it.

b) Draw a figure that represents 50%. Label it 50%.

c) Draw a figure that represents 25%. Label it 25%.

d) Draw a figure that represents 150%. Label it 150%.

e) Look at the four drawings. Re-label each figure using fractions. Write the fractions under the matching percent.

f) Now re-label the four drawings using decimals.

g) Each drawing has a percent, fraction, and decimal because these are three different ways to describe the same amount. Explore with your calculator to see how to go back and forth between these equivalent numbers. What keys did you use?

STUDY SKILLS

When a problem has you stumped, it is a good idea to write a question in its place to ask your study team or teacher.

For example: "How do I go from 0.5 to $\frac{1}{2}$ on my calculator?" or
"What's the difference between area and perimeter?"

GS-10. Copy these Diamond Problems and use the pattern you discovered earlier to complete each of them.

a)   b)   c)   d)   e)

f)   g)   h)   i)   j)

GS-11. Review your work in problems GS-7 and GS-8.

a) Without drawing the rectangles, think about all the possible rectangles with an area of 36. Sketch the rectangle you think will have the largest perimeter. Describe its shape. Label its length, width, area, and perimeter.

b) Sketch the rectangle you think will have the smallest perimeter. Describe its shape. Label its length, width, area, and perimeter.

c) Predict the shape of any rectangle with whole number dimensions that has: the largest perimeter; the smallest perimeter.

ROLL AND WIN, PART 1: DATA COLLECTION

GS-12. Share your answers to GS-9. Have the team decide which diagrams you will share with the class. Have one member of your team put these diagrams on the table provided by your teacher. Compare your calculator techniques and questions with each other.

GS-13. ROLL AND WIN

Your team is going to play a dice game called Roll and Win. The game is played with two dice, each numbered 1 through 6. The two dice are rolled and the numbers that come up are <u>subtracted</u>. For example, when a 2 and 5 come up, the difference is 3 so the team member who chose the difference 3 will get a point. The person with the most points wins. Each person must choose a <u>different</u> difference as their winning value before beginning the game.

a) What differences are possible in Roll and Win?

b) Each person in your team should choose a difference from your list in part (a). Write down the name of each person and his/her number.

c) Break your team into partners to gather data. Each pair of people will toss two dice 20 times, subtract, and record the differences. Be sure to record the result of each roll, regardless of who wins. Before beginning, decide with your partner how you are going to record the differences.

d) Which difference occurred most often in your set of 20 tosses?

e) Combine your data with the rest of your team. Which difference occurred most often in your team's combined set of 40 tosses? Decide who won.

f) Draw a histogram displaying your team's data with the differences on the horizontal axis and the number of occurrences (the frequency) on the vertical axis.

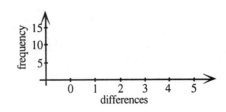

g) Is this a fair game? Why or why not? Before you answer, briefly state what you understand "fair game" to mean and give a simple example.

GS-14. Without drawing the rectangles, think about and then list all the possible rectangles with an area of 25 square units.

   a) Sketch the rectangle you think will have the largest perimeter. Label its length, width, area, and perimeter. Describe its shape.

   b) Sketch the rectangle you think will have the smallest perimeter. Describe its shape. Label its length, width, area, and perimeter.

GS-15. Copy these Diamond Problems and use the pattern you discovered earlier to complete each of them. Some of these may be challenging!

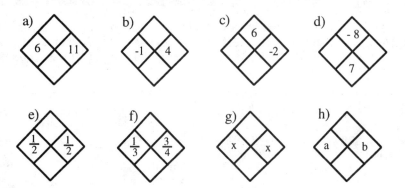

GS-16. The graph at right is called a bar graph.

   a) According to the graph, how many hours of television does the average 9th grader watch?

   b) Which grade level student watches the least amount of TV, on average? Why do you think this is so?

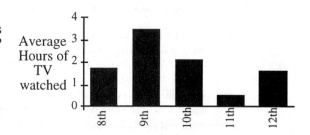

   c) Describe the amount of TV watched by the different grade levels.

   d) Based on this graph, how many hours does the average student watch?

GS-17. Mrs. Katz bought a bike for $50 and sold it to her neighbor for $60. She decided to buy it back, but had to pay $70 for it. If she then sold it for $80, what was her net profit or loss? Show your solution in enough detail so that a teammate could easily understand how you found your answer.

GS-18. Copy and complete each sequence. Using words, not numbers, describe how the patterns work (for example, "double the previous number").

a) 1, 3, 6, 10, ___, ___

b) 1, 3, 9, 27, ___, ___

c) 49, 47, 52, 50, 55, ___, ___

d) 1, $\frac{1}{2}$, $\frac{1}{4}$, $\frac{1}{8}$, ___, ___

e) 8, 7, 5, 2, ___, ___

ROLL AND WIN, PART 2: PROBABILITY

GS-19. Exchange notebooks with a member of your study team. Is your partner's notebook organized, as described in GS-2, or as required by your teacher? Work together to be sure that both notebooks are organized. When you are finished, read the definition in the box below.

> **PROBABILITY** is a mathematical way to predict how likely it is that an event will occur. If all the outcomes of an event are equally likely to occur, then the probability (or likelihood) that a desired result occurs is expressed by the fraction:
>
> $$P(\text{outcome}) = \frac{\text{number of ways that the outcome occurs}}{\text{total number of possible outcomes}}$$
>
> P(outcome) is the way we specify that we want the probability "P" of a certain outcome. For example, there are 6 faces on a die. Suppose the desired outcome is a 4. Since the number 4 appears once, we write the probability of getting a 4 as: $P(4) = \frac{1}{6}$.

GS-20. You are now going to calculate the mathematical probability of each difference which occurred in the Roll and Win Game, GS-13.

The first thing you need to find in calculating probability is the total number of possible outcomes. You can do this by making a **systematic list** to show all the possibilities that can occur. Here is one way to organize a systematic list:

Die #1	Die #2	Difference
1	1	0
1	2	1
1	3	2
etc.	etc.	etc.

Copy and complete this list. What is the total number of possible outcomes? Why is this list called "systematic?"

Getting Started: Working in Teams

GS-21. The 36 entries in the list represent the total number of outcomes. Using the definition above, find the following probabilities. You do not need to reduce the fractions.

Your answer should be written in this form: $P(3) = \frac{6}{36}$

a) P(0)

b) P(1)

c) P(2)

d) P(3)

e) P(4)

f) P(5)

g) What is the sum of all the probabilities? Does your sum make sense? Explain.

h) What number would you now choose to play Roll and Win? Explain why.

GS-22. Because you have two separate dice another method to organize your data is to **make a table** like the one at right. Copy and complete this chart. The numbers along the side and top represent the two dice faces. Fill in the chart with the differences.

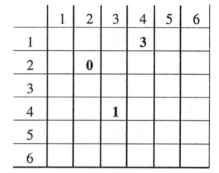

a) How many possible outcomes are shown in the chart?

b) Which method (list or chart) do you prefer?

GS-23. Use probability to answer the following questions:

a) If you roll a pair of dice 72 times, how many of those rolls do you expect to have a difference of 0? 1? 4? 6? Explain how you know.

b) If you roll a pair of dice 100 times, how many of those rolls do you expect to have a difference less than 7? Why?

GS-24. Compare your team results from problem GS-13 with the probabilities you calculated in problem GS-21. Describe how they are similar or different.

GS-25.  Measure and record your height to the nearest inch then record your shoe size. Next get the same two measurements for a family member. If possible, pick a family member with adult-sized shoes. Be sure to bring this information to class tomorrow.

GS-26. In a lottery, two spinners are spun to select your prize. One spinner determines if you win a boat or a car. The second spinner determines if your prize is red, blue or green.

    a) Make a systematic list of all the possibilities.

    b) What is the probability that you win a car?

    c) What is the probability that you win a green car?

 COMMUNICATING WITH YOUR TEACHER

Just a quick note about the importance of doing your work in this class. Homework and classwork are the best ways to communicate with your teacher. One purpose of assigned course work is that **your teacher can use it to evaluate your level of understanding**. If you do not do your work, the teacher can only assume you know nothing. You need to complete all assignments--especially tool kit entries, unit summaries, and portfolio work. Otherwise, exams are the only other way that your teacher has to evaluate your knowledge of algebra. We all have bad exam days, so this leaves assigned course work as the best way to regularly demonstrate your level of understanding to your teacher. I will even say that daily assignments are far more important than exams. The moral to this story is DO ASSIGNED WORK EACH DAY.

Now do GS-27 and you will have completed the first unit in this course!

GS-27. Reflect on the study team activities you experienced the last few days. Which activities were your favorites? Why? What about your team made you feel comfortable? What makes an effective team member?

Unit One
The Difference of Squares
Organizing Data

Unit 1 Objectives
Difference of Squares: ORGANIZING DATA

In this unit we continue to develop good work habits and the ability to work as a productive team member. Many of the problems in this course can be solved by using a variety of strategies and tools. In this unit you will look at ways to **organize and analyze data** in tables, graphs and charts. You will use models to review integer operations and combine like terms. You will also develop the problem solving strategy of Guess and Check and explore the use of your calculator when dealing with large numbers.

In this unit you will have the opportunity to:

- understand the rules for doing arithmetic with integers.
- combine algebraic terms using addition and subtraction.
- interpret graphs.
- organize data using a variety of methods, including systematic lists and tables.
- investigate patterns and use them to predict outcomes.
- solve word problems using the "Guess and Check" strategy as preparation to write and solve equations.

Read the problem below. Over the next few days you will gain the skills to solve it.
Do not try to complete it now.

SQ-0. **DIFFERENCE OF TWO SQUARES**

Many numbers can be written as a difference of two squares, such as the ones below. You will find ways to write numbers as a difference of two squares, organize your data and look for patterns. The difference of two squares will occur again in a later unit.

NUMBER	SOLUTIONS (Difference of Squares)
8	$3^2 - 1^2$
15	$8^2 - 7^2$; $4^2 - 1^2$

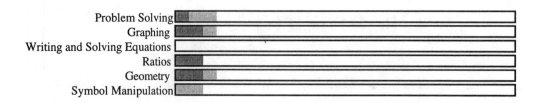

Unit 1
Difference of Squares: ORGANIZING DATA

 WHAT TO DO IN THIS COURSE

What is your primary responsibility as a student in this algebra class? I am going to answer that question with one simple phrase: **to think!** Throughout this course, the number one requirement will be to think about mathematics. This requires you to be active in your learning. The point of CPM and its style of teaching and learning is to help you think. **You will need to read the book and do your homework every night to practice thinking.** Your years of college prep mathematics and even beyond will require you to be an active thinker.

I will not kid you by saying this is easy, but it will get easier as you do it on a regular basis. You also have other responsibilities, most of which were mentioned in the welcome note in the "Getting Started" unit (PZL-1). Make sure you read this note again. Also, be sure you know all the requirements for this course. If you are still not sure what they are, ask your teacher as soon as possible.

MISSION POSSIBLE (something to do)

Re-read the second to last paragraph in the welcome note.

a) Do you agree with the statements about what it takes to learn how to play a musical instrument? Why or why not?

b) Read the box below, then begin today's lesson with SQ-1. By the way, what is your <u>primary</u> responsibility in this course?

REMINDER: Guidelines for Study Teams

1) Each member of the team is responsible for his or her own behavior.

2) Each member of the team must be willing to help any other team member who asks for help.

3) You should only ask the teacher for help when all team members have the same question.

4) Use your team voice.

SQ-1. On the poster paper on the wall, place a sticky dot above your height to indicate your shoe size.

Do the same for your family member's shoe size.

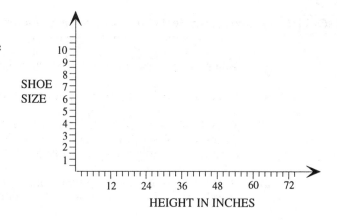

SQ-2. THE BEEBOPPER SHOE STORE

As the manager of the Beebopper Shoe Store, you are responsible for ordering the shoes to stock a brand new line of athletic shoes. You have no idea how many shoes of each size to order, but you have heard from a wise old woman (who does <u>not</u> live in a shoe) that you can probably figure out a person's shoe size by estimating his or her height. The woman believes that taller people wear larger shoe sizes and shorter people wear smaller sizes. You are not sure about this wisdom, so you have decided to conduct an investigation by asking the local algebra students to provide heights and shoe sizes.

Make a table that represents the data from the Beebopper Shoe Store graph on the wall. You should make your own data table and verify the entries with your team members before writing them down.

SQ-3. Use the table you made in problem SQ-2 to make your own graph of the class's height and shoe size data on graph paper. On the horizontal (height) axis, use a scale where one unit represents 2 inches of height. On the vertical (shoe size) axis, let each unit represent a half shoe size.

SQ-4. Use your own Beebopper Shoe Store graph (or the graph on the wall) to answer the following questions in complete sentences:

a) What are the smallest and largest shoe sizes represented?

b) What are the shortest and tallest heights represented?

c) Which shoe size would you stock the most of if you were the manager of the Beebopper Shoe Store? Which size would you stock the least? Would it be profitable to carry every possible size? Why or why not?

d) What difference, if any, do you notice between the data for males and the data for females?

e) Was the old woman correct? Can you predict a person's shoe size if you know his or her height? Why or why not? Do you think this relationship would be valid for adults? Why or why not?

f) Explain how you would use the graph to predict the shoe size for someone 5'6" tall. Explain how you would use the graph to predict the shoe size for someone 7' tall.

g) Look at the graph. Find the height represented most frequently. Calculate the average shoe size for classmates of this height. Does this shoe size make sense? Why or why not?

h) Is there any data point that does not seem to follow the trend?

SQ-5. Use the pattern you found in the Getting Started Unit to complete these Diamond Problems. Create two new Diamond Problems of your own.

a) diamond with $\frac{1}{2}$ and 1

b) diamond with 10 and 7

c) diamond with 5 and $\frac{1}{3}$

d) diamond with 8 and -6

SQ-6. Enrollment in math courses at a particular high school is shown in the pie chart at right. If there are 1,000 students enrolled in math courses, approximately how many students are enrolled in Algebra? In Geometry? In Calculus?

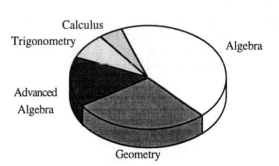

SQ-7.  A local radio station is giving away prizes to the ninth caller. After a listener successfully dials in, the radio station then spins two spinners. The first determines if you win a cassette or CD, while the second selects from five artists: two of whom you like.

a) Assuming that you are the ninth caller, what is the probability that you win a CD?

b) What is the probability of getting one of your favored artists on tape or CD?

c) What is the probability that you win a CD of one of your favored artists?

d) What is the relationship between the fractions in (a) and (b) to your answer in (c)?

ADDING AND SUBTRACTING INTEGERS

The number 3 can be represented by

+++ or +++++++ / ---- or +++++ / --

The number 0 can be represented by the same number of positive and negative numbers. An equal number of positive and negative tiles is called a **NEUTRAL FIELD**.

++ / -- or +++++ / ----- or +++ / ---

The number -1 can be represented by

- or +++ / ---- or ++++ / -----

SQ-8. Use a minimum number of tiles to represent the following integers. Build each model and then sketch it on your paper.

a) 2   b) -7   c) 0

Use at least 12 tiles to represent each of the integers below. Sketch each model after you build it.

d) 5   e) -1   f) 0

SQ-9. Represent each of the statements below using tiles. Sketch each model and write the problem with its answer below your drawing.

    a)   -7 + (-2)   c)   3 + (-4)   e)   -2 - 5

    b)   -1 + 5   d)   4 - (-2)   f)   -7 - (-8)

Problems that are especially important have a box around the problem number. They help you develop understanding and consolidate ideas. Pay careful attention to these problems, and be sure to revise your work if necessary. Complete, correct solutions should be in your notebook to serve as examples of why a concept or procedure works and as an example of how to do it.

[SQ-10.] Khalifa has never worked with integers before. Explain to her your rules for doing integer problems, including a diagram. Answer in complete sentences.

    a)   -4 + (-2)   b)   3 + (-7)   c)   2 - 5   d)   3 - (-2)

SQ-11. Lindsay thinks the problems 2 + 4 and -2 + (-4) are similar while Juan disagrees. Explain your position on this discussion.

SQ-12. Compute <u>without</u> a calculator:

    a)   427 - (-3)   b)   -50 + (-150)

SQ-13. Sam and Samantha put the problem 3 + 2 · 4 in their calculators. Sam got 20 for the answer while Samantha got 11. Ms. Speedi then knew Samantha had a scientific calculator and Sam did not. Sam had to return his and buy a scientific calculator for his algebra class.

    a)   How did Ms. Speedi know Samantha's answer was correct and Sam's was wrong? What did Samantha's calculator do that was different from Sam's?

    b)   Check your calculator to see if it is a scientific calculator by doing the problem. What kind of calculator do you have? If it is not scientific you will need to make a trip to the store and buy one.

SQ-14. Deondre was tired of playing Roll and Win. Instead of subtracting the two dice, he decided to add them. He called his new game Roll and Add. Use the strategy you used in problems GS-20 through GS-22 to decide which number Deondre should choose to win. Show all your work.

Difference of Squares: Organizing Data

SQ-15. Sketch at least 12 tiles that represent each of the integers below.

   a)   4   b)   -2   c)   -5

SQ-16. Represent each of the statements below using tiles. Sketch each model and write the problem with its answer below your drawing.

   a)   2 + (-4)   b)   -3 - (-5)

SQ-17. Compute without a calculator.

   a)   -15 + 7   e)   -50 - 30   i)   9 + (-14)

   b)   8 - (-21)   f)   3 - (-9)   j)   28 - (-2)

   c)   -12 - (-4)   g)   -75 - (-75)   k)   -3 + (-2) + (-5)

   d)   -9 + (-13)   h)   (-3) + 6   l)   3 + 2 + 5

INTERPRETING GRAPHS AND COMPARING AREAS

SQ-18. Today you received a sample page for your **ALGEBRA TOOL KIT**, essentially a summary of what has been covered so far in this course. YOU will be responsible for the creation of subsequent entries in your tool kit.

We suggest that you summarize each new idea, with an appropriate diagram or example, as it is presented. It is also a good idea to label each entry with the textbook page number or problem number where the idea appeared.

Your tool kit is a handy reference guide you will create during the year. As you identify and organize the important ideas, you will have a useful tool to make algebra more manageable.

Add to your tool kit what you know about adding and subtracting integers.

INTEGER ADDITION AND SUBTRACTION

In class we have been using integer tiles to model the operations of addition and subtraction. Here is an example that explains the process with words and a sketch of what you record.

Addition Example: -8 + 6

Start with a neutral field, that is an equal number of positive and negative tiles, that has a value of zero. You can review the definition of "neutral field" in the box after problem SQ-7.

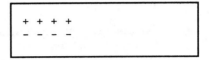

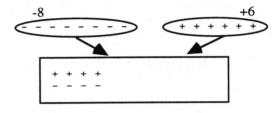

Display the two numbers using tiles.

Combine the two numbers with the neutral field.

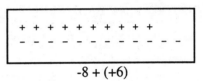

Circle the "zeros".

Record the sum.

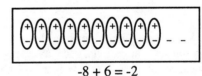

Subtraction Example: -2 - (-4)

Start with the first number displayed with the neutral field.

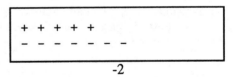

Circle the second number in your sketch and show with an arrow that it will be removed.

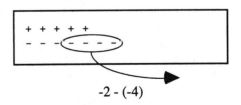

Remove the second number.

Circle the "zeros"
Record the difference.

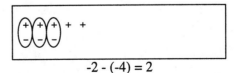

Difference of Squares: Organizing Data

SQ-19. This problem must be done with a scientific calculator. Use your study team to make sense of the calculator results.

a) Fill your calculator display with 9's. How many 9's does your calculator display? Copy the display on your paper with commas in the appropriate places.

b) Without using your calculator, add one to the number you just wrote down. Compare your answer with others. Count your zeros, put in commas and write that answer in words.

c) Now use your calculator to add one to your original number. Write down exactly what your calculator displays. This is the calculator's way to display those numbers that do not fit in its screen.

d) Compare your answers for parts (b) and (c).

e) Emil's calculator displays $\boxed{5.286^{12}}$. What does the "12" represent?

f) Extension: Can you get your calculator to display $\boxed{5.286^{12}}$? Write down your steps.

SQ-20. Latisha is determined to do well in school this year. Her goal is to maintain an at least 85% average in all her courses. You will need to help her keep track of her average each night. So far she has two scores in Algebra, 72 and 89. Find her average now. Show your work.

SQ-21. CAR COMPARISON

The following three graphs describe two cars, A and B.

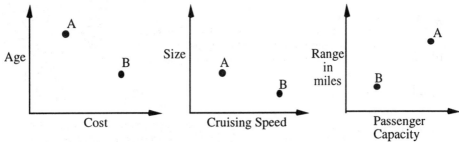

Decide whether each of the following statements is true or false. Explain your reasoning.

a) The newer car is more expensive.

b) The slower car is larger.

c) The larger car is newer.

d) The less expensive car carries more passengers.

SQ-22. Suppose we discover the island shown in the drawing at right and want to know its area. If we cover the island with a grid we can approximate the area.

a) What is the area of the grid?

b) The island's area is less than the area of the grid. Why?

c) Using the grid, approximate the area of the island, <u>including</u> the lake.

d) Does the coastline represent the area or the perimeter of the island? Explain the difference between area and perimeter.

SQ-23. a) Copy the two graphs at right. On each graph label two points that would represent cars A and B as described in the Car Comparison Problem, SQ-21.

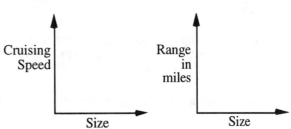

b) Which car would you buy? Why? Explain your answer in complete sentences.

Difference of Squares: Organizing Data

SQ-24. First try these problems without a calculator. Then check your answers on your calculator.

a) 32 + (-7)

b) -5 + (-10)

c) -13 + (-12)

d) (-8) + (-8) - (-8)

e) $\left(-1\frac{1}{2}\right) + \left(-2\frac{1}{3}\right)$

f) $\left(-4\frac{1}{5}\right) + (10)$

g) $\left(\frac{3}{4}\right) - \left(-2\frac{1}{3}\right)$

h) $\left(16\frac{1}{8}\right) + \left(-\frac{3}{4}\right)$

SQ-25. Copy and complete each of these Diamond Problems:

a) 15 / 8

b) 3, 1

c) 10 / 5

d) $\frac{1}{2}$, $\frac{1}{3}$

e) 4 / $\frac{1}{2}$

f) 0.3, 0.2

g) 11 / 12

h) -3, -4

SQ-26. Use the work from Deondre's Roll and Add game to answer the following questions. Refer to your work from SQ-14.

a) Which sums have the same probability of showing up? Explain why.

b) If Deondre rolled the dice 72 times, how many times would you expect the sum of 12 to come up? the sum of 7? the sum of 10?

SQ-27. Max, who is in the third grade, asked his older sister, Maxine, to explain what 3 · 4 means.

a) In complete sentences, explain how Maxine could help Max understand what "3 times 4" means. Draw a picture that Maxine could use to show Max what "3 times 4" means.

b) Give a real life example of when you might use 3 · 4.

MULTIPLYING AND DIVIDING INTEGERS

 UNDERSTANDING HOW TO USE THIS TEXTBOOK

By now you have noticed that this Algebra 1 course expects you to be an active learner. You work with other students in study teams, the problems guide you to learning the mathematical ideas, you are asked questions about what your work means, and your teacher works with you in several ways during class time.

Another key step to get the most out of this course is to understand the textbook. **Understanding the textbook** is quite a bit different from understanding the problems within it.

Quickly <u>glance</u> at the welcome note (PZL-1 in Unit 0) and see if you can find the three big ideas it contains about this course.

How did you find them? If you noticed that they are in **bold print**, you have just discovered **one way for the authors to relate big ideas to you**. Flip through a few pages and find a few other examples of using bold print for emphasis. Bold print is one of the most common ways to point out information. **Sentences in bold print contain valuable information.** Always make sure you <u>add</u> <u>information</u> <u>in</u> <u>bold</u> <u>print</u> <u>to</u> <u>your</u> <u>tool</u> <u>kit</u> <u>and</u> <u>know</u> <u>how</u> <u>the</u> <u>surrounding</u> <u>paragraphs</u> <u>relate</u> <u>to</u> <u>the</u> <u>idea(s)</u>.

There are some other signs to help you find key information in the text. Every unit of this text contains several main ideas. These ideas are the fundamental concepts of algebra. **You will absolutely need to know these ideas to be successful in this course.**

- The mathematical ideas and definitions are in **single** and **double line boxes**. These boxes sometimes contain **examples** of how to use an idea or do a procedure.

- Important problems--usually ones that develop an idea or help pull it together--have a **box around the problem number.**

The information you find in boxes and important problems should be in your tool kit.

There are a few more ways to locate key information in this course. In the front of your book there is a **table of contents** that lists what you will study unit by unit. In the back there is an **index** that helps you locate where a term or concept is introduced. At the beginning of each unit there is a **list of what you will learn** in that unit. At the end of each unit, everything you should have learned and added to your algebra tool kit is listed in a **"Tool Kit Check-up."** In addition to your textbook, there is a **"Parent's Guide to Math 1"** that will help your parent(s) understand how to help you with this course. Finally, there is a **"Math 1 Supplement"** that has additional examples and lots more practice problems. Ask you teacher how to obtain these resources or use the ordering information in the "Note to the Parents" at the beginning of the book.

MISSION POSSIBLE (something to do)
Make a list of all the locations in the text that will help you find the important information it contains.

Difference of Squares: Organizing Data

SQ-28. Padraig has never multiplied integers before. Explain to him your rules for doing integer problems. Include a diagram. Then compute the following products:
   a) 3(2)
   b) 5(-3)
   c) -3(5)
   d) -4(-2)

SQ-29. Compute:
   a) -12 ÷ 3
   b) $\frac{6}{-2}$
   c) $\frac{-8}{-4}$

SQ-30. What do the rules you use for multiplying and dividing integers have in common? Be specific enough that a student who transfers into class tomorrow would easily understand your response.

SQ-31. Try these problems without a calculator first. Then use a calculator to check your answer.
   a) -16 + 7
   b) 10 - (-24)
   c) (3)(-9)
   d) -9 + (-11)
   e) -49 - 36
   f) -56 ÷ (-7)
   g) 15 ÷ -3
   h) -7 ÷ 7
   i) (-6) · 9
   j) 27 - (-3) - 12
   k) (-5)(-5)
   l) -6 · (43)

SQ-32. Show <u>all</u> your work to prove that your answer is correct.

   a) If you spin both spinners and add the two numbers, what is the probability that the answer will be positive? Make a list of the possible outcomes.

   b) If you multiply the results of the spins, what is the probability that the answer will be positive?

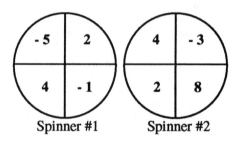

Spinner #1   Spinner #2

SQ-33. Latisha's newest score is a 90. Use her scores from SQ-20 to figure out her average now. Be sure to show your work.

SQ-34. Draw a diagram to represent $\frac{2}{3}$. Label your diagram with the fraction. Would 0.66 or 0.67 be a more appropriate decimal equivalent?

SQ-35. Write what you know about how to round decimals. Include several examples.

INTERPRETING GRAPHS

SQ-36.* The diagram below shows a map surrounding the local Community Center. Five of Tara's friends are at the locations shown and are all traveling to meet her there to play basketball. Kelly and Ricardo are each riding their bikes. Elizabeth is walking. One of the remaining friends walked while the other drove. Assume that all the same modes of transportation travel at the same rate. Analyze the map and determine each person's placement on the graph below.

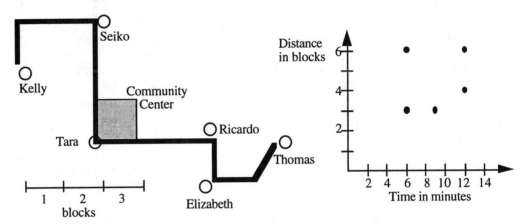

a) Trace the graph above on your paper. Label each point with the name of its corresponding person.

b) How long did it take Ricardo to get to the Community Center?

c) How did Seiko get to the Community Center?

d) How did Thomas get to the Community Center?

---

* Adapted from *The Language of Functions and Graphs*, Joint Matriculation Board, Shell Centre for Mathematical Education, 1985

PZL-7 HUMAN RESOURCES

In PZL-6, I listed all the printed resources that are available for help. I forgot to mention another kind of resource -- the most important one -- YOU and your study team members! Parts of the textbook are written for you to work with other students. The authors <u>assume</u> that you will have other students with you so that you can talk about what you are doing, ask questions, and get immediate help from your partner or teammates. They also believe (and have read research that agrees) that talking about the math you are doing will help you learn it--even if you are the so-called "smartest" person in the class.

GET THE PHONE NUMBERS OF STUDENTS IN YOUR TEAM so that you have someone to talk to when you are doing work out of class. You may want to form study pairs with a friend who is also taking this course or study teams to work after school, in the evening, and/or on the weekend. The point is that there is plenty of help available if you take advantage of it. Caution! Be careful not to fall into any of the traps that some students think is the way to use their study team:

- letting the smart one do all the work and then copying his or her answers.
- doing the work on their own because they don't like working with others.
- visiting with their friends and then asking the teacher for help.

If you think like one of these students, you will have to change how you study in a team. Teams are there to use as a resource. Let your friends help you understand things that you do not, and in turn you can help them with things they need. EVERYONE SHOULD CONTRIBUTE TO THE WORK OF THE TEAM.

MISSION POSSIBLE (something to do)

Discuss with your partner or teammates several ways for everyone to contribute to the daily classwork. Write a brief summary of your conclusions. When you are finished, continue with today's lesson.

SQ-37. Suppose you toss three coins, a penny, a nickel and a dime. They might come up "heads, heads, heads" (H, H, H).

a) What other possible outcomes might occur? List all the possibilities you can find.

b) How can you be sure that you listed all possibilities in part (a)? Describe a strategy for checking your list.

c) Find the probability that <u>exactly</u> (only) one coin comes up "heads."

d) Find the probability that <u>at least</u> one (or more) coin(s) comes up "heads."

SQ-38.   You roll a die and it comes up a "6" three times in a row. What is the probability of rolling a "6" on the next toss?

SQ-39.   Add to your tool kit what you know about how to multiply and divide integers.

INTEGER MULTIPLICATION

Multiplication is repeated addition or subtraction in a problem with two factors: The first factor tells us how many groups we are adding (+) or subtracting (–). The second factor tells us how many are in each group and whether they are positive (+) or negative (–).

For example: (2)(3) means add 2 groups of 3 positive tiles.
(2)(–3) means add 2 groups of 3 negative tiles.
(–2)(3) means remove 2 groups of 3 positive tiles.
(–2)(–3) means remove 2 groups of 3 negative tiles.

Below are examples of the diagrams used to record multiplication.

Example: (2)(–5)

Start with a neutral field. Since the first factor is positive groups will be added to the neutral field. Build 2 groups of 5 negative tiles.

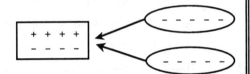

Physically push the groups into the neutral field lining up the positive and negative tiles.

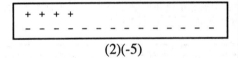

(2)(–5)

Circle the "zeros".

Record the product.

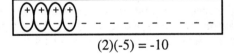

(2)(–5) = –10

Example: (–2)(–5)

Start with a neutral field.

Since the first factor is negative, circle the two groups of negatives that will be removed. Use arrows to indicate removal.

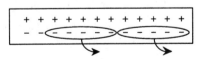

Remove the groups.
Circle the "zeros" and record the product.

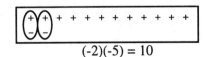

(–2)(–5) = 10

Difference of Squares: Organizing Data

SQ-40. Try these first, then check your answer with a calculator.
- a) -50 - 30
- b) 4(-15)
- c) (-13)(-2)
- d) -178 - (-3)
- e) $-\frac{1}{2} - 2\frac{3}{4}$
- f) $-\frac{1}{2} + (-\frac{1}{6})$
- g) (2.35)(-4.01)
- h) (0.005) ÷ (-0.021)

SQ-41. Today, Latisha earned a score of 67. Figure out her average now. How much did her grade drop? Show your work and use your previous work from SQ-20 and SQ-33 to help you.

SQ-42. Write another fraction that is equivalent to $\frac{4}{5}$. Draw diagrams to show that they are equal.

a) Find the equivalent decimal for both fractions. Was rounding your answer necessary?

b) Find the equivalent percent for both fractions.

SQ-43. Trace the number line below on your paper. Locate the following numbers by placing the lower case letters a through e on the number line corresponding to the values given below. Part (a) is done for you.

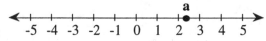

- a) $2\frac{1}{3}$
- b) 2.7
- c) $-1\frac{1}{2}$
- d) -0.2
- e) $33\frac{1}{3}$ % of 1

SQ-44. Estimate the area of Montana and California using the grid. Which state is larger? Compare the area of Montana to the area of California.

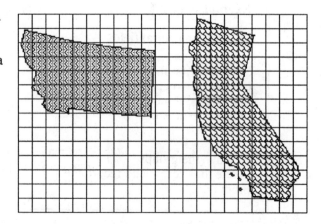

PROBLEM SOLVING WITH GUESS AND CHECK TABLES

SQ-45. In the past, you have guessed at the answer to a problem and then figured out if you were right or wrong. In this algebra course, a method called "Guess and Check" is a first step to mastering the power of algebra and solving problems. The problems you do in class should be well organized and readable. THE ORGANIZATION IS IMPORTANT. It tells you how you checked your guess. Be neat and clear in your work, like the examples you will see in class. Otherwise, your work may not be acceptable, and it will certainly be less useful when we connect the process to algebraic methods. Show four guesses and checks for each problem, even if you get the answer sooner. Copy this example, starting with Step 5, in your tool kit.

GUESS AND CHECK TABLES: AN EXAMPLE

The base of a rectangle is three centimeters more than twice the height. The perimeter is 60 centimeters. Use a Guess and Check table to find the base and height of the rectangle.

STEP 1 Draw a diagram and start a table. Why is the height a good choice to guess?

Guess Height	

STEP 2 Make a Guess.

Guess Height	
10	

STEP 3 Calculate the base.

Guess Height	2(height) + 3 Base
10	

STEP 4 Find the perimeter. Remember, rectangles have two "base" sides and two "height" sides.

Guess Height	2(height) + 3 Base	Perimeter = 2(Height) + 2(Base)
10	2(10) + 3 = 23	2·10 + 2·23 = 66

STEP 5 Check the perimeter against 60 and identify it as correct, too high, or too low.

Guess Height	2(height) + 3 Base	Perimeter = 2(Height) + 2(Base)	Check 60?
10	2(10) + 3 = 23	2·10 + 2·23 = 66	too high

STEP 6 Go back to Step 2 and make a new Guess. Use the previous guess to make a more accurate guess. Repeat this process until you find the solution, using a minimum of four guesses.

STEP 7 Write your answer to the problem in a complete sentence.

Difference of Squares: Organizing Data

Solve each of the following problems by making a Guess and Check table with at least four guesses as modeled in the example above. The table for SQ-46 is started for you.

SQ-46. Find two consecutive odd numbers whose sum is 376.

First Odd Number	Second Odd Number	Sum	Check 376?
101	103	204	too low

SQ-47. One number is five more than a second number. The product of the numbers is 3,300. Find the two numbers.

SQ-48. The perimeter of a triangle is 76 centimeters. The second side is twice as long as the first side. The third side is four centimeters shorter than the second side. How long is each side?

SQ-49. Find four consecutive odd integers such that the sum of the second integer and twice the fourth integer is 65.

SQ-50. Copy the axes at right and put a dot for:

a) Student A, who studies hard but gets only average grades.

b) Student B, who studies little but gets good grades.

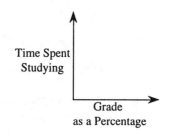

SQ-51. Copy and complete the drawing at right to show -3 + (-1).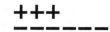

SQ-52. Latisha got a score of 95 today. Calculate her new average. How much did her average improve? Show your work. You may wish to refer back to SQ-41.

SQ-53. Compute:

a) -15 + 7

b) -50 - 30

c) 5.8 + (-6.37)

d) (-2)(-25)

e) (-6 + 17) - 20

f) 0.0001 ÷ 10

g) [-12+ (-18)] - 15

h) (4982)(-556)(0)

i) 0.0001 · 10

SQ-54. Draw a diagram that represents the quantity $\frac{1}{2} \cdot 3 \cdot 5$.

PATTERNS AND SPECIAL NUMBERS

SQ-55. Copy and complete the table.

10^1 10   ten
10^2 100   one hundred
10^3
10^4
10^6
10^9

SQ-56. A scientific calculator is necessary for this problem. Do not attempt it with a graphing calculator.

a) Using a scientific calculator, enter a nine followed by all zeros. Write that number carefully on your paper, with commas. This is called **standard notation**. How many zeros does your number contain?

b) Write its name. For example, 1,000 would be called "one thousand".

c) On paper, multiply this number by two.

d) On the calculator, multiply the original number by two. Copy the display on your paper. Why do you think your calculator displayed the answer that way?

e) Compare the number of digits to the right of the one (1) on your paper with the calculator display. Write down your observations. The table in SQ-55 may help with your conclusion.

f) What is the biggest number your calculator can hold? Compare your result with your team members.

SQ-57. In the pattern at right, each figure is composed of tiles.

a) Copy the first three figures in the pattern and draw the fourth.

figure 1   figure 2   figure 3

b) Copy and complete the table:

figure #	1	2	3	4	5	6	10	23
# of tiles	1	4						

c) Describe with words what pattern you used to fill in the table. Describe any other patterns you see.

d) The sequence you have developed is a famous pattern called "Square Numbers". Why do you think the pattern has that name?

Difference of Squares: Organizing Data

SQ-58. In the pattern at right, each figure is composed of tiles.

a) Copy the first three figures in the pattern and draw the fourth.

figure 1   figure 2   figure 3

b) Copy and complete the table:

figure #	1	2	3	4	5	6	10	23
# of tiles	2	6						

c) Describe with words what pattern you used to fill in the table. Describe any other patterns you see.

d) The sequence you have developed is a famous pattern called "Rectangular Numbers". Why do you think the pattern has that name?

SQ-59. In the pattern at right, each figure is composed of tiles.

a) Copy the first three figures in the pattern and draw the fourth.

figure 1   figure 2   figure 3

b) Copy and complete the table:

figure #	1	2	3	4	5	6	10	23
# of tiles	1	3						

c) Describe with words what pattern you used to fill in the table. Describe any other patterns you see.

d) The sequence you have developed is a famous pattern called "Triangular Numbers." Why do you think the pattern has that name?

SQ-60. Copy the tables below into your tool kit, listing at least ten numbers in each pattern.

FAMOUS PATTERNS										
	1	2	3	4	5	6	7	8	9	10
The Square Numbers:	1	4	9							
The Rectangular Numbers:	2	6	12							
The Triangular Numbers:	1	3	6							

SQ-61. Now examine your tool kit entry with the three famous number patterns. Do you see any relationships between them? Describe what you have found.

SQ-62. Latisha earned an 85 today. Her previous scores were 72, 89, 90, 67, and 95. Calculate her new average.

SQ-63. Use a Guess and Check table to find the solution and state it in a sentence.

Jabari is thinking of three numbers. The greatest is twice the least. The middle is three more than the least. The numbers total 75. Find the numbers.

SQ-64. Use a Guess and Check table to find the solution and state it in a sentence.

The total cost for a chair, a desk, and a lamp is $562. The desk costs four times as much as the lamp and the chair costs $23 less than the desk. Find the cost of the chair and the desk.

INTRODUCTION TO VARIABLES AND COMBINING LIKE TERMS

 UNDERSTANDING THE ROLE OF THE TEACHER

Earlier I mentioned that your teacher would use several methods to help you learn algebra. The idea is to use several approaches to give you more opportunities to understand algebra. In the past, you may have had math classes where the teacher almost always stood in front of the class and gave you instructions on how to complete different math problems.

In other words, the teacher **told you** exactly what to write, what to say, and what to think. This method of teaching leaves little room for you to think and understand. Being told what to do before understanding the reason(s) behind it makes concepts easy to forget. Past experience by students who learned through reasoning and discovery showed that it helped eliminate the tendency to merely memorize information and then see it quickly slip away.

Your teacher will usually give step by step instructions at key points in the course, frequently when tying together several days of class exploration.

<u>**In most cases, your teacher will ask questions designed to help YOU develop the steps yourself.**</u> The teacher will not stand up in front of the class and lecture you very often. Rather, you will usually work with a partner or in study teams and help each other. Therefore, try not to get mad at your teacher when you do not get a step by step explanation to your question. **Your instructor wants you to become a strong thinker, not a tape recorder.** You may get frustrated with this style at first, but over time you will see your confidence and success with math grow. The teacher will make sure your frustration does not become overwhelming. Open your mind and relax. Start thinking about the advantages this approach may offer you.

MISSION POSSIBLE (something to do)

a) Summarize what you can expect from your teacher in this algebra course.

b) Compare and contrast the above description (list similarities and differences) to the way class time has been used so far this year.

SQ-65. You have made or have been provided with sets of tiles of three sizes. We will call these "algebra tiles". Suppose the big square has a side length of x and the small square has a side length of 1. What is the area of:

a) the big square?   b) the rectangle?   c) the small square?

d) Trace one of each of the tiles in your tool kit. Mark the dimensions along the sides, then write the area of each tile in the center of the tile and circle it.

From now on we will name each tile by its area.

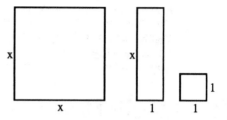

SQ-66. Check your results with your team members.

a) Find the areas of each tile in SQ-65 if x = 4. Find the areas of each tile if x = 6.

b) Why do you think x is called a variable?

SQ-67. Summarize the idea of Combining Like Terms in your tool kit. Then represent the following situations with an algebraic expression.

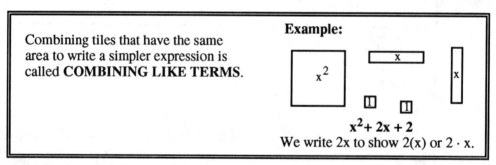

Represent each of the following situations with an algebraic expression.

a)   b)   c) 38 small squares
20 rectangles
5 large squares

d)

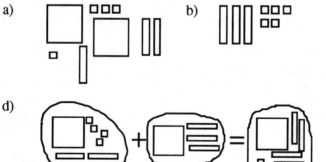

SQ-68. You put your rectangle and two small squares with another pile of three rectangles and five small squares. What is in this new pile?

SQ-69. Suppose one person in your team has two big squares, three rectangles, and one small square on his desk and another person has one big square, five rectangles, and eight small squares on her desk. You decide to put all the tiles together on one desk. Write an algebraic equation that represents this situation. Your equation should look like
(   ) + (   ) = _____ .

SQ-70. You are busy working on a problem with algebra tiles. You have three big squares, five rectangles, and 10 small squares on your desk when your friend leans over and borrows two big squares, two rectangles, and four small squares. Write an algebraic equation that represents the tiles you had, what your friend took, and the tiles you have left.

SQ-71.
Example: To show that 2x does not usually equal x^2, you need two rectangles and one big square.

$2x \neq x^2$

a)   Show that $3x + x \neq 3x^2$.

b)   Show that $2x - x \neq 2$.

SQ-72. Bob, Kris, Janelle, and Pat are in a study team. Bob, Kris, and Janelle have algebra tiles on their desks. Bob has two big squares, four rectangles, and seven small squares; Kris has one big square and five small squares; and Janelle has 10 rectangles and three small squares. Pat's desk is empty. The team decides to put all of the tiles from the three desks onto Pat's desk. Write an algebraic equation that represents this situation.

SQ-73. Another time you were working with one big square, seven rectangles, and six small squares when the teacher walked by and knocked the big square on the floor and the wind blew through and scattered five of the small squares all over the room. Write an equation to represent the tiles you had, what you lost, and the tiles that remained.

Difference of Squares: Organizing Data

SQ-74. Use a Guess and Check table to solve the problem below. Then state your solution in a sentence.

Mairé is thinking of two numbers. When she adds them, she gets 40. When she multiplies them, she gets 351. Help her younger sister, Enya, figure out the numbers.

SQ-75. Latisha is really motivated. She earned a 92 on her test this time.

a) Calculate her grade now. (Recall her past scores were 72, 89, 90, 67, 95, 85.)

b) She has one more grade before the progress report grades are calculated. Figure out what grade she needs to earn to have at least an 85% average. Show your work.

c) Add what you know about **averaging numbers** to your tool kit.

GUESS AND CHECK TABLES: MORE PRACTICE

Make a Guess and Check table to solve problems SQ-76 through SQ-79. Then state your solution in a sentence.

SQ-76. Duke cleared the cash register of nickels and dimes. There were twenty coins in all, and the total value of the coins was $1.35. How many of each type of coin were in the cash register?

The table has been created for you. Copy it and add guesses of your own to solve the problem.

Number of Nickels	Number of Dimes	Value of Nickels	Value of Dimes	Number of Coins	Value of Coins	Check $1.35?
10	10	$0.50	$1.00	20	$1.50	too high

SQ-77. The drama department at Galileo High is having a production. Tickets cost $3 for members of the student body and $5 for anyone else. A total of 515 tickets were sold, bringing in $1785. How many student body members attended?

SQ-78. A cable 84 meters long is cut into two pieces so that one piece is 18 meters longer than the other. Find the length of each piece of cable.

SQ-79. Janelle has only quarters, dimes and pennies in her pocket. There are eight coins in all and their total value is 83 cents.

a) How many of each kind of coin does Janelle have?

b) If she takes one coin out of her pocket, what is the probability that it is a dime?

SQ-80. Latisha's friend Brandee forgot to make up a test and had these scores: 70, 92, 91, 71, 89, 84, 0, 85.

a) Calculate Brandee's average. Does this average score really represent her abilities? Why or why not?

b) Brandee persuaded Ms. Speedi to allow her to make up the missed test. However, it had been awhile since the original test date. Brandee received a 60. Calculate her new average.

c) What difference did the 0 score make? Does this new average better represent Brandee's ability?

SQ-81. Caitlin was minding her own business using her algebra tiles -- one big square, four rectangles, and five small squares -- when Sean came in and added four rectangles and two of the small squares. Write an equation to represent the tiles Caitlin started with, what Sean added, and the tiles Caitlin had after Sean left.

SQ-82. Copy and solve these Diamond Problems:

SQ-83. Use a Guess and Check table to find the solution and state it in a sentence.

The sum of two numbers is 111. The greater exceeds the lesser by 17. What are the two numbers?

SQ-84. Latisha earned an 88 as her last grade before progress reports were calculated. Calculate her average. You may wish to refer to SQ-75. Explain what percent grade Latisha has now.

DIFFERENCE OF SQUARES

SQ-85. **DIFFERENCE OF TWO SQUARES**

In this unit you have organized data, described patterns with diagrams and words, and talked and worked together in a study team. You now have a problem to investigate that will require all of those skills. This is a big problem that should need the entire team's effort in order to complete it successfully.

a) Study the examples in the table.

NUMBER	SOLUTIONS (Difference of Squares)
8	$3^2 - 1^2$
15	$8^2 - 7^2$; $4^2 - 1^2$

b) Express the numbers 1 through 25 as a difference of two squares. Some numbers have more than one solution as shown with the number 15. Some numbers might have no solution. Find as many solutions for each number as your team can and organize them neatly and clearly on a chart.

c) Describe as many patterns as you can find.

d) Write an explanation of how your study team went about solving this problem. How did you start? Did someone find a technique that helped you find more solutions? Did you use patterns or guess and check to help? How did you decide what your chart would look like?

e) Prepare a presentation to share your results with the class. Everyone in your study team must be part of the presentation. Your presentation should include an introduction, your chart(s), explanations of your patterns, and an explanation of how your team solved the problem. (Hint: it is important for the class to be able to read your results and see your patterns as you speak. Therefore be aware that your chart needs to be written large enough to be seen easily by everyone.)

Make sure each person in your study team understands what he or she needs to complete before coming to the next class. Your team will need to present your results then.

f) Extension: Can you describe a way to find solutions for any given number? For example, can you write 43, 99, or 60 as a difference of two squares? If so, explain how you found this general solution and include it in your presentation.

SQ-86. Compute:

a) $-3 + 8$

b) $-12 - 8$

c) $17 + (-19)$

d) $(-2)(-12)$

e) $\frac{1}{3}(-6)$

f) $(-6)(14)$

g) $-4 + 9 - 11$

h) $(-1)(-1)(-1)$

i) $-15 + 15$

SQ-87. You were minding your own business using your algebra tiles: two big squares, four rectangles, and three small squares. Tom came in and took the four rectangles and two of the small squares. Write an equation to represent the tiles you started with, what Tom took, and the tiles you had after Tom left.

SQ-88. Copy and complete the drawing at right to show -3 + (-4).   +++ ------

SQ-89. Make a Guess and Check table to solve and state your solution in a sentence.

Todd is 10 years older than Jamal. The sum of their ages is 64. How old are Todd and Jamal?

SQ-90. Write a team and learning reflection for Unit One. Use complete sentences and write as honestly as you can. Use the questions below as a guide. Include any other information you like.

a) **Study Team Reflection:** Who was in your study team and how did your team work? Be as detailed as possible. What exactly did you do to help someone? Who helped you and how did they do so? What talents did you find you had? Describe a different talent someone else had. What was disappointing about your team? What are important team skills that help a team work effectively? What team skills do you personally want to work on in your next team?

b) **Learning Reflection:** What part of your learning do you feel good about? What is still difficult for you? What learning are you proudest of? Did you do all your assignments? Did you participate in class discussions? What study skill do you want to improve in the next unit? Describe your plan to do so.

PZL-9   **REMINDERS**:   WHERE TO GET INFORMATION and
                          WHAT TO DO IF YOU FALL BEHIND

Before you start the next problem, go back and review PZL-6. Use the list you made to help assemble a COMPLETE took kit for Unit 1 as directed in the next problem.

By the way, some of you may have had a hard time getting started with this course and are beginning to fall behind. I have good news for you! The information contained in PZL-6 tells you where to look to catch up quickly. **Now I have to warn you: the number one rule for studying CPM math is DO NOT GET BEHIND!** You will have to study every day. In particular, homework is an essential part of your studies. This means that if you are behind, **use your time wisely**. Do not spend hours doing all the problems when you can choose a select few to get the main ideas. After you understand all the main ideas you can go back for more practice if you have time. Be honest about your level of understanding. If you understand, move on. If you do not, go back and practice.

Difference of Squares: Organizing Data

SQ-91. **TOOL KIT CHECK-UP**

Your tool kit contains reference tools for algebra. Return to your tool kit entries. You may need to revise or add entries.

Be sure that your tool kit contains entries for all of the items listed below. Add any topics that are missing to your tool kit NOW, as well as any other items that will help you in your study of algebra.

- Definition of Probability
- Diamond Problems
- Integer Multiplication and Division
- Famous Patterns
- Averaging Numbers

- Area and Perimeter of a Rectangle
- Addition and Subtraction of Integers
- Guess and Check
- Algebra Tiles with Labels
- Combining Like Terms

To The Students

Gold Medal problems present you with an opportunity to investigate complex, interesting problems over several days. The purpose is to focus on the process of solving complex problems. **You will be evaluated on your ability to show, explain, and justify your work and thoughts.** Save **all** your work, including what does not work in order to write about the processes you used to reach your answer.

Completion of a Gold Medal Problem includes four parts:

- **Problem Statement**: State the problem clearly in your own words so that anyone reading your paper will understand the problem you intend to solve.

- **Process and Solutions**: Describe in detail your thinking and reasoning as you worked from start to finish. Explain your solution and how you know it is correct. Add diagrams when it helps your explanation. Include things that did not work and changes you made along the way. If you did not complete this problem, describe what you <u>do</u> know and where and why you are stuck.

- **Reflection**: Reflect on your learning and your reaction to the problem. What mathematics did you learn from it? What did you learn about your math problem solving strategies? Is this problem similar to any other problems you have done before? If yes, how?

- **Attached work**: Include <u>all</u> your work and notes. Your scratch work is important because it is a record of your thinking. Do not throw anything away.

GM-1. **GOING BANANAS!**

Cleopatra ("Cleo") the Camel works for the owner of a small, remote banana plantation. This year's harvest consists of three thousand bananas. Cleo can carry up to one thousand bananas at a time. The market place where the bananas are sold is one thousand miles away. Unfortunately, Cleo eats one banana each and every mile she walks.

Your Task:

Of the three thousand bananas harvested, what is the largest number of bananas Cleo can get to market?

P.S. This problem is not impossible!

GM-2. **RUMORS, RUMORS, RUMORS!**

Part 1

Burton High School has 1,500 students. During first period, a rumor is started when Susan tells three friends a secret. Each of Susan's three friends tell three of their friends during second period, who in turn tell three different friends during third period. Assume each person tells the rumor to only three others.

Part 2

In an attempt to discredit the President, a member of the opposite political party wishes to start a scandalous rumor about him. He plans to tell people this rumor one week before election day and wants the entire population of the United States (that's 250,000,000 people!) to know the rumor. He needs to figure out how many people he should tell (and they each will tell the same number of people) so that everyone will have heard the rumor by the seventh day.

Your Task:

- Find the number of students who have heard the rumor at the end of the day. (Burton has 6 periods.)

- Find the number of people the evil politician needs to tell so everyone in the United States will have heard the rumor by the seventh day. Remember: each person needs to tell the same number of people.

Unit 2
The Kitchen Floor
Area and Subproblems

Unit 2 Objectives
Tiling the Kitchen Floor: AREA AND SUBPROBLEMS

This unit will use geometry as an introduction to the problem-solving technique of using **subproblems** (i.e., breaking problems into smaller parts). The ability to identify and solve subproblems will be one of the most useful skills you learn all year.

Algebra tiles will continue to be a handy tool for figuring out many problems in algebra. We will use them to review the order of operations, find the area and the perimeter of complicated figures, as well as investigate the Distributive Property.

In this unit we will continue to organize data, develop arithmetic skills, and use problem solving strategies to develop ideas and find solutions to problems.

In this unit you will have the opportunity to:

- learn how to break large problems into smaller parts that you know how to solve.
- review and consolidate the Order of Operations with integers, decimals, and fractions.
- develop area formulas for basic geometric figures and understand their origins.
- continue working with Algebra Tiles to learn the Distributive Property.
- explore your scientific calculator to manage large numbers.

Read the following problem carefully. **Do not try to solve it now**. During this unit, you will gain many new skills to help you solve it.

KF-0.   TILING THE KITCHEN FLOOR

The kitchen in Susan's apartment has an ugly floor covering and she wants to replace it with various sized ceramic tiles. There will be a rectangular design made of 3" square tiles set in one foot from the edge of the kitchen. You will need to find the number of tiles needed for this rectangle, as well as the square footage of the kitchen.

Graph paper and a scientific calculator will be used <u>daily</u>. Be sure to have the supplies that you need.

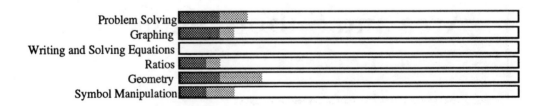

Unit 2
Tiling the Kitchen Floor: **AREA AND SUBPROBLEMS**

 HOW TO PREVIEW A UNIT

Your literature or social studies teacher may have suggested that before you read a story or a passage in your history text, you skim it quickly, looking for main ideas. You can do the same thing with your algebra textbook by using the ways the authors have highlighted important information. Take a few minutes--no more than five--to read the list of what you will learn in this unit (see previous page), then flip through the unit and skim each single and double line box. Finally, each person in the team should take a topic in bold print from one of the boxes, look it up in the index, and verify that the problem number with the box matches the problem number listed in the index. Once you have done all of this--remember, no more than five minutes, so work quickly--go on with today's lesson.

KF-1. THE PAINT JOB: INTRODUCTION TO SUBPROBLEMS

The inside of Susan's apartment needs a new coat of paint. She called a few contractors in her town and the lowest bid was $670. She noticed a paint sale at Home Emporium and wondered if she could save money by painting the walls and ceilings herself. She drew the floor plan shown below (which was easy since each wall measurement was a multiple of five feet) and measured the height of the ceiling (it was 8 feet tall). She will not paint the doors. Use the data with the scale drawing to determine whether Susan should hire the contractor or paint the apartment herself.

SCALE: —— 5 feet

Price of Paint:

$24.95 per gallon for approx. 300 sq. ft.

Tiling the Kitchen Floor: Area and Subproblems

KF-2. Describe the strategy your team used to find the total area of KF-1. Use complete sentences to list and describe the all steps you used.

KF-3. Write a description of how your team worked together to solve KF-1. How did your team collaborate? What did you contribute? How did your team help you? What would your team do differently next time you do an involved problem like this?

KF-4. While she's improving her apartment, Susan has decided to carpet her bedrooms. Home Improvement Emporium sells carpet at $20 per square yard. Roughly how much will Susan pay to carpet her bedrooms?

KF-5. The picture at right is a simple diagram of the apartment of Susan's friend, Randy. All numbers represent lengths in feet. Use the picture and answer the following questions:

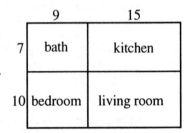

a) What are the dimensions of his living room?

b) Randy's friends are going to visit soon. He wants to keep them out of his bedroom. Make a list of the individual room areas he will have to clean.

c) What is the total area of the rooms he will have to clean?

d) Add this definition to your tool kit:

SUBPROBLEMS

Calculating the individual areas of each room is an example of finding **subproblems**. You need to solve smaller subproblems in order to answer a larger question such as part (c) above.

Breaking problems into smaller parts (subproblems) that you know how to solve is the first part of using this problem solving strategy. The second part of using **SUBPROBLEMS** is putting the results of the smaller problems together to answer the larger, often more complicated, question or problem.

KF-6. Approximate the area of each of the shaded regions below.

a)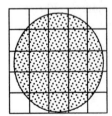

b)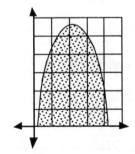

54   UNIT 2

KF-7. Make a Guess and Check table to solve the following problem. Write you answer as a complete sentence.

The length of a rug is one foot less than twice the width. If it takes 34 feet of fringes to wrap around its perimeter, find the length and width of the rug.

KF-8. Copy and simplify the following expressions by combining like terms. Using or drawing Algebra Tiles may be helpful.

a) $2x + 3x - 3 + 4x^2 + 10 - x$

b) $2x^2 + 30 - 3x^2 + 4x^2 - 14 - x$

c) $4x + 4x^2 - x^2 - 9 + 10 - x + 3x$

d) $20 + 3x - 3x + 4x^2 + 10 - x^2$

KF-9. Without a calculator, compute:

a) $-14 + (-31)$

b) $\dfrac{-16}{-8}$

c) $\dfrac{1}{2} - \dfrac{3}{4}$

d) $-(-8) - (-2)$

e) $-11 \cdot (24)$

f) $46 \div (-23)$

KF-10. CALCULATOR CHECK

Use your scientific calculator to compute the following and match the result to an answer. You may want to use a key marked $\boxed{+/-}$. This key changes the sign of a number.

a) $-3 + 16 - (-5)$ 1) -16

b) $(3 - 5)(6 + 2)$ 2) 327

c) $17(-23) + 2$ 3) 0.5

d) $5 - (3 - 17)(-2 + 25)$ 4) 18

e) $(-4)(-2.25)(-10)$ 5) -90

f) $-1.5 - 2.25 - (-4.5)$ 6) 0.75

g) $\dfrac{4 - 5}{-2}$ 7) -389

Tiling the Kitchen Floor: Area and Subproblems

AREA OF TRIANGLES AND SUBPROBLEMS

 PZL-11   HOW TO ASK YOUR TEACHER QUESTIONS

In Unit 1 I talked about the many ways you would learn algebra this year. Notice that I did not mention the teacher as the first resource. In CPM you should not need to go to your teacher first. The teacher is one resource, but you have some resources of your own. If, after checking your own resources, you need to ask the teacher for assistance, here is how to get the most out of your teacher. **BE PREPARED**. When you ask the teacher a question, make it a good one. Do not ask, "Can you help me?" **BE SPECIFIC**. A better question will come after you work on a problem with your partner or team and use the resources I described previously. Ask questions like, "Mrs. Cho, I am having a problem with SQ-79 part (a). My study team is having the same problem. We have tried using variables and someone thought about Guess and Check tables, but we are not sure where to go next. What else could we try?"

Your teacher will probably respond to your question by asking you a leading question. A leading question is one that makes you think, but at the same time gives you a hint. **Pay close attention to your teacher's questions. They contain hints and sometimes information** that you can use to move forward.

As you work on today's and future lessons, try to improve the quality of the questions you ask your teacher. The more specific you are about what confuses you, the better questions and hints you will get in return. Now start your work on KF-11.

KF-11.   The countertop in Susan's kitchen is shown at right. Find the amount of workspace (area) she has. Label each step (subproblem) in your solution.

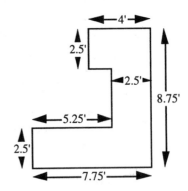

KF-12.   Compute the area of each figure below. With your team. Then describe two ways (using different subproblems) to find each area.

a)   b) 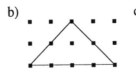   c) 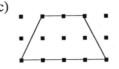   d)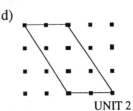

UNIT 2

KF-13. At Home-Floor-and-Tile, Susan encountered many different tile shapes and sizes. Several were not labeled with their prices. She did know that the triangular tile at right cost 75¢.

Assuming that the price per square inch is the same, determine the price of the tiles below. Explain how you found each amount.

a) 3" □ 3"   b) 6" □ 6"   c) 6" △ 6"

KF-14. AREA OF A TRIANGLE INVESTIGATION

You will be working with your study team to discover more about triangles. Each member of your study team needs scissors and each team needs one resource page.

a) How do you find the area of a rectangle? Does this work for all rectangles?

b) Examine the triangle at right. What fraction of the rectangle is △ABE?

c) On the resource page provided by your teacher, △ABC lies inside of rectangle ABDE in each example. Examine each rectangle with your study team. In each case, what fraction of the rectangle is the triangle?

d) Each member of your team should choose one of the figures on the resource page. Using scissors, cut out triangle ABC. Using all three pieces, compare the areas and test your idea from part (c).

e) If the area of a rectangle can be found by multiplying base times height, then how can you find the area of a triangle? Enter this formula, found in the book on the next page, and an explanation of how to find the area of a triangle in your tool kit.

>>Information to complete part (e) is on the next page.>>

Tiling the Kitchen Floor: Area and Subproblems

> To find the **AREA OF A TRIANGLE**, multiply the base and the height and divide by two.
>
> $$\text{Area} = \frac{1}{2} \text{base} \cdot \text{height} = \frac{\text{base} \cdot \text{height}}{2}$$
>
>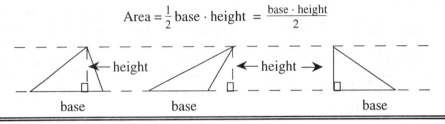

KF-15.   Construct this figure on graph paper, a geoboard, or dot paper: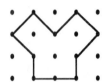

   a)   Separate the figure into squares, triangles, and rectangles and find the area of each. These are subproblems.

   b)   What is the area of the entire figure?

KF-16.   Susan wants to buy tulips for her flower garden. The planting manual suggests that tulips need $\frac{1}{4}$ sq. feet of space to grow properly. If her flower garden has 82 square feet, how many tulips should she buy?

KF-17.   CALCULATOR CHECK

Use your scientific calculator to compute the following and match the result to an answer. A fraction button marked $\boxed{a^b/c}$ can be useful when doing arithmetic with fractions. Consult your instruction manual for your calculator, or tomorrow ask your study team if you do not know how to use it.

   a) $\frac{(6)(1.2)}{2}$ 1)   0.375

   b) $\frac{2}{3}(-6) + 4$ 2)   2

   c) $\frac{3+5}{4}$ 3)   -1

   d) $\frac{1}{4}(5.25 - 3.75)$ 4) $7\frac{1}{2}$

   e) $-2(\frac{2}{3}) - 4$ 5)   3.6

   f) $\frac{1}{2} \cdot 3 \cdot 5$ 6) $-5\frac{1}{3}$

   g) $\frac{-6 - (-3)}{3}$ 7)   0

KF-18.  Make a Guess and Check table to solve the following problem. Write your answer in a complete sentence.

Susan is buying three different colors of tiles for her kitchen floor. She is buying 25 more red tiles than beige tiles, and three times as many navy blue tiles than beige tiles. If Susan is buying 435 tiles in total, how many tiles of each color did she buy?

KF-19. Carefully examine the area of the rectangles below. Be sure to explain your answers clearly.

a) Make a diagram and explain why you can use Figure C to find the sum of the areas of Figures A and B.

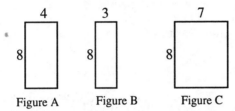

b) Sketch a different rectangle with the same area as the square at right. Label the sides appropriately. How does the rectangle's perimeter compare to the square?

Tiling the Kitchen Floor: Area and Subproblems

KF-20. Each of the four rectangles below has an area of 36 square units. Examine their dimensions (length and width) and answer the following questions.

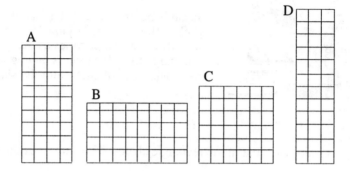

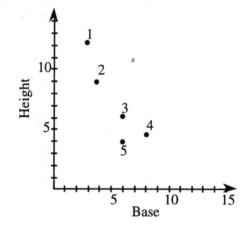

a) Copy the graph on your paper. Then match four of the points on the graph with the letters A, B, C, and D to correspond with the rectangles above.

b) Which point at right you leave unlabeled?

c) What is the base, height and area for that point?

d) Why should that point not be on this graph?

e) Use your calculator to find the dimensions of three more rectangles with areas of 36 square units. At least one of your examples should have dimensions which are not integers. Place the points corresponding to the three rectangles you found on the graph.

f) Connect all the points with an area of 36 and describe the resulting shape.

KF-21. Randy is ordering new carpet (for the apartment shown in KF-5). If he does not want to carpet the kitchen or the bathroom, find the area of carpet he will need to order. How much will his order cost if carpet is $3.50 per square foot? List the subproblems you use and solve them.

AREA AND CIRCUMFERENCE OF CIRCLES

KF-22. In your own words, add these definitions to your tool kit:

> A **CIRCLE** is the set of all points that are the same distance from a fixed point, G. The fixed point is called the **CENTER** of the circle and the distance from the center to the points on the circle is called the **RADIUS** (usually denoted r). A line segment drawn through the center of the circle with both endpoints on the circle is called a **DIAMETER** (denoted d). Note: d = 2r.
>
> You can think of a circle as the rim of a bicycle wheel. The center of the circle is the hub where the wheel is bolted to the bicycle's frame. The radius is a spoke of the wheel.

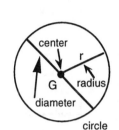

KF-23. Copy the table at right.

Item	Diameter	Circumference	Ratio

For each of the circular items provided by your teacher:

a) Use a ruler to measure the diameter of the circle. Enter this information in the table.

b) Use the string provided by your teacher to measure the **circumference** (the perimeter of a circle). Enter this information in the table.

c) In the last column of your table, calculate the value of the $\frac{\text{circumference}}{\text{diameter}}$.

d) Work with the other teams in your class to find a class average for part (c).

KF-24. Write a formula to find the circumference of a circle. Explain how to find the circumference of a circle if you know the diameter or radius.

Tiling the Kitchen Floor: Area and Subproblems

KF-25. You will need scissors and a resource page for this investigation. We will simulate a rectangle using a circle by following the steps below:

> **TO CREATE OUR "RECTANGLE"**
> - Cut the circle out of the resource page. Then cut this circle in half by cutting long the solid line.
> - With each semicircle, carefully cut on the dotted lines from the center to roughly $\frac{1}{2}$ cm to the outer edge. Be sure to not cut the entire length of the radius.
> - Open the semicircles and fit the "teeth" together. Keep these pieces together by gluing them down on another piece of paper.

a) Use your formula from KF-24 to determine the base of our shape.

b) Determine the height of the shape.

c) If our circle had been cut into smaller sections, called **sectors**, our shape would closely resemble a rectangle. Use the formula for the area of a rectangle to find the area of our new shape.

KF-26. Use a complete sentence to describe how to find the area of a circle given its radius or diameter.

KF-27. Roses need a lot of area to be healthy, and Susan has two different pots in which to plant hers. One is a circular pot with a radius of 8 inches, and the other is a flower box with a length 12 inches and width 16 inches. Which pot would give the roses the most surface area?

KF-28. The table in Susan's dining room is shaped like an octagon and has measurements as shown at right. All given measurements are in feet. Find the area of its surface. Clearly show each subproblem you use in your solution.

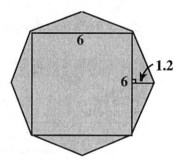

KF-29. Shatara is trying to picture a triangle with an area of 18 square units. Help her by using graph paper to draw at least five different triangles with area 18 square units.

KF-30. Add this definition and example to your tool kit:

> One type of **SUBSTITUTION** is replacing a variable with a number.
>
> **Example:** For the expression 2x + 4, when x = 3, substituting for x gives you:
> 2(3) + 4 = 10

KF-31. Use substitution to solve for y. Show your work like the example above.

a) y = 3x - 5, when x = 4.

b) y = 4 - 9x, when x = -3.

c) y - 2x = 5, when x = -4.

d) y + 6 = 2x - 4, when x = 9.

KF-32. Simplify the following expressions by combining like terms, if possible.

a) $x + x - 3 + 4x^2 + 2x - x$

b) $8x^2 + 3x - 13x^2 + 10x^2 - 25x - x$

c) $4x + 3c$

d) $20 + 3c - 3 + 4c^2 + 10 - 2c^2$

CIRCLES AND SUBPROBLEMS

KF-33. Add these definitions to your tool kit:

> The **CIRCUMFERENCE** of a circle (C) is its perimeter, or distance around the outside of the circle. Length is <u>one</u> dimensional. We can calculate the circumference using the formula $C = 2\pi r$ or $C = \pi d$, where "r" is the radius and "d" is the diameter.
>
> The **AREA** of a circle (A) is the amount of space the circular region covers. Unlike length, area is <u>two</u> dimensional. It can be found using $A = \pi r^2$.
>
> As a decimal, π never ends and never repeats. The first twenty-eight digits of π are:
> **3.1415926535897932384626433383.**
>
> It is impossible to use all the digits of pi in calculations (since the digits go on forever and it cannot be written as a fraction), so often 3.14 or $\frac{22}{7}$ are used as an <u>approximation</u> of π. Unless you are told otherwise in a problem or by your teacher, use the π key on your calculator.

KF-34. The box above contains two formulas for the circumference of a circle: $2\pi r$ and πd. Are they the same? Explain.

KF-35. In the kitchen, Susan's stove is 36 inches wide by 30 inches deep. She has 4 burners, each with a diameter of 10 inches. Find the area of the stovetop around the burners (shaded in the diagram). Show and label all your subproblems.

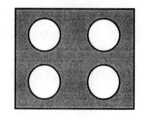

KF-36. North High School will have its first graduating class in the year 2001. To celebrate, it has designed watches with a gold leaf sector between the 12 and the 1 as shown at right.

Next year's graduating class will have gold between the 12 and the 2 as shown. Each of the first twelve graduating classes will have a similar design. This way, when graduates meet other alumni many years from now, they will instantly be able to determine which year they graduated.

As student body president, Susan is concerned with the cost of the watches and needs to tell the jeweler the amount of gold each watch will require. If the radius of the face of the watch is 2 cm, help Susan out by making a table with the areas of gold leaf for each of the first 12 graduating classes.

KF-37. A 1994 Almanac lists the diameter of the Earth to be 7,926 miles. Find the length of the Equator, rounded to the hundreds place.

KF-38.  If telephone poles (18 feet tall) suspend a telephone wire off the ground along the equator, how much longer than the equator would the wire be? (Ignore the sag between poles.)

UNIT 2

KF-39. Examine the tiles representing the following problems. Summarize why tiles are eliminated in some problems and not in others.

a) -3 + -1 = -4

b) -2 + 5 = 3

KF-40. After a party, there were still two pieces of pizza left! The original pizza, with a diameter of 12 inches, was evenly cut into 16 pieces. Find the area of the leftover pizza.

KF-41. Make a Guess and Check table to solve the following problem. Write your solution as a sentence.

The length of a rectangle is six more than the width. If the perimeter is 160, find the length and width of the rectangle.

KF-42. You have two dice.

a) If you roll one die, how many outcomes are possible?

b) If you roll one die, what is the probability of getting a 5 ?

c) If you roll both dice, how many outcomes are possible? (Hint: the answer is not 12. Refer to Unit 0, problem GS-21, Roll and Win.)

d) If you roll both dice, what is the probability of getting a sum of 5 ?

KF-43. Copy the following figures and divide them into parts that will help you to find the total area of each figure. Show the subproblems you use to find each area.

a)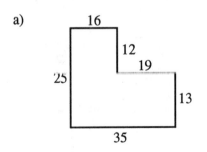

b)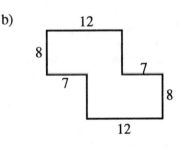

KF-44. Look for special keys $\boxed{\pi}$ and $\boxed{x^2}$ on your calculator that make these calculations easier. Find the area and circumference of a circle, to the nearest hundredth, with:

a) radius of 5

b) radius of 3.75

c) diameter of 8. Be careful! What subproblem must you do first?

KF-45. SUBPROBLEMS IN APPLICATIONS: THE BROKEN WINDOW

The window in Susan's living room needs new glass. The glass shop prices depend on the area of each pane of glass, and their price chart is shown below:

Area	Price	Area	Price
0 - 399 cm²	$25	1000 - 1999 cm²	$139
400 - 599 cm²	$49	2000 - 2499 cm²	$219
600 - 799 cm²	$69	2500 - 2999 cm²	$299
800 - 999 cm²	$99	3000 cm² and up	$0.10 per cm²

Use the design of her window to determine the cost of replacing all panes of glass. Assume that the four panes of glass making up the outer circular region are equal in area.

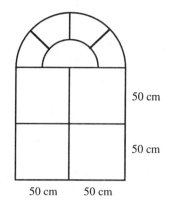

KF-46. How much money would Susan save if she replaced the entire window with one solid piece of glass?

KF-47. Find the perimeter and area of Susan's swimming pool shown in the diagram at right. Be sure to show and label your subproblems.

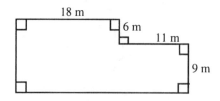

KF-48. If the area of a rectangle is 12, what could the base and height be? Find two different possibilities.

a) If the base is 24, what is the height?

b) Generally, if you know the area and base of a rectangle, how can you determine the height?

KF-49. Copy and solve these Diamond Problems:

   a)   b)   c)   d)

KF-50. Evaluate each of the following expressions when $x = -3$ and then again when $x = \frac{1}{2}$. Show all of your subproblems. Each problem has two answers.

a) $4x + x^2$

b) $3(x + 3)^2$

c) $x^2 - 3x + 5(-2x + 3)$

KF-51. A bag of Algebra Tiles has two big squares, twelve rectangles and twenty small squares. Susan and her partner split the tiles between them. Write an expression to represent how many tiles Susan receives. Describe how you determined your answer.

KF-52. Combine the following integers without a calculator. Check your result with your calculator.

a) $-16 + 42$

b) $-8 - 15$

c) $-22 - (-15)$

d) $41 + 13$

GROUPING AND THE DISTRIBUTIVE PROPERTY

MULTIPLYING WITH ALGEBRA TILES

Example: The dimensions of this rectangle are

x by 2x

Since two large squares cover the area, the area is $2x^2$.

We can write the area as a multiplication problem using its dimensions:

$$x(2x) = 2x^2$$

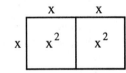

KF-53. Use the figure at right to answer these questions.

a) What are the dimensions of the rectangle?

b) What is the area of this rectangle?

c) Write the area as a multiplication problem.

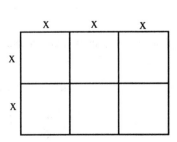

Tiling the Kitchen Floor: Area and Subproblems

KF-54. Use what you learned in the previous problem to multiply the following expressions.

a) (2x)(5x)

c) 3(2x)

b) (6x)(4x)

d) c(3c)

KF-55. GROUPING WITH ALGEBRA TILES

In order to develop good algebraic skills, we must first establish how to work with our Algebra Tiles. When we group rectangles and small squares together, as in the examples below, we read and write the number of rows first, and the contents of the row second. "3x" means three rows of x.

For example, all three figures below contain three rectangles and twelve small squares. The total area is 3x + 12, as shown in Figure A. In Figure B, the rectangles are grouped, forming 3 rows of x, written as 3(x). Three rows of four small squares are also a group, written as 3(4).

In Figure C, notice that each row contains a rectangle and four small squares, (x + 4). Since three of these rows are represented, we write this as 3(x + 4).

Figure A	Figure B	Figure C
3x + 12	3(x) + 3(4)	3(x + 4)
Total Area	3 rows of x and 3 rows of 4	3 rows of (x + 4)

Write down your observations of the different ways to group 3x + 12.

KF-56. Match each geometric figure below with an algebraic expression that describes it. Note: "3 · x" means "3 times x" and is often written 3x. This represents 3 rows of x.

a)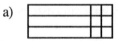

b)

c)

d)

e)

f)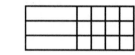

1. 2(x) + 2(2)
2. 3(x + 2)
3. 5(x + 1)
4. 3(x) + 3(1)
5. 2(x + 5)
6. 3(x) + 3(4)

UNIT 2

KF-57. Sketch the geometric figure represented by each of the algebraic expressions below.

   a) 4(x + 3)

   b) 4(x) + 4(3)

   c) Compare the diagrams. How do their areas compare? Write an algebraic equation that states this relationship. This relationship is known as **The Distributive Property.**

KF-58. Use the Distributive Property (from KF-57) to rewrite the following expressions. Use Algebra Tiles if necessary.

   a) 6(x + 2)

   c) 2(3x + 1)

   b) 3(x + 4)

   d) 5(x - 3)

KF-59. The picture at right has a rectangle inside a square. If the shaded region has an area of 44 sq. units, find x.

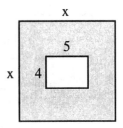

KF-60. Use a Guess and Check table to solve this problem.

In the football game, Rocky gained 3 times as many yards as Bullwinkle. Rocky also gained 10 yards more than Boris. The 3 players gained a total of 410 yards. How many yards did Boris gain?

KF-61.  Today is Home Emporium's 25th anniversary and their entire stock is 15% off. Susan approached the sales clerk with a large purchase of tiles, paint, and other odds and ends. The sales clerk started to mark 15% off of each price. Susan suggested that the sales clerk instead add up the entire order and then take off 15%. Which method will correctly compute the final price, and why?

Tiling the Kitchen Floor: Area and Subproblems

KF-62. For a class party, four 3' by 8' tables were assembled as shown at right.

a) In order to buy tablecloths, Susan needs the total surface area of the tables. Find the total area to be covered.

b) If the class is seated one student per two feet along the outside and inside of this configuration, how many students can be accommodated?

KF-63. Translate the following sentences into an expression with numbers, then find the net result.

a) Jasmine borrowed $20 dollars from her parents. She spent $12, received a paycheck for $55 and repaid her parents. What is her new balance?

b) Caesar walks to school each day. This morning, he left home and walked 9 blocks. Then he realized he forgot his Calculus homework. He turned around and walked back 4 blocks but then discovered his homework in his back pocket. He then turned around again and walked 11 blocks to school. How far away from school does Caesar live?

c) Mr. Dickinson refills the soda machine several times a day. One morning, after finding only 39 sodas in the machine, he added four 6-packs. After 32 sodas were sold, he added nine 6-packs. How many sodas are in the machine now?

KF-64. Simplify the following expressions by combining like terms, if possible.

a) $2x^2 - 5 + 2x + 10 - x^2$

b) $5x^2 + 3x - 6x^2 + 4x^2 - 3x - 1$

c) $14 + 7c - 2 + 4c^2 + 9 - 18c^2$

d) $6 + 2x$

KF-65. Ms. Speedi is going to have jewels, which cost $100 apiece, set along the outer border of the face of her watch. The jeweler explained that she could place two jewels per centimeter, but to estimate the price of the jewels, she would need to know the circumference of the watch. The radius of the face of the watch is 2 cm. Help Ms. Speedi by finding the approximate cost of the jewels (including 8% tax). Show all your subproblems.

DEVELOPING THE DISTRIBUTIVE PROPERTY

KF-66. Draw or use Algebra Tiles to represent the following expressions. Find the products.
a) (2)(2x) b) (2x)(2x) c) (2)(3x + 1)

KF-67. Use the Distributive Property to match algebraic expressions in the left column with equivalent expressions in the right column. Note that at least one of the expressions on the left cannot be matched with an expression on the right.

a) 4(x + 5)   1) x(x + 3)
b) 2(x + 3)   2) -4(x) - 4(2)
c) 5(2x) + 5(4)   3) (3 + x) · 3
d) x(x) + x(3)   4) 4(x) + 4(5)
e) 2x(x + 4)   5) 5(x + 4)
f) 3(3) + 3(x)   6) 2(x) + 2(3)
g) -4(x + 2)   7) 2x(x) + 2x(4)

KF-68.   THE GAME SHOW

Susan has had an incredible streak of good fortune as a guest on the exciting game show, "THE MATH IS RIGHT." So far, she has amassed winnings of $12,500, a sports car, two round-trip airline tickets and five pieces of furniture.

In an amazing finish, Susan landed on a "Double Your Prize" square and answered the corresponding math question correctly! She instantly became the show's biggest winner ever.

A week later, $25,000, a sports car, four round-trip airline tickets and five pieces of furniture arrived at her house. Susan feels cheated. What is wrong?

KF-69. How does Susan's game show experience relate to the Distributive Property?

Tiling the Kitchen Floor: Area and Subproblems

KF-70. Use the Distributive Property to rewrite each expression in parts (a) through (d).

a) 7(n + 3)
b) 4(g - 5)
c) p(p + 8)
d) 2x(3x + 4)
e) Reverse the process: 4x + 8 = 4(__ + __)
f) $5x^2 + 10x = 5x($__ + __$)$
g) 4x + 5x = __(__ + __)

KF-71. Ms. Speedi used the Distributive Property on a problem and got an answer of 4x - 8. When Janelle and Rhonda tried to figure out the original question, they got two different answers. Work with your study team to figure out what answers they got and why there is more than one answer.

KF-72. Use the figure at right to answer the questions that follow.

a) What is the area of the <u>unshaded</u> region? (We will call the unshaded region a "three-step" region.)

1) What is the area of the entire figure?
2) Write the unshaded area as a fraction of the whole figure.
3) Write this fraction as a percent.

b) Draw a four-step and five-step region and find the area of the unshaded regions.

c) Copy and complete the table below. Look for patterns.

Number of steps	Base	Height	Area of unshaded region
3-step figure	4	3	6
4-step figure			
5-step figure			

d) Without drawing the figure, find the unshaded area of a ten-step region. Explain how you found your answer.

e) **Extension:** Find the unshaded area of the region that has 157 steps.

KF-73. We are using subproblems in this unit to make complex problems more manageable. You may not be aware that you have used subproblems before. In the last unit you described the fraction $\frac{2}{3}$. Fractions allow you to represent the parts of a whole. In the past, you have been asked to add fractions. Let's examine the subproblems involved in adding fractions. Consider the sum of $\frac{2}{3} + \frac{1}{4}$. (You may want to add 'sum' to your tool kit -- the sum is the result of combining by adding.)

a) What must you do first to add these fractions?

$\frac{2}{3} + \frac{1}{4} =$

b) What is the common denominator?

c) The first subproblem is to find the common denominator and then rewrite each of the fractions with the same denominator. Show how to do this.

d) Now add the fractions.

e) Calculator Check: Find the fraction key on your calculator. Use your calculator to add the fractions. Does your result match your answer from part (d)?

KF-74. Show two methods of finding the area of the rectangles below. Be sure to write down and label your subproblems.

a) 7 9
    3

b) 15 20
    8

KF-75. Use the Distributive Property to match algebraic expressions in the left-hand column with equivalent expressions in the right-hand column. Not all expressions on the left can be matched with an expression on the right. The expressions in the right column have been simplified.

a) 4(x + 7)   1) $x^2 + xy$

b) x(x + 1)   2) $x^2 + x$

c) 2(x + 9)   3) $y^2 - 3y$

d) x(x + y)   4) 4x + 28

e) 3(2x + 5)   5) 10xy

f) y(y - 3)   6) 6x + 15

g) 10(x + y)   7) 2x + 18

Tiling the Kitchen Floor: Area and Subproblems

KF-76. Find the area of the figure at right.

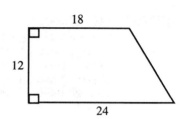

KF-77. We will use letters to represent variables for unknown numbers throughout this course. Use the correct order of operations to find the value of each part below. Show your substitution and simplifying steps.

a) For $y = 2 + 3x$, when $x = 4$, what does y equal?

b) For $a = 4 - 5c$ when $c = -\frac{1}{2}$, what does a equal?

c) For $n = 3d^2 - 1$ when $d = -5$, what does n equal?

d) For $v = -4(r - 2)$ when $r = -1$, what does v equal?

e) For $3 + k = t$, when $t = 14$, solve for k.

USING THE DISTRIBUTIVE PROPERTY

KF-78. In your own words, add the following information to your tool kit.

> ### THE DISTRIBUTIVE PROPERTY
>
> With two rectangles (2x) and eight small squares (8), we can represent the product $2(x + 4)$ by grouping the tiles together. Below right, a **GENERIC RECTANGLE** represents the same product as if its parts were the dimensions (length and width) of a rectangle. Study these representations and find their similarities and differences. Then record both examples in your tool kit.
>
>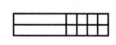
>
> $2(x + 4) = 2 \cdot x + 2 \cdot 4 = 2x + 8$
>
> In general, when you use tiles and rectangles to multiply two or more terms by the same number, variable, or mixed expression, you can see the multiplying pattern we have been using. The pattern is called the **DISTRIBUTIVE PROPERTY** and can be written as a rule:
>
> $a(b + c) = ab + ac$.

KF-79. KF-78 shows how to rewrite a product using Algebra Tiles and a generic rectangle. With your team, describe in complete sentences the similarities and differences between these two approaches.

KF-80. Ms. Speedi and her two sisters shared an apartment that was just down the street from Susan. Once all three of them were established in their professions, they decided that each of them would get her own apartment. As moving day approached, they began dividing most of what they had accumulated over the years. Among their possessions were three TV sets, 216 books, and six rugs. How should these items be dispersed?

KF-81. Explain in complete sentences the relationship of Ms. Speedi's situation and the Distributive Property.

KF-82. Use the Distributive Property to rewrite the expression and combine like terms (add) to simplify.

a) $3(x + 4) + 8(x + 1)$ c) $6x(9 - x) + 2(2x + 3)$
b) $2x(x - 3) + 5(x + 4)$

KF-83. Each of the expressions below was found using the Distributive Property. Can you find their original expressions? Some may have more than one solution.

a) $8x - 10$ c) $16x - 20$
b) $15y + 25$ d) $-2x - 6x^2$

KF-84. With your team members, create a problem using the Distributive Property that would have more than one solution.

KF-85. Throughout this unit you worked in your study team daily. Write a detailed response to each of these questions:

- What role did you play in your study team?
- What types of questions did you ask?
- How often did you just listen and copy down answers?
- How well did your team work?

KF-86. Find the area of the shaded region of the dart board at right. Show all your subproblems!

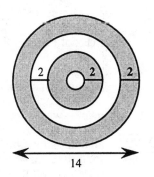

Tiling the Kitchen Floor: Area and Subproblems

KF-87. Find the total area of the rectangle at right in as many different ways as you can. Explain your methods. Are there more than two ways to solve this problem?

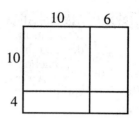

KF-88. Draw integer tiles to represent the following problems. Then combine the tiles and find a solution.

a) $-8 + (-2)$
b) $-3 + 7$
c) $14 - (-3)$
d) $-11 - (-4)$

KF-89. Use the Distributive Property to rewrite the following.

a) $3(x + 4)$
b) $3x(x + 4)$
c) $3(x - 4)$
d) $3x(4x + 5)$

KF-90. AREA OF A SQUARE

a) Draw a diagram of a square with edge length 6. Find its area.

b) Describe in words how to find the area of any square given the length of one edge.

c) If the area of a square is 64 square units, how long are its edges? Explain how you know.

d) If the area of a square is 20 square units, approximately how long are its edges? Explain how you arrived at your solution.

e) Is there a key on your calculator that can help you find the solution to part (d)?

KF-91. Use a Guess and Check table to solve the following problem. Write your solution as a sentence.

In science class, Hector decided to measure the mass of his math, science and Spanish textbooks. He discovered that the mass of his Spanish text is five more than four times the mass of his math book, while the mass of his science text is 110 grams less than his Spanish book. Combined, the mass of the texts is 1,178 grams. Find the mass of each book.

KF-92. In San Francisco, there are many one-way streets. Each block along these streets has a "one-way" sign posted to prevent accidents. The measurements of a typical sign are shown in the diagram at right. Find the area of this sign.

ORDER OF OPERATIONS AND SCIENTIFIC NOTATION

As you continue studying mathematics, the ability to identify and solve subproblems will be very useful. You used this problem solving technique in previous math classes when you simplified complex arithmetic expressions.

KF-93. Add this definition and example to your tool kit:

> We use the **ORDER OF OPERATIONS** to simplify complex arithmetic expressions like the one below. When a problem involves many operations, we perform them in the following order:
>
> | **Example:** | $12 \div 2^2 - 4 + 3(1 + 2)^3$ |
> |---|---|
> | Parentheses (or other grouping symbols) | $12 \div 2^2 - 4 + 3(3)^3$ |
> | Exponents (powers or roots) | $12 \div \mathbf{4} - 4 + 3 \cdot \mathbf{27}$ |
> | Multiplication and Division (from left to right) | $\mathbf{3} - 4 + \mathbf{81}$ |
> | Addition and Subtraction (from left to right) | $\mathbf{80}$ |

KF-94. Chun found the area of the shaded region at right.

a) Examine his work and explain how he found the area.

b) Chun used three operations in his solution. List the order of operations used in this solution.

c) Did Chun's method follow the Order of Operations?

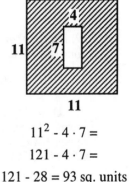

$11^2 - 4 \cdot 7 =$
$121 - 4 \cdot 7 =$
$121 - 28 = 93$ sq. units

KF-95. We have been using subproblems to make a bigger problem more manageable. You have also used subproblems when simplifying more complex arithmetic problems. Remember that you will have to write down and clearly show each subproblem as demonstrated in the example in KF-93. Simplify. Show all steps.

a) $5 + 3 \cdot 4$

b) $5(3 + 4)$

c) $2 - 5 \cdot 3$

d) $5 + 4^2$

e) $5^2 - (2 \cdot 4 + 6)$

Tiling the Kitchen Floor: Area and Subproblems

KF-96. Add Scientific Notation and Standard Form to your tool kit. Then answer the question that follows in your notebook.

> In Unit One we had situations in which our calculator jumped into scientific notation. Scientific notation is how our calculators display extremely large numbers. When your calculator shows $\boxed{1.3 \ \ 03}$ it means $1.3 \cdot 10^3 = 1,300$. Writing 1,300 in the form of "$1.3 \cdot 10^3$" is called **SCIENTIFIC NOTATION.** "1,300" is written in **STANDARD FORM**.
>
> To enter "$1.3 \cdot 10^3$" into your calculator, we will use the $\boxed{x^y}$ button, which raises a number (x) to any power (y). (Your calculator might have $\boxed{y^x}$ or $\boxed{\wedge}$ button instead) Use the following keystrokes:
>
> $$1.3 \quad \boxed{x} \quad 10 \quad \boxed{x^y} \quad 3$$
>
> Follow these key to verify that your calculator converts 1.3×10^3 into 1,300.

Predict what it means when your calculator shows $\boxed{3.9 \ \ 05}$. Write your answer in scientific notation and standard form.

KF-97. Which number is bigger: 5^{20} or 10^{15}? How can you tell? Enter both expressions into your scientific calculator and compare the results.

KF-98. The United States, with a population of 200 million people, has a national debt of roughly $4 trillion. How much money would each citizen need to pay in order to pay off this debt? Note that one million is 1,000,000 and one trillion is 1,000,000,000,000.

KF-99. Neil A. Armstrong was the first person ever to walk on the moon. After his historic landing on July 20, 1969, he stepped onto the Moon's surface and announced, "That's one small step for a man, one giant leap for mankind."

a) His craft, Apollo 11, traveled 238,900 miles from the Earth to reach the moon. How many feet was this? (Don't forget there are 5,280 feet in each mile.)

b) Extension: How many inches was this?

KF-100. The following pattern is composed of nested squares.

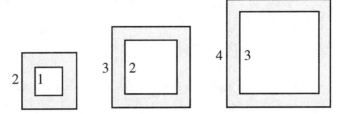

a) Draw the next figure in the pattern. Then find the area of the shaded region.

b) What is the shaded area of the fifth and sixth figures? Be sure to show your subproblems for each figure.

c) Predict the area of the 20th figure.

KF-101. Use your knowledge of integers and Order of Operations to compute the following:

a) $7 - 2 \cdot (-5)$

b) $5 \cdot (-3)^2$

c) $-3 \cdot 4 + 5 \cdot (-2)$

d) $6 + 3(7 - 3 \cdot 2)^2$

e) $35 \div (16 - 3^2) \cdot 2$

f) $7 - 6(10 - 4 \cdot 2) \div 4$

KF-102. Use the Distributive Property to rewrite the following expressions.

a) $2(x + 5)$

b) $4(x + 1)$

c) $5(x + 3)$

d) $10x(x - 2)$

KF-103. **CALCULATOR CHECK**

Use your scientific calculator to compute the following and match the result to an answer. A key marked $\boxed{\sqrt{}}$ will help you take square roots. A key marked $\boxed{x^2}$ will help you take square numbers. Consult your instruction manual for your calculator, or tomorrow ask your study team if you do not know how to use it. Decimal answers below are rounded to the nearest hundredth.

a) $\sqrt{12}$

b) $(-2 + \sqrt{3})^2$

c) $\dfrac{\sqrt{3}}{2}$

d) $\sqrt{5^2 - 4^2}$

e) $(-3)^2 - 2(-3) + 1$

f) $2 \cdot (\tfrac{2}{3})^2 - 1$

g) $\left(\sqrt{\dfrac{3}{4}} - \dfrac{1}{2}\right)^2$

1) 0.07

2) $-\dfrac{1}{9}$

3) 3.46

4) $\dfrac{1}{4}$

5) 3

6) 0.87

7) 16

Tiling the Kitchen Floor: Area and Subproblems

KF-104. Gary wanted to know how long the edge of a square would have to be in order to have an area of 52 square units. Ramon, a member of his study team, explained it had to be a little more than 7 since a square with edge 7 has an area of $7 \cdot 7 = 49$ square units. Shu Fei thinks the edge has to be 2,704 units since $52 \cdot 52 = 2{,}704$. Find the length of the edge and write a note to Gary explaining who is correct.

KF-105. Write each of these sets of numbers in order from smallest to largest.

a) $2, \sqrt{5}, 1.5$

b) $3, \sqrt{2}, 1.6$

c) $\sqrt{7}, 2, \frac{5}{2}$

d) $10, \sqrt{102}, 10.8$

e) $\sqrt{121}, 10.9, 11.1$

f) $-\sqrt{3}, -1.62, -1.83$

KF-106. TILING THE KITCHEN FLOOR: AREA AND SUBPROBLEMS

The kitchen in Susan's apartment has an ugly floor covering and she wants to replace it with various sized ceramic tiles. There will be a rectangular design made of 3" teal tiles (Susan's signature color!) set in one foot from the edge of the walls as shown with dashed lines on the diagram below. Your task is to find the number of tiles needed for this rectangle, as well as the square footage of the kitchen.

a) Use the floor plan of the kitchen from your resource page to determine the area in square feet of her kitchen floor. Write the subproblems you use.

b) The tiles are currently selling for $2.45 per square foot. How much will the tiles for the whole floor cost? Be sure to include 7% sales tax!

c) The teal tiles are 3" by 3" and are positioned in a rectangle one foot away from the walls. How many teal tiles will she need to purchase? Remember that the dashed lines on your resource page indicate the position of this rectangle.

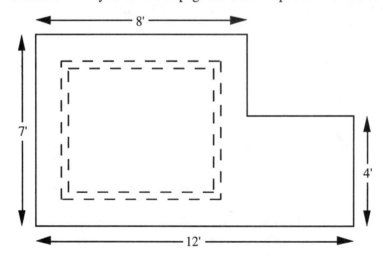

KF-107. The floor plan for Susan's bathroom is shown below. She wants to pull up the linoleum and replace it with 3" square tiles. While the tile does not go under the bathtub, it does cover the floor below the sink and toilet. If you ignore the grout lines, how many tiles will she need to buy?

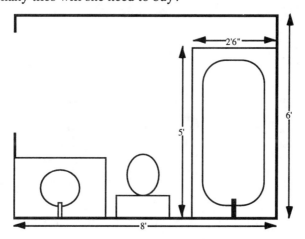

 PZL-12 REMINDER: THE ROLE OF THE TEACHER

You've almost completed another unit of your algebra course. I've found that even after students work with these materials for several weeks, some students are still confused about the teacher's role in the class. Sometimes this confusion--even anger and frustration--is because the student still expects the teacher to tell the class what to do for each problem. This often means the student wants the teacher to use most of the class time telling students how to work examples and homework problems.

Your teacher <u>should</u> do some brief lecturing and demonstrate <u>selected</u> problems. However, we've already established that the authors of the book want you to understand mathematics. Their experience has been that, for most students, telling them what to know is not enough for them to learn <u>and</u> retain the ideas. If you are still uncomfortable with the structure of the class, go back and read PZL-8, just after problem SQ-64 in Unit 1. You should also take some time to meet with your teacher--probably outside of class time--to discuss your feelings. I hope at least two things will happen: you will understand more about why the class is run the way it is, and your teacher will learn more about what you need to learn algebra successfully.

Now continue the lesson by doing KF-108.

KF-108. Use a Guess and Check table to solve the following. Write your answer in a complete sentence.

Latisha is thrilled with her grade! With four tests behind her, so far she has a 93% average and has only one test left! She wants to keep her "A" which means she needs to keep her percentage above 90%. Help her find the minimum score for her fifth test.

KF-109. Use the Order of Operations to simplify the following expressions.

a) $5 - 2 \cdot 3^2$

b) $18 \div 3 \cdot 6$

c) $(5 - 3)(5 + 3)$

d) $(-2)^2$

e) -2^2

f) $24 \cdot \frac{1}{4} \div -2$

KF-110. A local restaurant offers a Dim Sum lunch special which includes two dumplings, three egg rolls, a sweet bun and a drink. Susan and her friends order four Dim Sum lunch specials.

a) How many of each item should they receive?

b) Compare the situation in part (a) with the Distributive Property. What do they have in common?

KF-111. In this problem you will investigate the relative position of some square roots on the number line. Use the resource page provided by your teacher.

x	Exact Value of $\sqrt{x}$	Approximate Value of $\sqrt{x}$
1	$\sqrt{1} = 1$	
2	$\sqrt{2}$	1.4
3		
4	$\sqrt{4} = 2$	
5		
6		
7		
8		
9		
10		

a) Use your calculator and complete the table. Write each decimal approximation to the nearest tenth.

b) Use the decimal value for each $\sqrt{x}$ to locate its position on the number line. Label the position using square root notation, <u>not</u> a decimal.

c) 4 and 9 are square numbers. Why do $\sqrt{5}, \sqrt{6}, \sqrt{7}$, and $\sqrt{8}$ fall between 2 and 3 on your number line?

d) Locate the positions of $-\sqrt{10}, -\sqrt{9}, -\sqrt{8}, -\sqrt{7}, -\sqrt{6}, -\sqrt{5}, -\sqrt{4}, -\sqrt{3}, -\sqrt{2}$, and $-\sqrt{1}$ on the number line, and label each point using square root notation.

KF-112. Susan had some leftover tiles and has decided to use them to tile around her hot tub. This hot tub, located at the center of her 15 foot by 18 foot deck, has a diameter of 6 feet. Find the area around her hot tub that would need tile.

KF-113. Think about how your team did today. Choose a problem that your team worked on in class and focus on how your group worked together. Use complete sentences to describe what happened in your study team while you were solving the problem. Include in your description:

- What worked well? What could you do to help the team work more effectively?
- When your team was stuck, what happened?
- Did you offer your team your best participation? If not, why not?

UNIT SUMMARY AND REVIEW

KF-114. SUMMARY POSTERS

We have been introduced to several big topics in Units 0, 1, and 2. Today, as a class, we are going to create posters similar to the example at right to represent these main ideas. The posters will serve as models of the material we have studied so far. They will include an example problem, solution, and an explanation of how to solve the problem.

	Group Names
Mathematical Topic	
Representative Problem: KF__	
Solution:	
Explanation:	

a) Identify the mathematical theme of the problem that was assigned to your team. Use this idea to title your poster. Put the names of your team members on the poster. Remember to make the contents of your poster large enough to be read and clear enough to be understood by everyone in the class.

b) Copy the problem on your poster, leaving enough room for your solution and explanation. Abbreviations may help. Make sure all important parts of the problem are shown. Be sure to include any given diagrams or tables.

c) Next, solve your team's problem on paper. Work together as a team to determine the answer. Verify that all members agree with the solution. On your poster, copy your team's <u>solution</u>, showing all steps involved in the solution.

d) Write an explanation so that others can follow what steps you took to solve your problem. Be very detailed. Statements such as "We used our calculators" and "Just set up a table" are not helpful, whereas "We multiplied first, then added" or "We decided to put the width in the guess column because..." are much clearer.

KF-115. For parts (a) and (b) below, demonstrate how to rewrite the following products using Algebra Tiles and generic rectangles.

a) 4(x + 5)

b) x(2x + 5)

c) At the end of each Unit, Ms. Speedi puts her Algebra Tiles together in a tub. Each bag of Algebra Tiles has 4 large squares, 12 rectangles and 20 small squares. If she has 32 bags, how many of each size tiles does she have? Write an equation to represent this situation.

d) Mr. Ettlin's tub has 72 large squares, 240 rectangles and 432 small squares. He wants to separate them into the most bags possible so that each bag contains the same number of each size of tile. How many bags can he create and how many of each size tile will a bag contain? Write an equation to represent this situation.

KF-116. Analyze the two graphs below and write a sentence for each one describing their information.

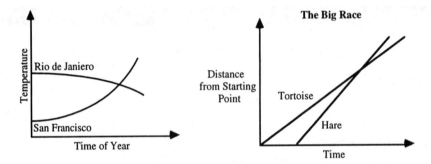

a) What is the meaning of the point of intersection (where the two curves meet) for the Temperature graph?

b) What is the meaning of the point of intersection (where the two curves meet) for The Big Race graph?

c) Create your own scenario and make a graph. Have at least two lines or curves intersect, as in the graphs above. Explain what the intersection represents in your situation.

KF-117. The wooden deck off Susan's bedroom is shown at right. She plans to apply a sealer to its top surface to protect the wood from sun and rain. In order to get the proper amount of sealant, she needs to know the total surface area. Help her find the area. Show all subproblems.

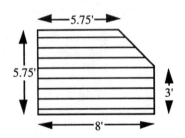

84   UNIT 2

KF-118. Solve the following problems using Order of Operations. Verify your answers with your calculator.

a) 12 - 18.5 + 15 + 6.3 - 1 + 28

b) (-4)(-2) - 6(2 - 5)

c) $23 - (17 - 3 \cdot 4)^2 + 6$

d) $14(2 + 3 - 2 \cdot 2) \div (4^2 - 3^2)$

KF-119. Sketch tiles to illustrate the solution to each part below.

a) Represent the number 3 in two different ways.

b) Solve the problems below. Remember: sketch tiles to support your solution.

1) 3 - 8

2) -4 - 3

3) -5 + 2

4) 2 · 3

5) (-1)(3)

6) (-4)(-2)

KF-120. Each figure below is composed of rectangles of various sizes.

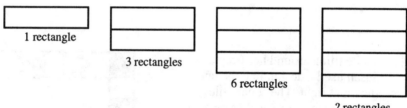

a) How many rectangles are in the fourth figure?

b) How many rectangles would be in the fifth figure? (Hint: a table is helpful.)

c) How many rectangles would be in the sixth figure?

d) Without drawing any more figures, determine how many rectangles are in the tenth figure and explain how you solved the problem.

KF-121. Patty's Hamburgers advertises that the company's franchises have sold an average of 405,693 hamburgers per day for the past 15 years.

a) About how many hamburgers is this? Round your answer to the nearest 100,000.

b) Express your answer from part (a) in scientific notation.

Tiling the Kitchen Floor: Area and Subproblems

KF-122. Use the process of "combining like terms" in each part below.

a) Liha has three large squares, two rectangles and eight small squares, while Makulata has five large squares and two small squares. At the end of class, they put their pieces together to give to Ms. Speedi. Write an algebraic equation for each student and find the sum of their pieces.

b) Write the length of the line below as a sum. Then combine the like terms.

c) Simplify the expression $4x + 6x^2 - 11x + 2 + x^2 - 19$.

KF-123. Determine the fraction and percent of the circle shaded at right. Then find the area of the shaded region. Show all subproblems.

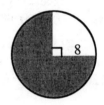

KF-124. While tiling around her fireplace, Susan used a design like the one shown at right. The square tiles (shaded) measure 8 inches on a side, and the rectangular tiles (unshaded) are cut from the 8" square tiles. The grout line between the tiles will be $\frac{1}{8}$" thick. She needs to figure out how wide to cut the rectangular tiles.

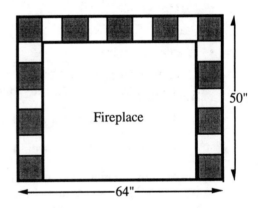

KF-125. There are two or more ways to find the area of the figures below. Find the areas of each figure using at least two methods. Show all subproblems.

a) 4 + 7 / 6 + 2

b) 2 2 2 / 5 / 2

KF-126. Using your scientific calculator, find the result of $8 \div 2 \cdot 6$.

a) Nina tried to compute $8 \div 2 \cdot 6$ by multiplying the 2 and 6 first, then dividing that product into 8. What will she get?

b) Explain to Nina what she did wrong.

KF-127. TOOL KIT CHECK-UP

Your tool kit contains reference tools for algebra. Return to your tool kit entries. You may need to revise or add entries.

How many entries do you have? Examine the list of tool kit entries from this unit below and add any topics that are missing.

- Circle vocabulary
- Area of a Triangle
- Area of a Circle
- Circumference of a Circle
- Substitution

- Standard Form
- Scientific Notation
- Distributive Property
- Order of Operations
- Subproblems

To The Students:

Gold Medal problems present you with an opportunity to investigate complex, interesting problems over several days. The purpose is to focus on the process of solving complex problems. **You will be evaluated on your ability to show, explain, and justify your work and thoughts**. *Save **all** your work, including what does not work in order to write about the processes you used to reach your answer.*

Completion of a Gold Medal Problem includes four parts:

- **Problem Statement**: *State the problem clearly in your own words so that anyone reading your paper will understand the problem you intend to solve.*

- **Process and Solutions**: *Describe in detail your thinking and reasoning as you worked from start to finish. Explain your solution and how you know it is correct. Add diagrams when it helps your explanation. Include things that did not work and changes you made along the way. If you did not complete this problem, describe what you <u>do</u> know and where and why you are stuck.*

- **Reflection**: *Reflect on your learning and your reaction to the problem. What mathematics did you learn from it? What did you learn about your math problem solving strategies? Is this problem similar to any other problems you have done before? If yes, how?*

- **Attached work**: *Include <u>all</u> your work and notes. Your scratch work is important because it is a record of your thinking. Do not throw anything away.*

GM-3. **GRAPHING MADNESS**

On graph paper starting at (0,0), carry out the following moves:

Move Number	Directions
1	Right 1 unit
2	Up 2 units
3	Left 3 units
4	Down 4 units
5	Right 5 units

Continue moving counter clockwise using this pattern.

Your Task:

- There are visual patterns, as well as quadrant, coordinate, and move patterns. Describe <u>each</u> of them.
- Name which quadrant the 79th move will be in. What will its coordinates be? How do you know this?
- For **any** move, name which quadrant it will be in and what its coordinates will be. Explain the method you are using.

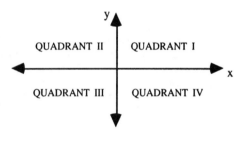

Tiling the Kitchen Floor: Area and Subproblems

GM-4. FIBONACCI RECTANGLES

The Fibonacci Numbers are the following sequence:

1, 1, 2, 3, 5, 8, 13, ...

Describe this sequence. Continue it until you have the first 15 Fibonacci numbers.

If the measures of the sides of a rectangle are consecutive Fibonacci numbers, it is called a "Fibonacci rectangle".

Here are the first four Fibonacci rectangles:

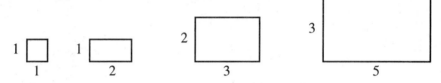

The sum of the areas of the first two Fibonacci rectangles is 3 square units.

The sum of the first three Fibonacci rectangles is 9 square units.

When you find the **sum of the areas** of the first two Fibonacci rectangles, the first three Fibonacci rectangles, the first four Fibonacci rectangles, etc., a pattern exists between the **area sums** and the original Fibonacci numbers.

Your Task:

- Find the sums of the areas of the Fibonacci rectangles, starting with the first two, then the first three, up through the first 15 rectangles.
- Describe the pattern that exists between the area sums and the original Fibonacci sequence.

Unit Three
The Burning Candle

Graphs and Patterns

Unit 3 Objectives
The Burning Candle: PATTERNS AND GRAPHS

In previous units, we have organized data and patterns with the use of tables. Now we will discover how to represent this information on graphs.

We will continue to use subproblems and Guess and Check to explore ideas and solve problems. We will also continue to practice algebraic manipulation of numbers and variables using the Order of Operations, Distributive Property, and Combining Like Terms.

During this unit, you will **create graphs from rules, words and tables**. These graphs will form predictable patterns that you will be able to describe with words and variables.

In this unit you will have the opportunity to:

- use patterns and organized data tables to draw graphs and solve problems.

- explore and use the xy-coordinate system.

- explore families of equations and their graphs, with a primary focus on linear and quadratic functions.

- begin writing algebraic expressions to describe the rule that governs tables of input and output values.

Read the following problem. **Do not try to solve it now.** Throughout this unit, you will develop skills and understanding to help you solve this problem.

BC-0. It is your best friend's birthday. You want to surprise her by walking into the room carrying a piece of cake with a lighted candle. Naturally, she wants to make a wish before blowing out the candle, and she can be awfully slow about making wishes. Can you predict when the candle will go out? What do you need to know in order to do so?

Problem Solving
Graphing
Writing and Solving Equations
Ratios
Geometry
Symbol Manipulation

Unit 3
The Burning Candle: PATTERNS AND GRAPHS

BC-1. ALGEBRA WALK: UNDERSTANDING INPUT/OUTPUT RELATIONSHIPS

The Algebra Walk is an exercise in "human graphing" where people represent points on a graph. Your teacher will give the class instructions on how to form human graphs. Then you will work in study teams to complete the problems below.

For each of the following rules, copy and complete the table. Then neatly graph each point. Use the resource page provided by your teacher.

The x-values in the table are sometimes referred to as **input** values, since they are the values used with the rule for x. The y-values are the **output** values, since they are the result of what happens to the input (x) value.

IN (x)	-6	-5	-4	-3	-2	-1	0	1	2	3	4	5	6
OUT (y)													

a) $y = 2x + 1$

b) $y = -2x$

c) $y = x + 4$

d) $y = -x + 4$

e) $y = x^2$

f) Compare the graphs in parts (a) through (c). How are they similar? How are they different?

g) Express each symbolic rule in parts (a) through (e) in words.

BC-2. Write a paragraph that describes what you did and what you observed in today's classwork. Did you notice any patterns? Why was it easy to spot someone who was out of place?

The Burning Candle: Patterns and Graphs

BC-3. Use your graphs from problem BC-1 to answer the questions below.

a) How can the graph for the rule $y = 2x + 1$ be used to predict the result for an input (x-value) of 7? How can the graph be used to predict the output (y-value) associated with an input of $3\frac{1}{2}$?

b) If you wanted an output of 7 for the rule $y = -x + 4$, what would you need as an input?

c) For each of the rules in parts (a) through (e) of problem BC-1, where does the graph cross the y-axis? Describe any patterns you notice.

PZL-13 NOTEBOOKS: YOUR COLLECTION OF EXAMPLES

I just wanted to remind you that you will be much more successful in this course if you keep a good notebook. This is critical, because many of the **examples** in this book are the problems you solve with your partner or study team. Keep track of the problems that you solve. Organize them neatly in a notebook. Keep them in a safe place.

Make sure they are correct and use them as a resource. CORRECT PROBLEMS ARE YOUR EXAMPLES. BE SURE TO REVISE AND CORRECT YOUR PROBLEMS AND KEEP THEM IN A NOTEBOOK FOR EASY REFERENCE.

Now continue with today's lesson -- and keep a good notebook and tool kit!

BC-4. The two lines at right represent the growing profits of companies A and B in thousands of dollars.

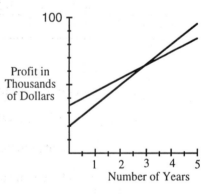

a) Sketch this graph on your paper. If company A started out with more profit than company B, determine which line represents A and which represents B. Label the lines appropriately.

b) In how many years will both companies have the same profit?

c) What will that profit be?

d) Which company has faster growing profits? How can you tell?

BC-5. Use substitution to find y.

   a) y = 2 + 4.3x, when x = -6

   b) y = x - 2, when x = 3.5

   c) y = (x - 3)², when x = 9

   d) y - 5x = -4, when x = -2

BC-6. Use the exponent feature of your calculator to evaluate the following expressions.

   a) $10^4 \cdot 10^{-7}$

   b) $(5^2)^3$

   c) $(5 + 2)^2 - 20$

   d) $(2^5 - 8)^2$

   e) $(5 + 2^2) - 20$

   f) $(10^5)^4$

   g) $3^{-4} \cdot 3^{-2}$

   h) $(2^5 + 2)^2$

   i) $(10^3)^4$

BC-7. In part (i) of problem BC-6, your calculator display was 1 ^{12}. Write this answer in both scientific notation and standard notation. Why did your calculator use scientific notation? When is it better to use scientific notation?

TABLES AND GRAPHS

BC-8. Complete the table below using y = 2x - 1. Plot and connect the points on a graph, then label the graph. Be sure to label the axes and to include the scale.

IN (x)	-6	-5	-4	-3	-2	-3/2	-1	0	0.5	1	2	3	4	5	6
OUT (y)															

BC-9. Use your graph of the equation y = 2x - 1 to complete parts (a) through (d) below.

   a) Approximate the value of x when y = 8.

   b) Approximate the value of x when y = 0.

   c) Explain how you can use the graph to find this value.

   d) Copy a model of a 'complete graph' into your tool kit or notes. Use the box on the next page as a reference.

>>Information to complete part (d) is on the next page.>>

TABLES AND GRAPHS

A complete graph has the following components:

- **x-axis** and **y-axis** labeled, clearly showing your scale.
- Equation of the graph near the line or curve.
- Line or curve extended as far as possible on graph.
- **x-** and **y-intercepts** labeled.
- **Coordinates** of points are stated in (x, y) format.

x	-1	0	1	2	3
y	6	4	2	0	-2

Tables can be formatted horizontally, like the one above, or vertically, as shown below:

x	y
-1	6
0	4
1	2
2	0
3	-2
4	-4

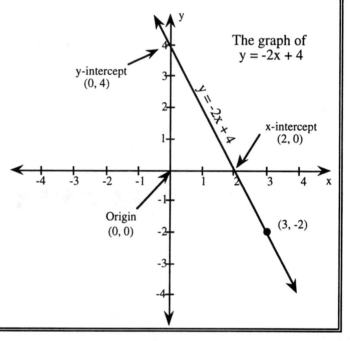

Throughout this course, we will continue to graph lines and other curves. Be sure to label your graphs appropriately.

BC-10. Make a table and complete it for the rule, " to find y, triple x, then add 1." Plot and connect the points on a set of axes, then label the graph. Be sure to label the axes and include the scale.

BC-11. Make a table and complete it for the rule $y = x^2 + 1$. Plot and connect the points on a set of axes, then label the graph. Be sure to label the axes and include the scale. Graphs of this shape are called **parabolas**.

    a) Use your graph to approximate the value of y when x = 5.

    b) Approximate the value of y when x = 0.

    c) Write the rule $y = x^2 + 1$ in everyday words.

BC-12.   ONE OF THESE POINTS IS NOT LIKE THE OTHERS, Part One

a) Plot and connect the points below.

IN(x)	-2	4	1	6	-5	0
OUT(y)	-6	-2	-3	2	-9	-4

b) Identify the point(s) that do not appear to fit the pattern.

c) Correct the point(s) found in part (b) above so it fits the pattern.

d) Write the coordinates of the point where the graph crosses the y-axis, called the **y-intercept**.

BC-13.   Translate the following sentences into an expression with numbers and find the net result.

a) Maurice dug a 5 foot hole and later filled it in with 3 feet of dirt. What was the net result?

b) Liha had $56 in her bank account. She spent $18 at a cafe and put a $75 pair of shoes on her charge card. What has the net result?

c) Mai works on the 38th floor in the World Trade Center in New York City. She took the elevator down 11 floors to copy a document and went up 43 floors to hand deliver it to her boss. She then went down 73 floors to the parking garage to get her car. On which floor is the parking garage?

BC-14.   Think about 2^5. What does it mean? Without your calculator, write its meaning (not value) without an exponent. Use your calculator to compute its value.

BC-15.   Plot the points A = (5, 3), B = (-4, 3), C = (-4, -6), and D = (5, -6) on a set of axes. Use a ruler to connect them in order, including D back to A, to form a **quadrilateral** (a shape with four sides).

a) What kind of quadrilateral was formed?

b) How long is each side of the quadrilateral?

c) What is the area of the quadrilateral?

d) What is the perimeter?

BC-16.   In Unit Two, we used Algebra Tiles to represent algebraic expressions.

a)   Draw a rectangle using Algebra Tiles with area 2x + 4. Write an algebraic expression for the dimensions of the rectangle formed with the tiles.

b)   Draw a rectangle using Algebra Tiles with area $2x^2 + 4x$ in two different ways. Write the algebraic expression for each rectangle.

c)   How are the sketches for parts (a) and (b) different?

BC-17.   Use the Distributive property to rewrite the following expressions:

a)   4(x - 5)

b)   6(x - 3)

c)   -2(4x - 5)

d)   3y + 9 hint: __ (__ + __)

BC-18.   The area of the big rectangle is $2x^2 + 6x - 10$ and the area of the small rectangle is $x^2 + 5x + 6$. Find the area of the shaded region.

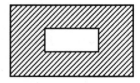

GRAPHING INPUT/OUTPUT REALTIONSHIPS

In each of problems BC-19 through BC-21:
a)   Make a table like the one below and complete using the rule provided.
b)   Plot the points on a graph, then label the graph. (Be sure to label the axes and include the scale.)

IN (x)	-4	-3	-2	-1	0	1	2	3	4
OUT (y)									

BC-19.   y = 3x + 3

BC-20.   y = -x + 1

BC-21. $y = x^2 - 2$

BC-22. Examine your three previous graphs.

a) How are the graphs different?

b) Write the coordinates (x, y) on the graph for the point where each line or curve crosses the y-axis.

c) Describe two ways that the graphs of $y = 3x + 3$ and $y = -x + 1$ are different. How are they the same?

PZL-14 UNDERSTANDING THE SPIRALING OF SKILLS AND CONCEPTS

The ideas and skills of algebra are spiraled throughout your text and over your additional years of CPM or any college prep math courses you take. What is spiraling? Put a dot on a piece of paper, then draw a curve that is somewhat like a circle but that keeps getting a little wider with each revolution. Suppose the dot represents the idea of "graphing." As you learn more and more about graphing, your understanding of it expands. Each new aspect of graphing could be placed along the expanding bands of the spiral. Each time you add a new band it is close to the previous band (what you already know) but takes the idea a little further. By the end of four years of college preparatory mathematics you would have a spiral perhaps the size of a full sheet of paper with everything you learned about graphing. You would have several more pages for the other big ideas in the courses. You could even do a spiral that ties all the big ideas together.

The authors wrote this course so that you could build your understanding of the big ideas of mathematics throughout the four years of your college preparatory math studies. **Clear understanding and mastery take time.** The authors wrote this text knowing you may not grasp all the information the first time you see it. So they give you an opportunity to use the ideas again and again, gradually adding more depth or something new to your understanding. If you do not fully understand a concept the first time, don't give up. **You will have the opportunity to see it and practice it again.** However, it is important to learn **as much as you can the first time** so you have a place to start the next time you see it.

When the textbook introduces an idea, the problems that surround the information are there to help you understand it. The **problems that come later offer you practice**. This practice is just as important as the initial work to develop understanding, because you will use these skills to solve new problems later. By the end of the program you should have a complete understanding of the big ideas. But be careful! Because you will practice skills and ideas over several weeks and months does not mean you can let things slide. Students who frequently skip assignments or spend class time doing other homework or chatting with friends fool themselves by thinking that they will learn ideas when they see them later.

Do not wait until the last minute. A full explanation of ideas does not occur later in the course. Remember: problems that show up later are for practice and smoothing rough edges and to get you ready to move on to something bigger. They are not there for you to start from scratch. Learn as much as you can as early as you can. Use the clues we discussed earlier about how to read and use this textbook. Add information in boxes to your tool kit. Each new piece adds to your understanding. This is the big idea behind spiraling.

Knowing what spiraling is will help you use the book to your advantage. Here is how. Many teachers structure their exams so that 60% of the material comes from prior units. That 60% is all the material that spirals with you. This is a major reason why it is important to keep a good, up-to-date tool kit to use as a reference and for review prior to tests. It is also why you should do careful, complete summaries at the end of each unit.

What happens when you come across a problem after it has been introduced and you do not know how to do it? Keeping spiraling in mind. Then your response might be, **"To look back at earlier material in the book"** or **"To look in my tool kit."** You may want to check your list of resources you made for PZL-6.

MISSION POSSIBLE (something to do)

Use the problems listed below.

a) Quickly flip through the book to check them out, then briefly describe what they have in common and how they illustrate spiraling.

SQ-45 (46-49), 63-64, 74, 76-79, 83, 89.

KF-7, 18, 41, 60, 91, 108.

BC-26, 41, 50, 88.

b) Name and describe at least one other idea that was introduced earlier in the year that illustrates spiraling in this course.

BC-23. Find the area of this trapezoid. Clearly show each of your subproblems.

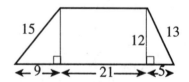

BC-24. If the sides of a polygon are given as indicated on the diagram,

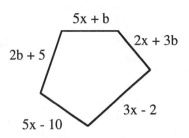

a) write the perimeter as a sum of the sides and simplify the expression.

b) find the perimeter if x = 2.5 and b = 1.25.

BC-25. ONE OF THESE POINTS IS NOT LIKE THE OTHERS, Part Two

a) Plot and connect the points below.

IN(x)	-2	4	1	6	-5	0	-3	2
OUT(y)	7	-5	1	-9	13	3	2	-1

b) Identify the point(s) that do not fit the pattern.

c) What shape does the graph appear to make?

d) Correct the point(s) found in (b) above so it (they) fit(s) the pattern. Name the correct point(s) in (x, y) notation.

e) Label the coordinates of the point where the graph crosses the y-axis.

BC-26. Use a Guess and Check table to solve this problem.

Melissa cut a 150-centimeter board into two pieces. One piece is 24 centimeters longer than the other piece. How long is each piece of board?

BC-27. Write the rule $y = \frac{1}{2}x + 3$ in everyday words.

WRITING RULES FOR INPUT/OUTPUT RELATIONSHIPS

BC-28. Copy and complete the table.

IN (x)	0	2	6	-8		-10		x
OUT (y)	0	1	3	-4	8		-14	

a) Determine the smallest and largest x-value in the table.

b) Determine the smallest and largest y-value in the table.

c) Use this information to plan the axes of your graph. Use the same scale for both axes. This means that if you let one square represent 2 units on the x-axis, you must also let one square represent 2 units on the y-axis.

d) Plot and connect the points.

The Burning Candle: Patterns and Graphs

For each of problems BC-29 through BC-30:

a) Copy and complete the table.

b) Write the process you described in part (a) in algebraic symbols.

c) <u>Explain</u> in <u>words</u> what is done to the input value, x, to produce the output value, y.

d) Graph the data. Use the same scale on both the x-axis and the y-axis. (You may not be able to fit all the points from the table on your graph because their x or y values are too large or too small. Graph enough points to see a pattern.)

BC-29.

IN (x)	9	-6	10		20	7	-2	x
OUT (y)	27	-18		15				

BC-30.

x	2	5		4.5	6		8	-3	0.1	-5	-1	x
y	4	25	9			100		9				

BC-31. After Cheryl completed her table and graph for the equation $y = -x - 1$, a member of your team thought her graph (shown at right) did not look correct. Cheryl needs your help to find her mistake.

x	y
-3	2
-2	1
-1	0
0	-1
1	-2
2	-3
3	-4

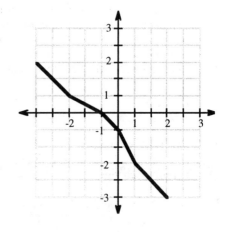

BC-32.

IN (x)	5	10	$6\frac{1}{2}$	$15\frac{1}{2}$	-9			x
OUT (y)	$7\frac{1}{2}$	$12\frac{1}{2}$				$4\frac{1}{2}$	12	

BC-33. Copy and complete each table. Write each rule in words.

a)

IN(x)	OUT(y)
3	-6
7	-14
6	-12
4	
10	
-1	
100	
x	

b)

IN(x)	OUT(y)
1	5
5	9
2	
	10
-10	
	21
1000	
x	

BC-34. Copy and complete the table.

IN (x)	2	10	6	7	-3		-10	100	x
OUT (y)	8	32	20			-13			

a) Explain in words what is done to the input value, x, to produce the output value, y.

b) Write the process you described in part (a) in algebraic symbols.

c) Graph the data. Use the same scale on both the x-axis and the y-axis. (You may not be able to fit all the points from the table on your graph because their x or y values are too large or too small.)

BC-35. Use the Distributive Property to write these products as a sum:

a) 3(-2x + 3)   b) -4(5b - 6)   c) -x(x + y)

BC-36. As Maisha ran a race, her distance from the starting line was graphed as shown in Figure 1 at right.

a) How long did it take her to finish her race?

b) How long was the race?

c) How fast was she running?

d) Why might the graph shown in Figure 2 be a better model?

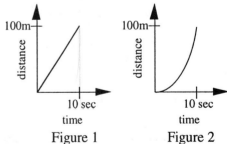

Figure 1   Figure 2

The Burning Candle: Patterns and Graphs

ESTIMATING X-INTERCEPTS

BC-37. Copy and complete this table. Describe in words the rule you used.

x	2	0	-3	$\frac{1}{2}$	-1	0.3	3	$-\frac{1}{3}$	45	-11	-2	x
y	5	1	-5	2		1.6	7	$\frac{1}{3}$	10	-21	-3	

a) State the rule using a variable, such as x.

b) Graph the **ordered pairs** (x, y). Find two more pairs of points that satisfy the rule without using any more <u>whole</u> numbers.

c) Estimate where the graph crosses the x-axis. This point is called the **x-intercept**. Notice that at this point, the value of y is always 0.

d) Find a value for x that makes the equation 2x + 1 = 0 true. If you are unsure how to solve the equation algebraically, use the Guess and Check strategy.

e) How do your answers to parts (c) and (d) compare?

f) Reading the line from left to right, would you describe this line as going uphill or downhill? Is it very steep?

g) What do all the points on the graph of the equation have in common?

h) Why is the point (4, 5) not on the graph?

i) What does the graph represent?

BC-38. Write an expression for the length of the given lines:

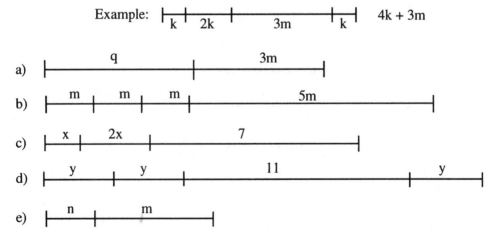

BC-39. The perimeter (distance around the outer edge) of the figure in Figure 1 is eight units and the perimeter of the figure in Figure 2 is 14 units.

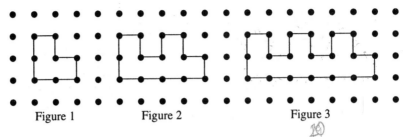

Figure 1   Figure 2   Figure 3

a) Find the perimeter of the figure in Figure 3. Find the perimeter of the figure you would draw for Figure 4.

b) Copy and complete the following table.

Figure Number	1	2	3	4	5
Perimeter	8	14			

c) Graph the information from your table. Label the x-axis "figure number" and the y-axis "perimeter."

d) How can you use the graph to find the perimeter of the sixth figure?

e) If the pattern of forming figures continues as shown above, what would the perimeter of the seventh figure be? What would the perimeter of the ninth figure be? (Hint: use the table.)

f) Explain in a sentence or two how to determine the perimeter of any figure in this pattern without the graph.

g) Write the pattern as a rule with x and y. Let x represent the figure number and y represent the perimeter.

h) In this problem there were many ways to represent the same information. List all the different ways.

BC-40. ONE OF THESE POINTS IS NOT LIKE THE OTHERS, Part Three

Plot and connect the points below.

IN (x)	-2	4	1	-4	0	3	-3	2
OUT (y)	0	12	-3	12	-4	5	-2	0

a) Identify the point(s) that does (do) not fit the pattern.

b) What shape does the graph appear to make?

c) Correct the point(s) identified in part (b) so it (they) fit(s) the pattern. Write the point(s) in (x, y) notation.

The Burning Candle: Patterns and Graphs

BC-41. Marla set up this Guess and Check table for a problem about a rectangle. Copy and complete the table, as it is shown, but do not solve the problem.

Guess Width	Length	Perimeter	Check 60 ?
5	11	32	too low
7	15		
10	21		

a) If the next Guess for the width is 9, describe how you would calculate the length.

b) Use your description from part (a) to write the length in symbols if the next guess for the width is x.

In problems BC-42 and BC-43,

a) Copy and complete each table.

b) Write a rule using the variable x.

c) Graph the rule and estimate the **X-INTERCEPT** (where the graph crosses the x-axis) and **Y-INTERCEPT** (where the graph crosses the y-axis). Write these coordinates (x, y). Be sure to label your graph and the axes!

BC-42.

x	2	-1	0	-2	$\frac{1}{2}$			x
y	-4	2	0			6	3	

BC-43.

x	10	0	1	-2	$\frac{1}{2}$	x
y	28	-2	1			

BC-44. Suppose that your grandmother gave you $1 on your first birthday, $2 on your second birthday, $3 on your third birthday, etc. Also suppose that you saved all the birthday money your grandmother gave you in a piggy bank.

a) Let your age in years be the input (x), and let the total savings from all your birthdays be the output (y). Organize your data in a table and draw a graph that includes at least ten birthdays.

b) Describe the graph and explain what it represents in two or three sentences.

c) Are there any points on your graph <u>between</u> years one and two? Explain.

d) How much birthday money will you have in your piggy bank on your sixteenth birthday?

e) How much birthday money will you have altogether on your twenty-first birthday?

COMPARING GRAPHS OF LINES AND PARABOLAS

BC-45. Complete this table for the rule $y = x^2$. Plot and connect the points.

x	-4	-3	-2	-1	-0.5	0	0.5	1	2	3	4
y											

a) Is this graph a straight line? Describe the result.

b) Have we seen this shape before? Look back on your previous graphs. What feature of this rule, $y = x^2$, determined this shape?

c) Can any value for x be used? Explain.

d) Can any y-value result from this rule?

e) Use the graph to approximate the x-value(s) when $y = 5$.

f) Use your calculator to solve part (e) more accurately.

BC-46. On the same set of axes in problem BC-45, complete a table and graph $y = 2x$.

a) How is the graph of $y = 2x$ different from the graph of $y = x^2$?

b) In Unit Two, we used the area of our Algebra Tiles to represent x^2 and x. Draw tiles to represent x^2 and 2x. How are they different?

The Burning Candle: Patterns and Graphs

BC-47. So far in this unit, most of the rules you have graphed were **LINEAR**, that is, straight lines. A few rules produced curved in a shape called **PARABOLIC**.

a) Using your work in the previous two problems, describe the difference between the rules that produce lines and those that produce parabolas.

b) Record an entry in your tool kit to describe in general a rule that produces lines for graphs and a rule that produces parabolas.

BC-48. We have studied both the area and circumference of circles. Now we will examine the relationship of the radius of a circle and its circumference and area.

a) Use the resource page provided by your teacher. Compute the area and circumference of circles with a radius of 1, 2, 3, 4, and 5. Place these values in your table.

b) Graph the radius compared to the circumference on one set of axes. Graph the radius compared to the area on another. Label the axes completely.

c) Compare the two graphs and write any similarities and differences on your paper.

d) Use your graph to approximate the area and circumference of a circle with radius of 2.5.

e) Use your graph to approximate the radius of a circle with a circumference of 10 units.

f) Use your graph to approximate the radius of a circle with an area of 20 square units.

BC-49. In the pattern at right, each figure is composed of squares.

1 square

5 squares

9 squares

a) Determine how many squares the fourth figure will have.

b) Record all of your information in a table. Use a pattern you see in your table to predict how many squares will be in the sixth and tenth figure. Write one or two sentences to explain how to find the number of squares for any figure in this pattern.

BC-50. Use a Guess and Check table to solve the following problem. Write your answer in a sentence.

The length of a rectangle is three times the width. If the area is 18.75, find the width and length.

BC-51. Without drawing a graph, describe the shape of the graph of each of the equations below. Explain how you know.

    a) $y = x^2 - 3$ b) $y = 2x - 3$

BC-52. Marina is trying to enter the expression -5^2 into her scientific calculator. She has entered a 5 and does not know what to do next. According to the Order of Operations, which operation should come first?

    a)   What result should she get?

    b)   Enter the expression $(-5)^2$ into your scientific calculator and record the result. What is the Order of Operations for this expression?

    c)   Does $-3^2 = (-3)^2$? Show why or why not.

BC-53. Use the Order of Operations to compute the following:

    a) $5 \cdot 4 - 3 \cdot (2 - 3^2)$ c) $1^2 - 2^2 + 3^2 - 4^2 + 5^2$

    b) $3(12 + 8.6 \cdot 2 - 5)$ d) $10 \div 5 \cdot 3 \div 2$

BC-54. Solve the problems below. Look for a pattern.

    a) $1^2 =$
            $1^2 - 2^2 =$
            $1^2 - 2^2 + 3^2 =$
            $1^2 - 2^2 + 3^2 - 4^2 =$
            $1^2 - 2^2 + 3^2 - 4^2 + 5^2 =$
            $1^2 - 2^2 + 3^2 - 4^2 + 5^2 - 6^2 =$

    b)   Examine your tool kit entry for famous number patterns from Unit One (from SQ-60). Which pattern does this resemble? How is it different?

EXPLORING PARABOLAS

BC-55. Consider the expressions $-x^2$ and $(-x)^2$.

    a)   Translate these expressions into words.

    b)   Is $-x^2$ the same as $(-x)^2$?

    c)   Evaluate $-x^2$ and $(-x)^2$ for the values $x = 2, 0, -1$, and -3.

    d)   What can you conclude about the expressions $-x^2$ and $(-x)^2$?

The Burning Candle: Patterns and Graphs

> **INEQUALITY SYMBOLS**: ≤ and ≥
>
> The symbol " ≤ " read from left to right means "less than or equal to."
> For example, $x \leq -5$ says "x is less than or equal to -5."
>
> The symbol " ≥ " read from left to right means "greater than or equal to."
> For example, $x \geq 7$ says "x is greater than or equal to 7."
>
> We can express the idea that "x is greater than or equal to -3 <u>and</u> x is less than or equal to 4" by writing
> $$-3 \leq x \leq 4$$
> This means that x can be any number between -3 and 4, including -3 and 4.

BC-56. Make a table and graph $y = x^2 - 4x$ for $-1 \leq x \leq 5$. Use several values for x, including fractions and decimals, in the given interval. Be sure to label the graph with the equation.

a) Use the graph to approximate the x-value that corresponds to a y-value of 1.

b) Determine the x-intercepts (where the graph crosses the x-axis).

> The symbol "..." is called an **ELLIPSIS**. It indicates that certain values in an established pattern have not been written, although they are part of the pattern.
>
> For example, in the pattern
> $$3, 3.5, 4, \ldots, 8.5, 9$$
> the ellipsis (...) means that the values 4.5, 5, 5.5, 6, 6.5, 7, 7.5, and 8 are part of the pattern even though they are not written.

BC-57. Make a table for $y = x^2 + 3x + 1$, using x = -5, -4, -3, ... , 1, 2.

a) Graph the equation and label your graph.

b) Locate the "bottom" of the parabola. We call this point of a parabola the **vertex**. Approximate the coordinates of the vertex

BC-58. Graph $y = 3(x - 2)$ and $y = 3x - 6$ using x = -2, -1, ... , 4, 5 on the same set of axes. What do you notice about these two equations? Did you expect this to happen? Why or why not?

For each of problems BC-59 and BC-60:

a) Make a table using several values, including fractions and decimals, using the given sets of input values, called the **domain**.

b) Plot and connect the points you found in part (a). Be sure to label each graph.

c) Use the graph to approximate the x-value that corresponds to a y-value of 1.

d) Find the x- and y-intercepts.

BC-59. $y = x^2 - 4$ for values $-3 \leq x \leq 3$.

BC-60. $y = 4 - x^2$ for values $-3 \leq x \leq 3$.

BC-61. Sketch a graph to represent each relationship described below. Label the axes. The <u>vertical</u> axis is given first.

a) the total cost of gasoline compared to the number of gallons of gasoline you buy.

b) the height of a burning candle compared to time.

c) the height of an unlit candle compared to time.

BC-62.
a) Write a description for what the graph at right might represent.

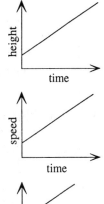

b) Write a description for what the graph at right might represent.

c) Create another situation that can be represented by this graph. Label the axes and explain what your graph represents.

BC-63. Use your calculator's $\boxed{y^x}$, $\boxed{x^y}$, or $\boxed{\wedge}$ key to compute each of the following values.

a) 2^3

b) 3^0

c) 2^{-1}

d) $(\sqrt{3})^4$

The Burning Candle: Patterns and Graphs 111

GRAPHS OF NON-LINEAR FUNCTIONS

BC-64   Use the $\frac{1}{x}$ key on your calculator to find the results for substituting each of the following values of x.

a) 2   d) -0.75   g) 0.01

b) 3   e) 100   h) $\frac{1}{2}$

c) 0.5   f) -100   i) $-2\frac{1}{2}$

BC-65. Examine your results from the previous problem. What happened as x increased? Decreased?

a) What do you notice about your results when x > 1 (when x is greater than one)?

b) What do you notice about your results when 0 < x < 1 (when x is between zero and one)?

c) What do you notice about your results when x < 0 (when x is less than zero)?

d) What happens if x = 1 ?

e) What happens if x = 0 ?

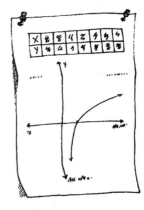

BC-66. On a large set of axes, graph the equation assigned to your team by your teacher. Pay attention to the given domain and scaling information. Be prepared to share your graph with the class.

	Equation	Domain and Axes Scaling
a)	$y = 2^x$	x = -2.0, -1.5, -1, ... , 2.5, 3.0 Scale the y-axis so that two marks on the graph paper represent one unit.
b)	$y = \sqrt{x}$	x = 0, 0.5, 1, 2, ... , 8, 9, 9.5, 10 Scale the axes so that two marks on the graph paper represent one unit.
c)	$y = 0.5^x$	x = -4, -3, ... , 3, 4 Scale the axes so that two marks on the graph paper represent one unit.
d)	$y = \frac{1}{x}$	x = -5, -3, -2, -1.5, -1, -0.5, -0.33, -0.2, 0, 0.2, 0.33, 0.5, 1, 1.5, 2, 3, and 5 Scale the axes so that two marks on the graph paper represent one unit.
e)	$y = x^3$	x = -3, -2, -1.5, -1, -0.5, -0.33, -0.2, 0, 0.2, 0.33, 0.5, 1, 1.5, 2, and 3 Scale the axes so that one mark on the graph paper represent one unit.

BC-67. On your Resource Page for problem BC-66, examine graph (a) of $y = 2^x$. Use the graph to answer the following questions:

a) What happens to y as x gets larger?

b) What happens to y as x gets smaller?

c) Where does the graph cross the x-axis?

d) Use the graph to estimate the x-value when the y-value, 2^x, is:

1) 5
2) 2
3) $\frac{1}{2}$
4) 3.5

e) Use the graph to find x so that:

1) $2^x = 7$
2) $2^x = 5$

BC-68. On your Resource Page for problem BC-66, examine graph (b) of $y = \sqrt{x}$.

a) Use the graph to estimate the values of x so that:

1) $\sqrt{x} = 1.7$
2) $\sqrt{x} = 2.2$
3) $\sqrt{x} = 0.5$

b) Can you find more accurate solutions using a calculator? If so, what are they?

1) $\sqrt{x} = 1.7$
2) $\sqrt{x} = 2.2$
3) $\sqrt{x} = 0.5$

c) Put the value x = -1 into the rule $y = \sqrt{x}$ using your calculator. Then try x = -3. What happens? Why are there no points on the graph of $y = \sqrt{x}$ to the left of the y-axis?

BC-69. On your Resource Page for problem BC-66, examine graph (c) of $y = 0.5^x$. Use the graph to answer the following questions:

a) What happens to y as x gets larger? What happens to y as x gets smaller?

b) Use the graph to estimate the value of x so that ...

1) $0.5^x = 2$
2) $0.5^x = 3$
3) $0.5^x = 1$
4) $0.5^x = \frac{1}{3}$

The Burning Candle: Patterns and Graphs

BC-70. On your Resource Page for problem BC-66, examine graph (d) of $y = \frac{1}{x}$. Use the graph to answer the following questions:

a) For positive values, what happens to y as x gets larger? What happens to y as x gets smaller?

b) What happens to y when x decreases from 1 to 0 ? What happens to y when x is increases from -1 to 0 ?

c) Use the graph to <u>estimate</u> a value for x so that $\frac{1}{x}$ is:

    1)   2   4) $\frac{-1}{4}$

    2) $\frac{1}{2}$ 5)   0.3

    3)   -2   6)   2.5

BC-71. On your Resource Page for problem BC-66, examine graph (e), of $y = x^3$. Use the graph to answer the following questions:

a) For positive values, what happens to y as x gets larger? What happens to y as x gets smaller?

b) Reading the graph from left to right, describe the path as "increasing" or "decreasing."

c) Use the graph to <u>estimate</u> a value for x so that x^3 is:

    1)   2   3)   -4

    2)   7

BC-72. THE BURNING CANDLE PROBLEM -- An Investigation

It is your friend's birthday. You want to surprise her by walking into the room carrying a piece of cake with a lighted candle. Can you predict when the candle will go out?

To answer this question, we will use a video presentation of a burning candle to collect data, make a graph, and look for a pattern.

a) Gather data from the "Burning Candle" video presentation and place it into a table. Note the mass of the candle at various times during the presentation and write down your observations. You should make at least five observations. Be sure to write down both the time and the associated candle mass.

b) Make a new column in your table to show elapsed time, that is, the total time since the candle was lit.

c) On axes (using equal intervals) with elapsed time on the horizontal axis, graph your data, comparing mass, y, to elapsed time, x. Sketch a line or curve by connecting your data points. Compare your graph with those of your team.

d) Use your graph to predict the mass of the candle at the elapsed time of 1 minute and 20 seconds. Predict the mass of the candle at the elapsed time of 2 minutes and 47 seconds.

e) If the candle continued to burn, when do you think it would go out? Discuss this question with your team and carefully explain your answer in complete sentences.

BC-73. Suppose your textbook gave you the answers to some Distributive Property problems but forgot to give you the questions! Can you figure out what the questions are?

a) $6x + 18 = \_\_(\_\_ + \_\_)$

b) $24y - 21$

c) $21x - 33y$

d) $12m^2 - 4m$

BC-74. For the following problems, combine like terms.

a) $9x + 6 - 4x - 2x + 1 - 15$

b) $5(x + 1) - 2(x + 1)$

c) $(6x^2 - 2x + 5) + (2x^2 - 7x - 11)$

d) $7x(x^2 - 5) - 2(x - 3) + 5$

BC-75. The treasurer of the Math Club made a graph to represent the club's monthly balance from the beginning of the school year, in August. Examine the graph and answer the questions below.

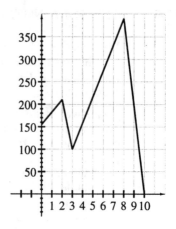

a) Find the balance in March, month 7.

b) When did the Math Club have the most money?

c) Explain what the x-intercept represents.

d) Explain what the y-intercept represents.

BC-76. Find the area of this figure. Show your work.

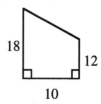

BC-77. Copy the table at right and continue the pattern for successive powers of 3.

a) Describe in a sentence or two a pattern formed by the units ("ones") digits of the numbers in part (a).

b) $3^1 = 3$. List the next three powers of 3 for which the ones place is a 3.

$3^1 = 3$
$3^2 = 9$
$3^3 = $ _____
$3^4 = $ _____
$\vdots$
$3^9 = $

BC-78. Find the area of the shaded region at right. Break the problem into subproblems and solve them separately.

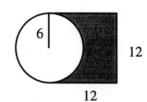

UNIT SUMMARY AND REVIEW

BC-79. UNIT THREE SUMMARY (BC-79 through BC-81)

With your study team, quickly list as many mathematical ideas or topics you remember from this unit. Do not flip back through the text or your notes! There will be time for that later. This exercise is to see what comes from memory. Be attentive to any suggestions offered by your teammates. Each member needs to write the list down.

BC-80. With the list generated from the class, discuss with your study team the relative importance of each item. Which ideas and skills were crucial to understand? Narrow the list to the most important four ideas.

BC-81. Obtain a Resource Page from your teacher.
   a) On the Resource Page, fill in your team's four big ideas as the titles of the four sections.
   b) For each topic, choose a problem from this unit that demonstrates the main idea. Be selective. While there are many problems to choose from, certain problems are better examples than others. Perhaps you would like to choose a problem you are most proud of solving. Check that the problems you choose are directly related to the main ideas of your summary.
   c) Solve each problem and show all work, especially the subproblems.
   d) Finally, explain the method you used to arrive at your solution. Use complete sentences and be as descriptive as you can.

BC-82. If a circle has a radius of 8.6 cm., find its area and circumference.

BC-83. Graph both $y = \frac{1}{3}x - 5$ and $y = \frac{1}{3}x + 1$ on the same set of axes. Use x-axis values of -6, -5, ...,5, 6. Do you notice anything in common about their equations? What about their graphs?

BC-84. Draw diagrams of three different triangles with an area of 7. Using graph paper may prove useful.

BC-85. Compute the numerical value of each of the expressions below.
   a) $3 + 3 \div 3 + 3$
   b) $3 + 3^3 \div 3$

BC-86. Use the graph below to ...

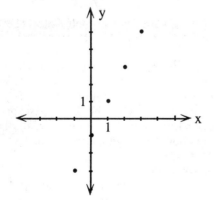

a) complete the table,

and

b) find the rule y = _____.

The Burning Candle: Patterns and Graphs

BC-87. For each of the following rules, create a table using the given domains. Then use the table to graph the rule.

	Equation	Domain
a)	$y = -x + 4$	$x = -4, -3, \frac{-1}{2}, 0, \frac{1}{2}, 0.3, 1, 2$
b)	$y = x^2 - 2$	$x = -3, -2, -1, -0.5, 0, 1, 1.3$
c)	$y = x^2 + 2x + 1$	$x = -3, -2, -1, 0, 1$
d)	$y = 2^x$	$x = -2, -1, -0.5, 0, 0.5, 1, 2$

BC-88. Solve this problem by using a Guess and Check table. Write your solution in a sentence.

West High School's population is 250 students less than twice the population of East High School. Combined, they serve 2,858 students. How many students attend West High School?

PZL-15 FOR THOSE WHO HAVE FALLEN BEHIND

NOTE: If you have been doing all your work and are up-to-date in all course requirements, you may skip this problem. Go ahead to problem BC-89.

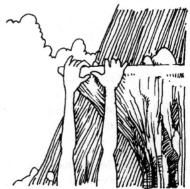

You may recall that I warned you about not falling behind at the end of Unit 1 in PZL-9. Now that you have finished a quarter of the course and know that the authors have spiraled ideas throughout the book, it should be absolutely clear that **YOU CANNOT FALL BEHIND!** As I said before, you have to study every day in this course. You still have to learn the material even if you fall behind, which could mean spending a lot of hours playing catch up. Eventually there are not enough hours left in the school year for you to catch up! Remember that you are responsible for learning all of the material in this course. Remember, too, that the next course in the college prep sequence--usually Geometry--assumes you know all the material in this course! Catch up NOW before it is too late!

If you find yourself behind, remember that you can find key ideas by looking for bold type, single and double line boxes, and problems that have rectangles around their numbers. If you are not sure what you are supposed to get out of a unit, read the list of topics to be studied on the first page of each unit. Also check to see when your teacher is available for extra help outside of class time. Finally, if you are really behind, struggling, or need more practice, it may be time to get a copy of the "Math 1 Supplement." Ask your teacher how to do this.

Remember -- start working to catch up. For now, please do the next problem.

BC-89. TOOL KIT CLEAN-UP

Tool kits often need to be reorganized to continue to be useful. Your tool kit spans entries from four different units: 0, 1, 2, and 3.

a) Examine the list of tool kit entries from this unit. Check to be sure you have all of these entries. Add any you are missing.

b) Identify which concepts you feel you understand.

c) Which concepts are still not clear to you?

d) Choose entries to create a Unit 0 - 3 tool kit that is shorter, clear, and useful. You may want to consolidate or shorten some entries.

e) How have you used your tool kit in the last two weeks?

- y-intercept
- Tables and Graphs
- x-intercept
- Linear and Parabolic Graphs

To The Students:

Gold Medal problems present you with an opportunity to investigate complex, interesting problems over several days. The purpose is to focus on the process of solving complex problems. ***You will be evaluated on your ability to show, explain, and justify your work and thoughts.*** *Save all work, including what does not work, in order to write about the processes you used to reach your answer. Completion of a Gold Medal Problem includes four parts:*

- ***Problem Statement***: *State the problem clearly in your own words so that anyone reading your paper will understand the problem you intend to solve.*

- ***Process and Solutions***: *Describe in detail your thinking and reasoning as you worked from start to finish. Explain your solution and how you know it is correct. Add diagrams when it helps your explanation. Include things that did not work and changes you made along the way. If you did not complete this problem, describe what you do know and where and why you are stuck.*

- ***Reflection***: *Reflect on your learning and your reaction to the problem. What mathematics did you learn from it? What did you learn about your math problem solving strategies? Is this problem similar to any other problems you have done before? If yes, how?*

- ***Attached work***: *Include all your work and notes. Your scratch work is important because it is a record of your thinking. Do not throw anything away.*

GM-5. TILING THE KITCHEN FLOOR

By using different sizes and shapes of tiles together, intricate geometric designs can be created. For example, using octagonal and square tiles, the popular design at right can be laid out.

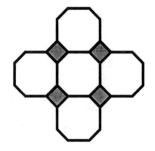

Using a variety of tile shapes and sizes allows for creative designs like the layout at left, which combines squares, triangles, and parallelograms.

After visiting a tile shop, Susan has chosen three different tiles for her kitchen floor, but would like your help in creating a design. Although Susan can not make the above patterns with the tiles she selected, many designs are possible. Use the floor plan provided in problem KF-106 to create a design. Then decide how many of each tile to buy.

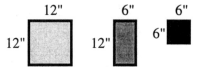

Your Task:

- Make an accurate scale drawing of the outside dimensions of the floor on graph paper. You can use as large a piece of graph paper as you want. Use the problem KF-106 resource page to get the dimensions and shape of the floor. Include your scale on the graph paper.

- Draw in the rectangular design of the 3 inch by 3 inch tiles one foot from the walls. Remember that this position is shown with the dotted line on the resource page.

- With your scale drawing, make a design using the three tiles Susan chose. Make the floor something you would enjoy owning. Use color and creativity that suits you.

- Create a shopping list for Susan's trip to the tile store.

GM-6. MOVE OVER, FRANK LLOYD WRIGHT

Ms. Speedi wants to build her dream house and would like you to design it. According to local zoning laws, she is limited to a one-story house with a maximum area of 1,000 square feet. She also has some personal requirements for this house:

- There must be two bedrooms with a combined area of at least 250 square feet.
- There should be one bathroom.
- The living room cannot be smaller than 200 square feet.
- The kitchen must be at least 220 square feet.
- The building costs are cheaper when the entire structure is rectangular. Therefore, make the overall shape of Ms. Speedi's house a rectangle.

Your Task:

- Create a blueprint of Ms. Speedi's house. Be sure to not only work within Ms. Speedi's requirements and the zoning law, but to also include elements essential to a house: a front and back door, windows, and interior doors. Add furniture if you like.
- Explain to Ms. Speedi the selling points of your design. What makes your design best? What assumptions did you make for your design? Did you need to make any difficult decisions?

Unit Four

Writing and Solving Equations

Unit 4 Objectives
Choosing a Phone Plan: WRITING AND SOLVING EQUATIONS

You already know you can solve problems by Guess and Check. In this unit you will learn to use a more efficient way to solve problems. The skills you have developed using Guess and Check tables will help you **write several equations** for the Phone Plan. Soon you will learn how to **solve these equations** and others like them.

In this unit you will have the opportunity to:

- represent word problems with algebraic equations.

- learn to solve linear equations with manipulatives and the fundamental laws of algebra.

- continue your examination of inverses by "undoing" mathematical operations; you will begin monomial factoring as "undoing" the Distributive Property.

- continue developing your ability to work with variables by solving literal equations.

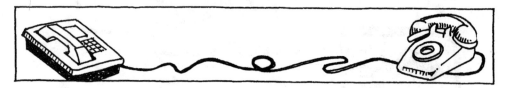

Read the following problem. **Do not try to solve it now**. During this unit, you will develop the skills to solve it.

CP-0. You have decided to try and convince your parents to let you get your own telephone line. After finding out that the monthly rate is $13.50, which you can afford, you offer to pay the cost yourself. However, just before you place your order, you see an advertisement that the local telephone companies are discontinuing the flat rate of $13.50 per month and will only offer metered service. AT^2 service is available for $3 per month plus 2¢ per minute, while PaBell is $8.50 per month and 1¢ per minute. Your parents think that you need to do a little research before making a decision. What would you do to help make your decision?

Obtain the CP-0 Resource Page: Choosing a Phone Plan Phone Log from your teacher today. Be sure to make an entry in your log each day of this unit, including non-school days.

Problem Solving
Graphing
Writing and Solving Equations
Ratios
Geometry
Symbol Manipulation

Unit 4
Choosing a Phone Plan: WRITING AND SOLVING EQUATIONS

 STARTING A MATH PROBLEM

How many times during a test have you come across a math problem and had no idea where to start? Before we start this unit, let's talk about ways to develop methods for beginning math problems.

Students always say to me that the hardest part of a math problem is the beginning. They say, "If I only knew where to start, I could do the rest." Does this sound familiar? If so, here is how we are going to beat this problem. The hard part of a math problem is trying to **figure out what the question wants from you**. There are several ways to do this. Let's look at an example.

SQ-48   The perimeter of a triangle is 76 centimeters. The second side is twice as long as the first side. The third side is four centimeters shorter than the second side. How long is each side?

This problem is from Unit 1 of your math textbook, and I must say it looks tough. Don't panic. We will figure out how to tame it. The first thing I am going to tell you is *I never read through the entire problem the first time I see it.* I know this is against everything you've heard about solving word problems, but I use another method.

I read the problem for information, stop when I find some, and write it down. The goal of this method is to **write the problem in your own words so that it makes more sense**. Notice I did not say copy the problem word for word. Try my method. Take notes from the problem and then read your notes back to yourself. Let's get going and I will show you how it is done.

Read the first part of the problem. It says "The *perimeter ...*" (Stop and write it down.) This is the first piece of information I found. What about you? Wait a minute--what does perimeter mean? Remember to look in your tool kit to understand any unknown math words.

The next piece of information is "triangle." When I did this problem, my notes looked like this:

    Perimeter (triangle) = 76 centimeters

    Side two = twice the first side.

    Side three = 4 cm shorter than side two

    <u>Find</u>: the length of each side.

Now read the notes. This looks a lot better than the way the book wrote it, and it makes sense to me when I read it. I also draw a figure based on my notes to help me see what the words and numbers mean. You may not like the way I write my notes, but that's fine. Write your notes any way you like, but make sure you organize them carefully. Do not leave out important information. **Label everything in your figure**. You will usually know how to continue from this point as long as you understand the vocabulary and have been participating in class.

This is one method of beginning problems. Another method for beginning problems is **drawing a picture** to visualize what the question wants. Notice that the method I use does both. Some people like using only the drawings to get them going. You choose which works best for you.

Still another method is talking to someone else about the problem. You may not be able to use this method all the time, but use it when you can. By trying to explain the problem to someone else, you will better understand what you have to do.

A modification of talking to someone else about the problem is talking to yourself. People might look at you funny, but it works. It helps me solve the problem correctly and that is the goal, isn't it?

Another variation of talking about a problem is reading it out loud. You hear things when you read something aloud that you don't always notice when you read silently. This is also a good idea when you are revising an essay for English or social studies.

Last but not least is something you have done for English classes. BRAINSTORM. Write down everything that comes to your mind about the problem. Don't think about it. Write it. Somewhere in the notes that you write you will find the cues you need to solve the problem. Look closely. Whatever method you choose, always try to relax. It is extremely difficult to start a problem when you are under stress. **Do not panic**. You can figure out anything as long as you participate in class and practice.

MISSION POSSIBLE (something to do)

Apply some of the ideas above as you solve the next set of problems, as well as the problems in the rest of this unit.

WRITING EQUATIONS FROM GUESS AND CHECK TABLES

CP-1. You know how to use a Guess and Check table to solve word problems. In this problem you will see how to extend the use of a Guess and Check table to include writing an algebraic equation to represent a word problem.

Add this example in your tool kit or notes. Make further notes from you teacher's presentation.

USING A GUESS AND CHECK TABLE TO WRITE AN EQUATION

The length of a rectangle is three centimeters more than twice the width. The perimeter is 60 centimeters. Use a Guess and Check table to find how long and how wide the rectangle is, and write an equation from the pattern developed in the table.

STEP 1 In Unit 1, we built the following Guess and Check table:

Guess Width	Length	Perimeter	Check 60 ?
10	2(10) + 3 = 23	2·10 + 2·23 = 66	too high
5	2(5) + 3 = 13	2·5 + 2·13 = 36	too low
8	2(8) + 3 = 19	2·8 + 2·19 = 54	too low
9	2(9) + 3 = 21	2·9 + 2·21 = 60	correct

STEP 2 What happens if we make x our next guess? What would the length be?

Guess Width	Length	Perimeter	Check 60 ?
x	2x + 3		

STEP 3 What would the perimeter be?

Guess Width	Length	Perimeter	Check 60 ?
x	2x + 3	2·x + 2·(2x + 3)	

STEP 4 How do we want the perimeter to relate to 60 ?

Guess Width	Length	Perimeter	Check 60 ?
x	2x + 3	2·x + 2·(2x + 3) =	60

STEP 5 Write the equation: $2 \cdot x + 2 \cdot (2x + 3) = 60$.

We will develop techniques for solving this equation later in this unit. The solution we find algebraically will correspond to a correct guess in the Guess and Check table.

Choosing a Phone Plan: Writing and Solving Equations

CP-2. Use your pattern-detection skills to write an equation represented by the Guess and Check table below.

Guess First number	Second number	Total	Check 149 ?
65	68	133	too low
71	74	145	too low
75	78	153	too high
73	76	149	just right

Answers: The numbers are 73 and 76.

Equation:

Solve each of problems CP-3 through CP-10 by using Guess and Check tables. Use the pattern you develop in the table to write an equation. Write your solution in a sentence.

CP-3. Heather has twice as many dimes as nickels and two more quarters than nickels. The value of the coins is $5.50. How many quarters does she have?

Guess # of Nickels	Number of Dimes	Number of Quarters	Value of Nickels	Value of Dimes	Value of Quarters	Total Money	Check $5.50?

CP-4. My new Saturn cost $14,000. Each month, it depreciates $100. In how many months will it be worth only $10,000?

CP-5. One number is five more than a second number. The product of the numbers is 3300. What are the numbers?

CP-6. On a 520-mile trip, Thu and Cleo shared the driving. Cleo drove 80 miles more than Thu drove. How far did each person drive?

CP-7. Chris is three years older than David. David is twice as old as Rick. The sum of Rick's age and David's age is 81. How old is Rick?

CP-8. Find three consecutive numbers whose sum is 57.

CP-9. Margaret is twice as old as Jenny, and Sarah is twice as old as Margaret. Their combined ages total 133 years. How old is each person?

CP-10. The State Market has 27 more apples than oranges. There are 301 apples and oranges altogether. How many apples are in the market?

CP-11. Use the Distributive Property to rewrite each expression.

  a) $4(2x + 3)$

  b) $-4(2x + 3)$

  c) $4x(2x + 3)$

  d) $-4(2x - 3)$

CP-12. Make a table with eight values for x between -3 and 3 and find the corresponding y-values. Then, graph the following equations on the <u>same set of axes</u>.

  1) $y = x$

  2) $y = x + 2$

  a) Where do the graphs of the two equations meet?

  b) List any similarities and differences you see in the graphs and equations.

CP-13. Find the value of each expression below for x = 2, then again for x = -3. Note: You will compute two different values for each expression.

  a) $x^2 - 3x + 8$

  b) $-3x^2 + x$

  c) $x^2 + x - 6$

CP-14. Log your phone time for today on the CP-0 Resource Page.

MORE WRITING EQUATIONS

Solve each of problems CP-15 through CP-22 by making a Guess and Check table and then write an equation from your table. Write your solution in a sentence.

CP-15. Mary sold 105 tickets for the basketball game. Each adult ticket costs $2.50 and each student ticket costs $1.10. Mary collected $221.90. How many of each kind of ticket did she sell?

# of Adult Tickets	# of Student Tickets	Value of Adult Tickets	Value of Student Tickets	Total Money	Check $221.90?

CP-16. Ms. Speedi keeps coins for paying the toll crossings on her commute to and from work. She presently has three more dimes than nickels and two fewer quarters than nickels. The total value is $5.40. Find the number of each type of coin she has.

Choosing a Phone Plan: Writing and Solving Equations

CP-17. Latisha and Maisha are twins. They have a brother who is eleven years younger than them and an older sister who is four years older. The sum of the ages of all four siblings is 69. Find Latisha's age.

CP-18. Antony joined a book club in which he received 5 books for a penny. After that, he received 2 books per month, for which he had to pay $8.95 each. So far, he has paid the book club $196.91. How many books has he received?

CP-19. A rectangular goat pen is enclosed by a barn on one side and by 100 feet of fence on the three other sides. The area of the pen is 912 square feet.

a) Draw a diagram of the goat pen and then find its dimensions.

b) Is there more than one answer? Does this change the area?

CP-20. For parts (a), (b), and (c), write an expression that represents each amount. Then generalize the process in part (d).

a) A three foot piece of string is cut. If one piece is x feet long, how long is the other piece?

b) Twenty raffle tickets were sold. If g girls bought raffle tickets, how many boys bought raffle tickets?

c) A pizza has twelve slices. If Karen eats p slices, how many slices are left?

d) Write a general method for expressing two parts of a whole using a single variable.

CP-21. The length of a rectangle is three times the width. The perimeter is 16 feet. What is the width of the rectangle?

CP-22. A notebook costs $0.15 more than a pen. The total cost of the pen and notebook is $2.25. How much does the pen cost?

CP-23. Simplify each expression by removing parenthesis and combining any like terms.

a) $3(2x + 7) - 10$

b) $-3(m + 2) + 4m + 7$

c) $-4 + 2(x - 7)$

d) $2x^2 + 3x - 7 + 5x^2 - 5x + 2$

CP-24. Suppose you have a standard deck of 52 playing cards. Cards have four suits: hearts, clubs, diamonds and spades. Each suit has 13 cards: Ace, 2 through 10, Jack, Queen and King.

a) What is the probability of drawing a five of spades?

b) What is the probability of drawing a five?

c) If you just drew a five and do not put the card back, what is the probability of drawing another five?

CP-25. Copy and solve each of the following Diamond Problems.

CP-26. Log your phone time for today on the CP-0 Resource Page.

Choosing a Phone Plan: Writing and Solving Equations

SOLVING SIMPLE EQUATIONS WITH CUPS AND TILES

CP-27. Build each equation with cups and tiles. Then solve it as we did in the class demonstration. Be sure to record each of the moves you make in solving the equation by making a sketch and writing what you did at each step. Show how you checked your solution.

a) $2x + 3 = 11$

b) $2x + 1 = 13$

c) $5x + 8 = 3x + 10$

d) $5 + 2x = 7 + x$

For each of the following word problems, you should:
a) Use a Guess and Check table to write the equation and use algebra to solve it. You may also solve the problem using Guess and Check, then write the equation.
b) Write a complete answer to the question, including units if needed; and
c) Check your answer.

CP-28. Last Friday night, 2000 tickets were sold at a local high school football game. Adult tickets sold for $7.50 and student tickets for $5.00. The total revenue was $11,625. How many student tickets were sold?

CP-29. José and Sachiko leave Sacramento driving in opposite directions. José drives five miles per hour faster than Sachiko. In four hours they are 476 miles apart. How fast is each person traveling?

CP-30. Alejandra cut a 40-inch long board into two pieces and painted one piece purple and the other piece orange. The purple board is four inches longer than the orange board. How long is each painted board?

CP-31. Raisa cut a string 112 centimeters long into two pieces so that one piece is three times as long as the other. How long is each piece?

CP-32. Fred, the Times newspaper distributor, collected all the dimes, nickels and quarters from one of his vending machines. He gathered twice as many dimes as quarters, and two more nickels than quarters. He collected a total of $7.60. How many coins did Fred take from the machine?

CP-33. Use the Order of Operations to determine the value of the following expressions.

a) $(2 \cdot 3)^2 + 4 \cdot 3 - 1$

b) $((2 \cdot 3)^2 + 4) \cdot (3 - 1)$

c) $(2 \cdot 3)^2 + 4 \cdot (3 - 1)$

d) $2 \cdot (3^2 + 4) \cdot (3 - 1)$

CP-34. Graph the line $y = -3x + 1$ for $-3 \leq x \leq 3$. Find the x- and y-intercepts.

CP-35. Find the area of each figure. Show all subproblems.

a)

b)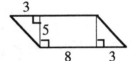

CP-36. Log your phone time for today on the CP-0 Resource Page.

CP-37. Solve each of the following equations by "inspection" (mental math).

a) $x + 7 = 2$

b) $3x = -45$

c) $-5 = \frac{x}{2}$

d) $-5 = \frac{1}{2}x$

e) $2 = -x$

Choosing a Phone Plan: Writing and Solving Equations

EXAMPLE OF USING CUPS AND TILES TO MODEL SOLVING EQUATIONS

Model with variables	Model with cups and tiles	Explanation
Step 1.	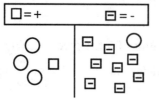	Here is the problem: The value of each side is the same and the same number of tiles must be in each of the cups. What can we do to simplify both sides of the equation and determine how many tiles are in each cup?
$3c + 1 = -8 + c$		Write the equation.
Step 2. $3c + 1 = -8 + c$ $\underline{-c -c}$ $2c + 1 = -8$	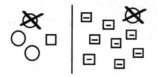	Remove one cup from each side.
Step 3. $2c + 1 = -8$ $\underline{-1 -1}$ $2c = -9$	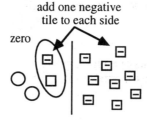 add one negative tile to each side; zero	"We cannot remove one positive tile from each side, but we can add one negative to each side."
Step 4. $\dfrac{2c}{2} = \dfrac{-9}{2}$	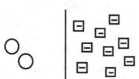	"If 2 cups balance 9 negative tiles, what does 1 cup balance?" We must divide the 9 negative tiles among the two cups.
Step 5. $\dfrac{2c}{2} = \dfrac{-9}{2}$, so $c = -4\dfrac{1}{2}$		Each cup has $4\dfrac{1}{2}$ of the negative tiles, so $c = -4\dfrac{1}{2}$.
Step 6. $3(-4\dfrac{1}{2}) + 1 = -8 + -4\dfrac{1}{2}$ $-12\dfrac{1}{2} = -12\dfrac{1}{2}$		Check the solution

CP-38. Use cups and tiles to solve each of the following equations. Show the check for each problem by substituting your answer for the variable and doing the arithmetic.
a) $2x - 1 = 13$
b) $-3 + 2b = -9$
c) $-5 = 3m + 1$
d) $-6 + 4y = 3$
e) $c - 8 = 2 + 5c$

CP-39. Try solving the following equations for the given variable <u>without</u> cups and tiles. Record each step.
a) $-15 = 2d - 25$
b) $p + 2p = -15$
c) $-6.2 = q - 4.4$

CP-40. After a home remodeling project, Susan paid $12,490 to a contractor who used the money to pay himself, a plumber, an electrician and a tiling expert. The plumber and the electrician received the same amount, while the tiling expert received $2,000 less. The contractor received three times as much as the tiling expert. How much did each person get?

CP-41. The two line segments at right have equal lengths. Write an equation that represents the diagram. How long is x?

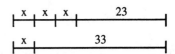

CP-42.  Hakeem has ten coins in his pocket, all quarters, nickels, and dimes, worth a total of $1.10. When he reaches in his pocket to grab one coin, the probability that he grabs a nickel is the same as the probability that he grabs a dime. What is the probability that he will grab a quarter?

CP-43. A fallen tree branch 94 inches long is cut into seven logs, all the same length. A ten inch piece is left over. How long is each log?

CP-44. Use cups and tiles, or a sketch to represent them, to solve each of the following equations. Show all parts of your work.
a) $x + 4 = 7$
b) $3x - 11 = 6$
c) $3 + 3x = 9 + 2x$
d) $2x - 7 = 3 + x$
e) $3x - 3 = 6 + x$

Choosing a Phone Plan: Writing and Solving Equations

CP-45. Right triangle ABC has an area of 45. Find the coordinates of B and C. Show all subproblems.

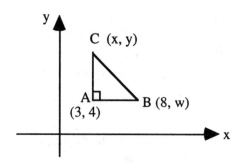

CP-46. Multiply. Sketch a generic rectangle as needed.

  a) 3(4x - 5)   b) 3x(4x - 5)   c) -5(4x - 5)

CP-47. Log your phone time for today on the CP-0 Resource Page.

COMMON TERM FACTORING AND MORE EQUATIONS

CP-48. Solve and check each of the following equations for x. Show your checks.

  a) -3x + 2 = 14   c) -2x - 4 = 4x + 8

  b) -2x + 6 = 10   d) 3x - 1 = 5x + 6

CP-49. For each equation below, first express the equation with words, and then describe what you will do to "undo" it. Finally, solve each equation for the given variable.

  a) -3x = 13   d) $\frac{2}{3}z - 2 = 8$

  b) $\frac{1}{2}y = -13$ e) $\frac{7}{2}x + 6 = x + 12$

  c) $\frac{2}{3}z = 8$

CP-50. Elizabeth correctly solved the following problems below. Copy her work. On each blank line explain what step she did.

EXAMPLE #1

$$-2x - 7 = -13$$
$$-2x = -6$$
$$x = 3$$

EXAMPLE #2

$$-x + 8 = 2x + 7x - 13$$
$$-x + 8 = 9x - 13$$
$$-x = 9x - 21$$
$$-10x = -21$$
$$x = \frac{-21}{-10} = 2.1$$

CP-51. Copy this definition into your tool kit.

> We have been using the Distributive Property to **MULTIPLY** and **FACTOR**.
>
> **MULTIPLYING** means removing parenthesis or changing products to sums.
> **COMMON TERM FACTORING** means writing with parenthesis or changing sums to products.
>
> For example: MULTIPLY $4(2x + 3) = 4 \cdot 2x + 4 \cdot 3 = 8x + 12$
> MULTIPLY $4x(2x + 3) = 4x \cdot 2x + 4x \cdot 3 = 8x^2 + 12x$
>
> For example: FACTOR $4m + 8 = 4m + 4 \cdot 2 = 4(m + 2)$
> FACTOR $3x + 6 = 3x + 3 \cdot 2 = 3(x + 2)$
> FACTOR $a^2 + ab = a \cdot a + a \cdot b = a(a + b)$

CP-52. Use the above information and the Distributive Property to rewrite each expression. Multiply in parts (a) and (b). Factor in parts (c) and (d).

a) $2(3x - 2)$

b) $2x(5x + 3)$

c) $5x + 15$

d) $x^2 + 4x$

CP-53. Write an equation and solve it. Be sure to identify what the variable represents if you do not show a Guess and Check table.

Find three consecutive even numbers whose sum is 54.

CP-54. Manuel's savings account has a return of 7.5%. After one year, during which Manuel did not add to or withdraw from his account, the account had $1410.40 in it. What amount was in the account at the beginning of the year?

CP-55. Part of $20,000 is invested at 12% interest. The rest is invested at 8% interest. The total interest from the two accounts is $1800. How much money was invested at each rate?

 HOW DO YOU STUDY FOR MATH TESTS?

Write a paragraph that describes what you do to study for tests in this class. I'll be back in a few days to talk more about studying for tests. See you then!

CP-56. Simplify.

a) $3x(2x - 4) + (7x)^2$ b) $(3 \cdot 10^4)(7.1 \cdot 10^3)$ c) $(0.75)^2 - 3(4 - 8)$

CP-57. Let x = the number of students in our school. Write an algebraic expression using x for the following amounts using a decimal and a fraction.

a) 50% of the students are female. How many students are female?

b) 25% of our students bring their lunch. How many students bring their lunch?

c) 10% of our students wear nail polish. How many students wear nail polish?

d) $\frac{2}{3}$ of our students wear pants. How many students wear pants?

e) 75% of our students wear a watch. How many students wear a watch?

f) 20% of our students had pizza for dinner last night. How many students had pizza?

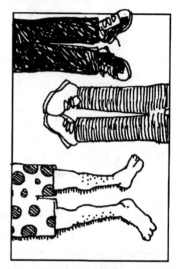

CP-58. The two line segments at right have equal lengths.

a) Write an equation to describe the relationship of the two lines.

b) Find the length x.

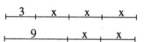

CP-59. Simplify:

a) $2x + \frac{1}{2}x$ c) $1y - 0.2y$

b) $x + 0.3x$ d) $m - \frac{1}{3}m$

CP-60. Solve for x.

a) $\frac{1}{2}x = -20$ c) $2x + \frac{1}{2}x = 20$

b) $\frac{4}{3}x = 9$ d) $\frac{3}{2}x - 8 = 16$

CP-61. Use your pattern-detection skills to write an equation and a word problem represented by the Guess and Check table below. There are many word problems you could write. Write only one, but be creative!

Guess First number	Second number	Product	Check 1974 ?
30	35	1050	too low
40	45	1800	too low
43	48	2064	too high
42	47	1974	correct

Answer: The number is 42.
Equation:
Your word problem:

CP-62. Log your phone time for today on the CP-0 Resource Page.

SOLVING EQUATIONS AS INVERSE OPERATIONS

CP-63. Solve for x and y given the information below. What subproblem must you do first?

$$3x + 2 = -4 \qquad \frac{1}{2}x - 5 = y$$

CP-64. Much of mathematics consists of the closely related processes of "doing" and "undoing." In many cases, to solve a problem we must "undo" something that has been "done." This is especially true when we solve equations. The notion of being able to "undo" something is part of the problem solving strategy of **WORKING BACKWARDS** and the mathematical concept called an **INVERSE**.

One way to think of this is to analyze the processes of wrapping and unwrapping a gift. The steps to wrap a gift will be reversed when unwrapping it. When we solve equations, we are also "undoing" operations on our variable.

What operation "undoes" addition? Subtraction? Multiplication? Division? Squaring?

Choosing a Phone Plan: Writing and Solving Equations

CP-65. Solve each of the following equations by "undoing" what has been done to the variable to find the "original number" represented by the variable.

a) $-8y + 35 = -21$

b) $-26 + 5 - 3t = 18t$

c) $9d - 1 = d - 25$

d) $-15v - 40 = 23 - 8v$

e) $-x - 1 = x - 21$

> For CP-66 and CP-68, write an equation and solve it. Be sure to define the variable(s) you use.

CP-66. The length of a rectangle is twice its width. Its perimeter is 36 centimeters. Find the area of the rectangle.

CP-67. Alicia is twice Barbara's age and Kamran is two years younger than Barbara. The total of their ages is 46. How old is each person?

CP-68. Jerry needs to raise his batting average. So far he has 20 hits in 100 times at bat for a 0.200 batting average. How many consecutive hits does Jerry need in order to raise his average to at least 0.250? Solve. Write an equation.

CP-69. Graph each of the following equations on the same set of axes. Remember to label the axes and each graph.

a) $y = x$

b) $y = x - 3$

c) $y = 3x$

d) Compare the graphs in parts (a) and (b). How did the "3" in part (b) change the graph from part (a)?

e) Compare the graphs in parts (a) and (c). How did the "3" in part (c) change the graph from part (a)?

CP-70. Solve each of the following equations.

a) $x + 8 = 2$

b) $5x + 0.5 = 0.2x + 1.5$

c) $\frac{2}{5}x = 5$

d) $0.3x + 1.2 = x + 8.4$

e) $x - 8\frac{1}{2} = -5\frac{1}{6}$

CP-71. A certain rectangle has a perimeter of 26 centimeters. Its length is 7.25 centimeters. Draw a diagram, write an equation, and find the width and the area of the rectangle.

CP-72. Use subproblems to find the area of the shaded region.

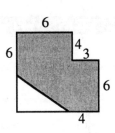

CP-73. Use your pattern-detection skills to write an equation and a word problem represented by the Guess and Check table below. There are many word problems you could write.

Guess First number	Second number	Total	Check 22 ?
4	9	13	too low
5	11	16	too low
9	19	28	too high
7	15	22	correct

Answers: The answers are 7 and 15.
Equation:
Your word problem:

CP-74. The answers in this problem are not numerical. Use a, b, c, and/or d to answer each question.

a) What are the coordinates of F?

b) What are the coordinates of G?

c) What is the area of triangle AHG?

d) Extension: What is the area of the shaded region?

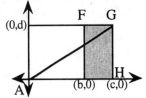

CP-75. Calculate to the nearest hundredth. Be careful with the Order of Operations.

a) $\dfrac{-5 + \sqrt{217}}{16}$

b) $\dfrac{-5 - \sqrt{217}}{16}$

CP-76. Log your phone time for today on the CP-0 Resource Page.

Choosing a Phone Plan: Writing and Solving Equations

EQUATIONS WITH PARENTHESES

CP-77. Add this information to your tool kit:

SOLVING EQUATIONS CONTAINING PARENTHESES

Example: Solve the equation $5(2x + 7) = 55$. $\qquad 5(2x + 7) = 55$

The feature that makes this equation different from the ones you have solved before is the parentheses. We will first need to remove the parentheses.

One way to remove the parentheses is to use the **Distributive Property** to rewrite the left side of the equation. $\qquad 10x + 35 = 55$

Now that the equation looks like many we have seen before, we can solve it in the usual way. First, we subtract 35 from each side. $\qquad 10x = 20$

Then we divide both sides of the resulting equation by 10. $\qquad x = 2$

CP-78. Here is one way to solve the equation $9(3x - 8) = 36$:

Copy the complete problem shown below on your paper. Fill in the blank labeled (b) to explain how the equation to its right was obtained from the equation above it. Then do the same for line (c).

Solve the equation $9(3x - 8) = 36$ $\qquad\qquad 9(3x - 8) = 36$
a) __Distribute the 9__ $\qquad\qquad\qquad\qquad 27x - 72 = 36$
b) _____ $\qquad\qquad\qquad\qquad 27x = 108$
c) _____ $\qquad\qquad\qquad\qquad x = 4$

CP-79. Here is another way to solve the equation $9(3x - 8) = 36$:

Copy the complete problem shown below on your paper. Fill in the blank labeled (b) to explain how the equation to its right was obtained from the equation above it. Then do the same for parts (b) and (c).

Solve the equation $9(3x - 8) = 36$ $\qquad\qquad 9(3x - 8) = 36$
a) _____ $\qquad\qquad\qquad\qquad 3x - 8 = 4$
b) _____ $\qquad\qquad\qquad\qquad 3x = 12$
c) _____ $\qquad\qquad\qquad\qquad x = 4$

CP-80. Compare the two methods in problems CP-78 and CP-79. Which would you prefer to use? Why? Which way is correct?

CP-81. Explain why the method shown in problem CP-79 would be inconvenient (but <u>not</u> incorrect) for solving the equation $3(2x - 1) = 7$.

CP-82. Solve each of the following equations for x. Show each step you use.

    a) $6(5x + 12) = 162$ c) $5(6x - 9) = -15$

    b) $2(7x + 15) = 128$ d) $3(9x - 14) = 16$

CP-83. Solve each of the following equations for x. Show the steps you use. Hint for part (f): Remember that $4 - 8(3x - 5)$ means $4 + (-8)(3x - 5)$, and the multiplication is done **first**.

    a) $5(x - 4) = 25$ d) $3(4x + 1) = 159$

    b) $x + 3 = 31$ e) $4(3x + 2) - 18 = 14$

    c) $7(2x - 4) + 3 = 31$ f) $4 - 8(3x - 5) = 92$

    g)   Can you use the method shown in problem CP-79 to solve parts (e) and (f)? Why or why not?

CP-84. If one side of a square is increased by 12 feet and an adjacent side is decreased by three feet, a rectangle is formed whose perimeter is 64 feet. Draw a diagram and find the length of the side of the original square. Write an equation and solve it. Be sure to define the variables you use.

CP-85. In honor of the 25th anniversary of Carla's Department Store, everything was 25% off. Jarlene bought a jacket at the sale for $45. What was the original price of the jacket? Write an equation and solve it. Be sure to define the variables you use.

CP-86. Solve for x:

    a) $3x - 5 = 5x - 9$ d) $5(2x - 3) - 4x = 8$

    b) $3(x - 4) = 15$ e) $4x - 8 + 2x = 6 + x$

    c) $-2(x - 2) = 11$ f) $\frac{2}{3}m = 9$

CP-87. Pat found a defective die: it had two 6's on it and no 1's. If Pat rolls the die, how likely is it to come up:

a) a 2 ?

c) either a 5 or a 6 ?

b) a 6 ?

CP-88. When Marla started her Algebra homework, she drew the diagram at right. Figure out what her original equation was, and solve for the variable (represented by the circle).

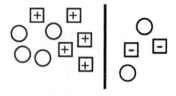

CP-89. Graph these equations on one set of axes. Label each equation clearly.

a) $y = 2x$

c) $y = x + 2$

b) $y = x^2$

CP-90. Log your phone time for today on the CP-0 Resource Page.

CP-91. Your phone log should have from six to eleven entries at this time.

a) Find the **mean** (or average) number of minutes you spend on the phone per day.

b) Assuming there are 30 days in a month, about how many minutes do you spend on the phone per month?

| CP-92. | CHOOSING A PHONE PLAN

INTRODUCTION: You have decided to try and convince your parents to let you get your own telephone line. After finding out that the monthly rate is $13.50, which you can afford, you offer to pay the cost yourself. However, just before you place your order, you see an advertisement that the local telephone companies are discontinuing the flat rate of $13.50 per month and will only offer metered service. Their ads are shown below:

PROBLEM: Your parents think that you need to do a little research before making a decision. Complete the series of questions below to help reach your decision.

a) About how many minutes of phone calls do <u>you</u> make per month? (Refer to CP-91) What is the largest number of minutes from your team?

b) For the flat rate and for each company make a table comparing total minutes to total cost. Choose at least four reasonable amounts to represent the number of minutes per month. It will be helpful to include 0 minutes and the largest number of minutes from your team. You will have three tables when you are done.

c) We are now going to plot our data. Carefully plan how to scale your axes. Label the vertical axis "cost per month" and the horizontal axis "minutes of calls" per month. Plot and connect the points for each table in part (b). Be sure to label the graph of each company plan.

d) When do both companies charge the same amount? When is AT^2 a better choice? When is Pa Bell a better choice? Discuss how your graphs relate to the solution of the problem.

e) Consider the options using the number of minutes you call per month. Which plan would you choose? Will you save money over the flat rate?

f) For each option, use the pattern in the table to write an equation which represents the cost in a given month in terms of the minutes of phone calls.

g) For each person in your study team, write down their name and how many minutes they estimate they spend on the phone. Which phone option should each person in your study team use and why?

CP-93. Solve each of the following equations.

a) $3(x + 1) = 4$

b) $4(y - 2) + 3 = 19$

c) $-2(x - 2) = 16$

d) $3(2x + 4) = 28$

CP-94. Solve each of the following equations by any methods you choose.

a) $\frac{1}{2}x + 4 = -\frac{1}{4}x + 7\frac{1}{2}$

b) $x + .63 = 1.56$

c) $0.78x - 2 = 0.8x + 8.4$

d) $0.38x = 1.82$

e) $1.2 - x = 0.8x - 1.2$

f) $2x - 1 = 0$

g) $\frac{x}{1.5} = 2$

> For each of problems CP-95 and CP-96, write an equation and solve it. Be sure to define the variables you use.

CP-95. Each side of a square garden is increased by three meters. The perimeter of the new garden is 50 meters. What was the length of the original square garden? Draw a diagram and label the sides.

CP-96. Mrs. Agnos keeps cats and canaries. The animals have a combined total of 30 heads and 80 legs. How many cats does Mrs. Agnos have?

CP-97. Solve each of the following equations. Leave answers as fractions, not decimals.

a) $9x = 3$

b) $5x = 11$

c) $14 = -3x$

d) $\frac{2}{5}x = \frac{1}{6}$

e) $\frac{9}{14} = -\frac{3}{7}x$

CP-98. Solve each of the following equations.

a) $5(4x + 3) = 75$

b) $-6(3x - 8) = -6$

c) $2 - 3(2x - 1) = 17$ (Hint: look back at CP-83 part (f).)

d) $-3(8x - 4) = 18$

e) $3 + 4(x + 1) = 159$

LITERAL EQUATIONS

 PZL-18 HOW TO STUDY FOR MATH TESTS

cl-ss

Everyone in the team should get out their response to PZL-17 (right after CP-57). I'll bet that at least one person in the team said, "I study the night before the test and memorize everything I

can." This method will not work in this course, and usually isn't very good for most courses. If your answer was even close to this, we need to talk. **The best way to study for an exam in this course is to participate in every class session**. In other words, you have to **learn the material as you go along**. Once you learn the material, then you can go back and **review several days ahead of the exam**. You can practice the night before to smooth any rough edges. *There is no way you will be able to learn the material for the first time the night before an exam!*

I do understand that it is difficult to review several days ahead, but the authors have built in some ways for you to do it. That is what the summary and tool kit assignments are for. Some teachers even give team quizzes and tests. That's another way to review. The real key is to **participate in every class session and ask good questions**. Then the night before a test is for polishing what you have learned, not a panicked cram session.

Here's a summary of the points I have made so far about studying for success in CPM Algebra 1. Call it the "Master Study Plan."

1. Do not fall behind.
2. Communicate with your partner, team, and teacher.
3. Build a complete tool kit.
4. Keep a good record of correct homework problems (notebook).
5. Know where to look for information (see PZL-6 in Unit 1).
6. Make the connections.

CP-99. Solve each of the following equations for x.

a) $2x + 3 = 13$ e) $5x + 4 = x + 20$

b) $2x + b = 13$ f) $5x + d = x + 20$

c) $3x - 2 = 10$ g) $\frac{x}{2} + 5 = 9$

d) $bx - 2 = 10$ h) $\frac{x}{2} + e = 9$

CP-100. Look back at problem CP-99. Compare how you solved the equation in part (b) with how you solved the equation in part (a). Write one or two sentences comparing the solutions.

CP-101. Solve as indicated in each part below.

a) Solve $3x + 91 = 43$ for x.

b) Solve $3x + c = 43$ for x.

c) Solve $3x + c = 43$ for c.

d) Solve $5x + 3c = 17$ for x.

CP-102. For each equation below, complete a table with the given domain (inputs). Graph each equation on a separate set of axes. Be sure everything is labeled.

a) $y = x^2 - 3x + 2$ for $-1 \le x \le 4$. Explain why it is important to use 1.5 as an input value.

b) $y = \frac{1}{3}x + 3$, for $x = -9, -6, -3, \ldots, 6, 9$.

c) $y = 8 - x$, for $-3 \le x \le 10$.

CP-103. Mrs. Sanchez needs to separate $385 into three parts to pay some debts. The second part must be five times as large as the first part. The third part must be $35 more than the first part. How much money must be in each part? Include an equation in your solution.

CP-104. Write an expression for the perimeter and the area of each of the rectangles below.

a)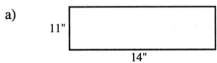

b)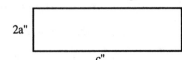

CP-105. All of the angles in the two figures at right are right angles.

a) Find the perimeter and area of the figure at right.

b) Find the lengths of the unlabeled sides in the second figure. Then find its perimeter and area.

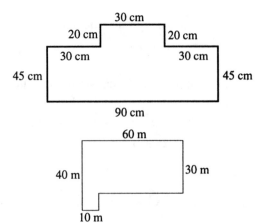

CP-106. Sunrise Sand and Gravel charges $4.50 for each cubic yard of sand plus $6.00 to deliver the sand. How much sand would they deliver for $100.00?

CP-107. The length of a rectangular picture frame is twice the width. The perimeter is 132 centimeters. What is the width of the frame? (Hint: draw a rectangle and label its sides.)

CP-108. Sketch a graph that fits the following description:

The height of a softball that was hit over a fence for a home run versus the distance the ball is from home plate.

CP-109. Find the area of the unshaded part. Show all subproblems in your work.

a) 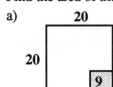   b) 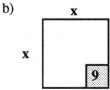   c)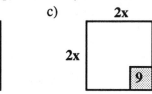

CP-110. Solve for the indicated variable.

a) $D = \frac{m}{v}$, solve for m.

b) $D = r \cdot t$, solve for t.

c) $y = mx + b$, solve for b.

d) $y = mx$, solve for m.

Choosing a Phone Plan: Writing and Solving Equations

CP-111. UNIT 4 SUMMARY

With your study team, quickly list as many mathematical ideas or topics you remember from this Unit. Do not flip back through the text or your notes! There will be time for that later. This exercise is to see what comes from memory. Be attentive to any suggestions offered by your teammates. Each member needs to write the list down.

a) With the list generated from the class, discuss with your study team the relative importance of each item. Which ones were crucial to understand? Narrow the list to the most important four main ideas. Each team member should be responsible for one idea. **Complete parts (b) through (d) as directed by your teacher.**

b) For each topic, choose a problem from the Unit that demonstrates each main idea. Be selective. While there are many problems to choose from, certain problems are better examples than others. Perhaps you would like to choose a problem you are most proud of solving. Check that the problems you choose are directly related to the main ideas of your summary.

c) Solve each problem and show all work, especially the subproblems.

d) Finally, explain the method you used to arrive at your solution. Use complete sentences and be as descriptive as you can.

CP-112.* Mr. Keller drove from Ukiah to Redwood City by the route shown on the map below. The graph depicts his trip.

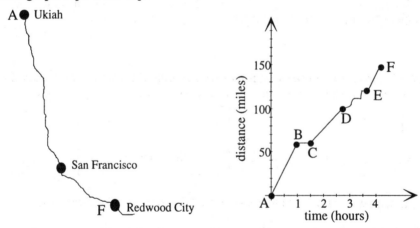

a) How far did Mr. Keller travel on his trip?

b) Describe Mr. Keller's speed during the interval from A to B.

c) What might have happened during the interval from B to C?

d) Something different happened as Mr. Keller traveled from D to E. Explain what you think might have happened.

e) In going from C to D, did Mr. Keller travel faster or more slowly than he did from A to B? Explain how you know.

---

* Adapted from *The Language of Functions and Graphs*, Joint Matriculation Board and the Shell Centre for Mathematical Education, University of Nottingham, England

For each of problems CP-113 through CP-115, write an equation and solve it. Be sure to define the variables you use.

CP-113. Mr. Keller traveled a total of 150 miles from Ukiah to Redwood City, as shown in the diagram in problem CP-112. If the distance between Ukiah and San Francisco is 18 miles more than twice the distance between San Francisco and Redwood City, what is the distance between Ukiah and San Francisco?

CP-114. The perimeter of a triangle is 33 centimeters. The second side is twice as long as the first side, and the third side is three centimeters longer than the second. How long is each side?

CP-115. The coins in a piggy bank are all dimes, nickels, and quarters. There are twice as many dimes as quarters, and two more nickels than quarters. If the total value of the coins is $7.60, how many coins are in the piggy bank?

CP-116. Solve each of the following equations for x.

a) $x + 1.5 = -3.25$

b) $8 = 4(x + 7) - 20$

c) $2x + 3 = 10$

d) $4x + 5 = 2x + 7$

e) $3x + c = 10$

f) $15 = 6x + 4x - 8$

CP-117. Use the Distributive Property to rewrite each of the following expressions. In (a) and (d) you must multiply. In (b) and (c) you must factor.

a) $4(x + 5)$

b) $x \cdot x + x \cdot 3$

c) $5 \cdot 2x + 5 \cdot 4$

d) $-3(2y - 5)$

CP-118. Write an equation and solve it. Be sure to define the variable(s) you use.

On an algebra test each question in Part A is worth three points and each question in Part B is worth five points. Latisha answered 19 questions correctly and had a score of 81. How many questions on each part of the test did she answer correctly?

CP-119. Solve each of the following equations for x.

a) $\frac{3x + 5}{4} = 5$

b) $\frac{3}{4}x - 6 = 12$

c) $4x + c = -x + 7$

d) $xc - 5 = y$

e) $14 - 3(2x + 1) = 15$

Choosing a Phone Plan: Writing and Solving Equations

CP-120. Copy and solve these Diamond Problems:

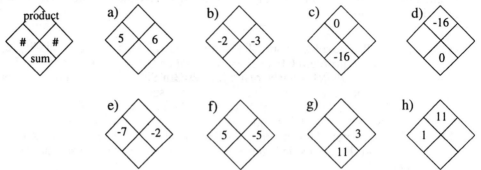

CP-121.* Look at each graph below, then write a story or description about what each graph shows. Your story may be different from what your study teammates will write.

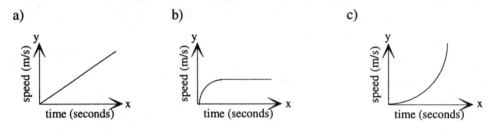

CP-122. Consider this number trick:

"Pick a number. Multiply it by 2. Add 6. Multiply the sum by 2.
Divide by 4, then subtract 3. The result is the same number you started with."

a) Show that the trick works if you start with a number of your choice.

b) Show that the trick works if you start with x.

CP-123. TOOL KIT CHECK-UP

Your tool kit contains reference tools for algebra. Return to your tool kit entries. You may need to revise or add entries.

Examine the list below. Each item should be included in your tool kit; add any that are missing now. You will find them in your text by skimming through the unit or by using the index in the back of the book.

- Using Guess and Check tables to write equations
- Multiply and factor
- Using cups and tiles to model solving equations
- Solving equations with parentheses

---

* Adapted from *The Language of Functions and Graphs*, Joint Matriculation Board and the Shell Centre for Mathematical Education, University of Nottingham, England

GM-7. MAKING DECISIONS, Part One

Carlos wanted to get a part-time job to have extra money for the holidays. After reading the advertisement at right, he applied. His interview went well, and he was offered the job. He had to choose between the following two pay scales:

Pay Scale # 1

He would make $10 per <u>minute</u>.

Pay Scale # 2

On day # 1, he would earn a total of 1¢
On day # 2, he would earn a total of 2¢
On day # 3, he would earn a total of 4¢

Each day, his salary rate would double from the day before.

Your Task:

- Before starting the problem guess which pay scale Carlos chose. Explain why you think this.

- If November 24th is a Monday, and the store is closed on Thanksgiving (November 27th), compare how much he would make for each day he works until December 24th using both scales.

- If you had a job offer like this one, which pay scale would you choose? Why?

- What number of days could you work to make the other pay scale attractive?

Choosing a Phone Plan: Writing and Solving Equations

GM-8. MAKING DECISIONS, Part Two

Carlos loves his job, so you decided to answer the advertisement shown below.

You were hired and given a choice of the following two pay scales:

Pay Scale # 1

You would make $20 per <u>second</u>.

Pay Scale # 2

On day # 1, you would earn a total of 1¢
On day # 2, you would earn a total of 8¢
On day # 3, you would earn a total of 64¢

Each day, your salary rate would multiply by a factor of eight from the day before.

Your Task:

- Since this job fewer days, compare your daily earnings using each pay scale. Remember that December 6th is a Monday.
- Which job would you choose? Why?

Summary:

- Describe what you learned by analyzing all 4 of the pay scales?
- Did the calculator do anything unusual? If so, describe it and explain why you think it happened.

Unit Five
Estimating Fish Populations
Numerical, Geometric, and Algebraic Ratios

Unit 5 Objectives
Estimating Fish Populations:
NUMERICAL, GEOMETRIC AND ALGEBRAIC RATIOS

In this unit you will be exploring the concept of **ratio**, an idea which is useful in many aspects of our lives. Ratios you have used so far include fractions, percents, and probabilities. You will further expand your knowledge of ratios, and will use the skills you develop to solve a problem similar to the one represented in the Estimating Fish Problem:

In this unit you will have the opportunity to:

- explore ratios from a numerical, geometric, and algebraic perspective.
- compare ratios of sides, perimeters, and areas for plane figures.
- explore percent as a ratio.
- use equivalent ratios in relation to graphs of lines.
- write and solve equations that involve ratios, including proportions.

We will continue to develop writing and solving linear equations, as well as solving word problems. Recognizing patterns will be an important skill used in our investigation of ratios.

EF-0.   ESTIMATING FISH POPULATIONS

Fish biologists need to keep track of fish populations in the waters they monitor. They may want to know, for example, how many striped bass there are in San Francisco Bay. This number changes, however, throughout the year as fish move in and out of the bay to spawn. Therefore biologists need a way to gather current data fairly quickly and inexpensively.

How do you think a fish biologist might estimate the number of fish in the Bay?

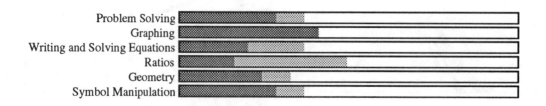

Unit 5
Estimating Fish Populations:
NUMERICAL, GEOMETRIC AND ALGEBRAIC RATIOS

EXPLORING RATIOS

EF-1.   What is the ratio of five days to two weeks (in days)?

EF-2.   Jill is 60 inches tall while her brother is 5 feet tall.

   a)   Is Jill 12 times as tall as her brother? Why or why not?

   b)   What is the ratio of Jill's height to her brother's?

EF-3.   What is the ratio of four inches to one foot (in inches)?

EF-4.   Jack Hammer does a job in 25 minutes. Peter Piper does the same job in half an hour. Write a ratio to:

   a)   compare Hammer's time to Piper's time.

   b)   compare Piper's time to Hammer's time.

EF-5.   What is the ratio of girls to boys in this math class?

EF-6.   Next let us look at the idea of ratios in a geometric setting. First we will examine the idea of **similar** geometric figures.

> ### SIMILAR GEOMETRIC FIGURES
>
> Here is an example of two figures that are similar:
>
>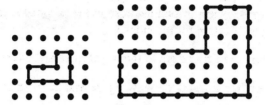
>
> Here is an example of two figures that are <u>not</u> similar:
>
>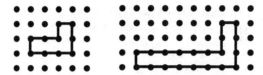
>
> a. Write one or two sentences to explain what the word **SIMILAR** means. Discuss your definition with your team and agree on a team definition. Have your teacher evaluate it with the team and, once everyone agrees you have an accurate definition, add it to your notebook.
>
> b. Discuss with your study team which figure below is similar to figure A. Write a sentence describing how you decided which figure is similar to figure A.
>
>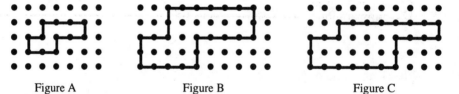
>
> Figure A   Figure B   Figure C

Before you work on the problem below, add these definitions to your tool kit.

>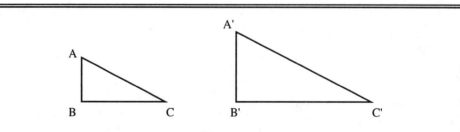
>
> If ΔABC is similar (has the same shape) to ΔA'B'C', then AB and A'B' are **CORRESPONDING SIDES**. However, AC and B'C' are <u>not</u> corresponding sides.

EF-7. Obtain a resource page for EF-7 from your teacher. As a study team, copy the figure at right onto dot paper.

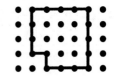

a) Use dot paper to enlarge the figure by making the corresponding sides of the new figure <u>twice</u> as long as those in the original figure. In this case, the enlargement ratio for the sides is $\frac{2}{1}$.

b) Compute the perimeters of the new figure and the original.

c) Compute the areas of the new figure and the original.

d) Find and reduce the following ratios.

$$\frac{\text{Length of any Side of new figure}}{\text{Length of corresponding Side of original figure}}$$

$$\frac{\text{Perimeter of new figure}}{\text{Perimeter of original figure}}$$

$$\frac{\text{Area of new figure}}{\text{Area of original figure}}$$

> In this course, **ratios** will always be listed in the order that compares "new" to "original". Therefore a ratio of $\frac{5}{1}$ or 5:1 means that the lengths of each side of the new figure are five times as large as the corresponding lengths in the original figure.

EF-8. Choose one of figures (a), (b), (c), or (d) below. Be sure that no two team members have the same shape. Enlarge your figure on dot paper by a factor of $\frac{2}{1}$. Then complete the table on your resource sheet with the information from your team members.

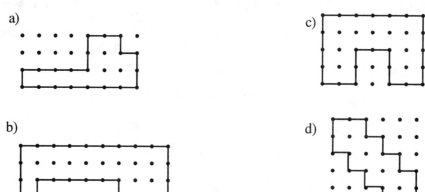

Team Names	Side Length of new figure / Corresponding Side Length of original figure	Perimeter of new figure / Perimeter of original figure	Area of new figure / Area of original figure

EF-9. Examine the tables from EF-7 and EF-8. What observations can you make about the ratios of the sides, perimeters, and areas? Do they have any connections? If so, what?

EF-10. Write an equation for the problem below, then solve it.

A bucket filled with water weighs 8.4 kilograms. If the water by itself weighs five times as much as the bucket, what is the weight of the bucket?

EF-11. Write a ratio of eight Choco-Nut cookies to two dozen cookies.

EF-12. What is the ratio of three ounces to three pounds (in ounces)?

EF-13. What is the ratio of two minutes to five seconds? Be careful!

EF-14. How are $\frac{1}{2}$, $\frac{3}{6}$, and $\frac{7}{14}$ related? Explain your reasoning. Then name three more fractions that are related in the same way.

EF-15. Rene collected information from her fellow basketball players and made this graph:

Describe the relationship between the height and weight of her fellow basketball players.

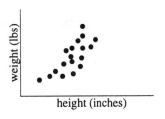

EF-16. Use your knowledge of percents to answer the following:

a) 14 is some percent of 24. Is that percent more or less than 100%? Estimate the percent.

b) 10% of a number is 50. Is that number more or less than 50? Estimate the number.

c) A number x is 20% of 40. Is x more or less than 40? Estimate the number.

d) 150% of a number is 28. Is the number more or less than 28? Estimate the number.

EF-17. Linda and Paul are recording artists. Last month their combined income was $13,000. Linda made $1,000 more than five times what Paul made. How much money did Paul earn?

ENLARGING AND REDUCING FIGURES

EF-18. Before you work on the problem below, add these definitions to your tool kit.

> Mathematicians call an educated guess (based on data, patterns and relationships) a **CONJECTURE**. Scientists use the term **hypothesis**. To **JUSTIFY** your conjecture means to explain the data, pattern, or relationship that led you to your educated guess and to convince your audience that your conjecture is true. Remember to continue onto the rest of this problem, after you have added the definitions of **conjecture** and **justify** to your tool kit

Based on the data from problems EF-7 and EF-8 and your observations from EF-9, predict the ratio of the sides, perimeters and areas when any figure is enlarged three times.

Test this conjecture by enlarging the shape at right by a factor of three and show the labeled ratios of sides, perimeters and areas.

Do these results lead you to confirm or reject your conjecture?

EF-19. Suppose a four-foot by five-foot rectangle is enlarged so that the ratio of corresponding sides is $\frac{4}{1}$.

a) What is the perimeter and area of the <u>original</u> rectangle?

b) What are the dimensions of the <u>new</u> rectangle?

c) Compute the perimeter and the area of the <u>new</u>, enlarged rectangle.

d) Write labeled ratios to compare the sides, perimeter and areas of the two rectangles.

e) Did these ratios confirm your conjecture from EF-18? Explain why or why not.

In order to save time and effort, we will introduce new notation.

For $\dfrac{\text{Length of any Side of new figure}}{\text{Length of corresponding Side of original figure}}$ we will use the notation $\dfrac{S_{new}}{S_{original}}$.

Similarly, $\dfrac{\text{Perimeter of new figure}}{\text{Perimeter of original figure}} \to \dfrac{P_{new}}{P_{original}}$, and $\dfrac{\text{Area of new figure}}{\text{Area of original figure}} \to \dfrac{A_{new}}{A_{original}}$.

EF-20. Choose a figure below. Verify that each figure was chosen by at least one member of your team. Copy your choice on dot paper.

a) Reduce the figure so that the corresponding sides are half the size of the original. In this case, the reduction ratio for the sides is $\frac{1}{2}$.

b) Compute the perimeters and areas of the original figures and their reductions.

c) Then compute and reduce the ratios $\dfrac{S_{new}}{S_{original}}$, $\dfrac{P_{new}}{P_{original}}$ and $\dfrac{A_{new}}{A_{original}}$ in the table on your resource page from EF-7 and EF-8.

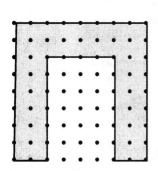 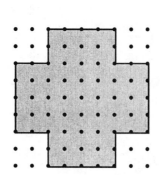 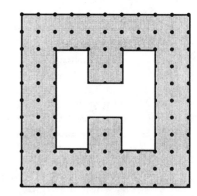

>>Problem continues on the next page.>>

Figure	$\dfrac{S_{new}}{S_{original}}$	$\dfrac{P_{new}}{P_{original}}$	$\dfrac{A_{new}}{A_{original}}$
1st			
2nd			
3rd			

EF-21. Susan's rectangular kitchen measures 12 feet by 15 feet. Her bathroom is similar (it has the same shape), but its area is only a quarter of the kitchen's area. Find the dimensions of her bathroom.

EF-22. Probability is a ratio comparison as well. A full deck of cards normally has 52 playing cards, but Cheryl's deck is missing two aces. Refer to CP-24 for basic information about a deck of cards.

a) Find the probability of drawing a King from this deck.

b) Find the probability of drawing an Ace from this deck.

EF-23. Solve each of the following equations for x.

a) $3x = b$

b) $2x + 7 = p$

c) $\dfrac{x}{3} = y$

d) $y = mx + b$

EF-24. Use the Distributive Property to multiply or factor.

a) $10(5 + 3)$

b) $10n + 30$

c) $4(x + 1)$

d) $-2(4x - 3)$

e) $24x + 6$

EF-25. Why does $-(2x + 7) = -2x - 7$?

EF-26. Solve each of the following equations for x.

a) $4(2x + 7) = 108$

b) $9(19x - 4) = 3x$

c) $15 - (2x + 7) = 14$

d) $14x - 53 - 32x = 73$

e) $\dfrac{x}{4} = 3$

f) $\dfrac{x}{3} = 13$

EF-27. Graph the curve $y = x^2 - 3$ for $-3 \leq x \leq 3$, THEN estimate the x- and y-intercepts.

EF-28. Look back to the ratios you obtained in problems EF-7, EF-8, and EF-20 when you enlarged and reduced figures. For each figure,

    a) compare the side length ratios to the perimeter ratios.

    b) compare the area ratios to the perimeter ratios.

    c) describe the patterns you see.

EF-29. A pizza has been cut into eight pieces and you get three pieces. Write a ratio to compare the number of pieces you get to the total number of pieces. What percent of the pizza did you consume?

RATIOS OF PERIMETERS AND AREAS

EF-30. Find your resource page from problems EF-7, 8, and 20.

    a) Enlarge the figure at right by ratios of 2:1, 3:1, 4:1, and 5:1. Divide the work evenly among your study team.

    b) Compute and reduce the ratios $\frac{S_{new}}{S_{original}}$, $\frac{P_{new}}{P_{original}}$ and $\frac{A_{new}}{A_{original}}$ for each ratio.

    c) Compare the ratios $\frac{S_{new}}{S_{original}}$ and $\frac{P_{new}}{P_{original}}$. How are the numerical values of the ratios related?

    d) Compare the ratios $\frac{P_{new}}{P_{original}}$ and $\frac{A_{new}}{A_{original}}$. How are the numerical values of the reduced ratios related?

EF-31. Suppose you enlarge (or reduce) a figure on dot paper of any size grid. Determine the ratios $\frac{P_{new}}{P_{original}}$ and $\frac{A_{new}}{A_{original}}$ if:

a) the ratio $\frac{S_{new}}{S_{original}}$ is $\frac{2}{1}$.

b) the new figure has corresponding side lengths three times the original.

c) the new figure has corresponding side lengths 10 times the original.

EF-32. a) Make a conjecture for what the ratios $\frac{P_{new}}{P_{original}}$ and $\frac{A_{new}}{A_{original}}$ will be if the ratio $\frac{S_{new}}{S_{original}} = \frac{N}{1}$ and N represents a positive integer.

b) Add the following information to your tool kit:

RATIOS OF SIMILAR FIGURES

If two figures are **similar**, such as the two triangles below, the **RATIO OF THEIR PERIMETERS** is the equal to the ratio of their sides. Also, the **RATIO OF THEIR AREAS** is equal to the square of the ratio of their sides.

If the **enlargement** or **reduction ratio** between a pair of corresponding sides is:

$\frac{S_{new}}{S_{original}}$, then: $\frac{P_{new}}{P_{original}} = \frac{S_{new}}{S_{original}}$, and

$$\frac{A_{new}}{A_{original}} = \left(\frac{S_{new}}{S_{original}}\right)^2$$

For example:

$\frac{S_{new}}{S_{original}} = \frac{3}{6} = \frac{2}{4} = \frac{2.5}{5} = \frac{1}{2}$

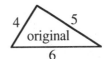

If $\frac{S_{new}}{S_{original}} = \frac{1}{2}$, then $\frac{P_{new}}{P_{original}} = \frac{1}{2}$ and $\frac{A_{new}}{A_{original}} = \left(\frac{1}{2}\right)^2 = \frac{1}{4}$

EF-33. Here are some equations to solve for x. They look a bit different from the equations you have already solved, but you can use the same process to solve them that you have used before.

a) $\frac{x}{3} = 10$

b) $\frac{2x}{5} = 9$

c) $\frac{x}{3} = \frac{6}{7}$

d) $\frac{3x}{2} = \frac{12}{5}$

e) $\frac{5x}{8} = \frac{x-4}{4}$

Estimating Fish Populations: Numerical, Geometric and Algebraic Ratios

EF-34. Pat created a figure on dot paper with a perimeter of 51 units and an area of 90 square units. Kim reduced the figure so that the side lengths are $\frac{1}{3}$ the original sides. Find the perimeter and area of the figure Kim made.

EF-35. On graph paper draw $\triangle ABC$ and $\triangle EDA$ with the points A(0, 0), B(4, 0), C(4, 3), D(8, 0), and E(8, 6). $\triangle ABC$ is similar to $\triangle EDA$. If AC = 5, find AE.

EF-36. Show your subproblems to find the shaded area of the figure. Write a ratio to compare the area of the shaded region to the whole region. If this figure were enlarged four times, what would the ratio be?

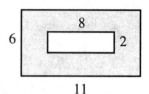

EF-37. Use the Distributive Property to rewrite each of the following expressions.
   a) 4(y - 7)
   b) 2y(y + 4)
   c) 3z(2z - 4)
   d) 5m + 10
   e) 7y + 49
   f) $m^2 + m$

EF-38. Copy and solve these Diamond Problems:

 a) b) c) d)

EF-39. Write a ratio for each of the following descriptions.
   a) eight minutes to one day (in minutes)
   b) one liter to one milliliter (in milliliters)
   c) forty-five minutes to two hours (in minutes)
   d) one pound to twelve ounces (in ounces)

EF-40. On the <u>same set of axes</u>, sketch:
   a) the speed of a ball compared to the time it takes to roll down a relatively flat ramp. Be sure to label the axes.
   b) the speed of the ball compared to the time it takes to roll down a steeper ramp.

EF-41. Multiply (1,048,570)(16,384). Write the result in both standard notation and scientific notation.

EF-42. A popular window design is shown at right.

a) Find the area of the glass if all measurements are in feet.

b) Wooden molding surrounds the outside of the window. Find the length of the molding.

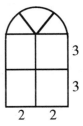

EF-43. Solve for x:

a) $9(x - 2) - 13 = 72.5$

b) $2 - 3(2 - x) = -13.75$

c) $9 + 3x = -14 + 5x$

d) $7 - 2x = 5(x + 2)$

EF-44. Next we want to look at ratios from a geometric point of view. Add this definition to your tool kit with an example.

RATIOS OF SIMILAR TRIANGLES

We say two triangles are **similar** if they are the same shape, but are not necessarily the same size. Their corresponding angles are the same size and their corresponding sides have the same (constant) ratio.

One way to demonstrate that two triangles are similar is to show (or be told) that all three pairs of corresponding angles are equal. This principle is illustrated below.

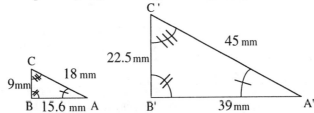

The symbol **A'** is read "A prime." It is a way to label corresponding parts of figures.

To denote the **LENGTH OF A LINE SEGMENT,** we will write the two letters which name the endpoints of the line segment in uppercase. "**AB**" means "the length of segment AB".

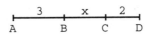

In the diagram above we see that the segment AB has length 3, so we write AB = 3. We also see that segment BD has length x + 2, so we write BD = x + 2.

Estimating Fish Populations: Numerical, Geometric and Algebraic Ratios

EF-45. RATIOS OF CORRESPONDING SIDES

In this problem you will use the measurements from problem EF-44 above to compare the ratios of a side length from one triangle △A'B'C' to the corresponding side of similar triangle △ABC. Be sure to record the labels of each ratio.

a) Calculate the following ratios:

1) $\dfrac{A'B'}{AB}$ 2) $\dfrac{A'C'}{AC}$ 3) $\dfrac{B'C'}{BC}$

b) What did you observe about the ratio of the corresponding sides of similar triangles?

c) What is the enlargement ratio of the sides of △A'B'C' compared to the corresponding sides of △ABC ?

EF-46. RATIO OF TWO SIDES WITHIN A TRIANGLE

Next you will investigate the relationship between the ratio of two sides within a single triangle and the ratio of the two corresponding sides within a similar triangle.

Use the measurements from problem EF-44 to calculate each of the following ratios. Round your results to the nearest **tenth**. (For example, the decimal 0.752 would be rounded to 0.8.) Compare the ratios.

	Ratios for △ABC	Ratios for △A'B'C'
a)	$\dfrac{AB}{BC}$	$\dfrac{A'B'}{B'C'}$
b)	$\dfrac{AB}{AC}$	$\dfrac{A'B'}{A'C'}$
c)	$\dfrac{BC}{AC}$	$\dfrac{B'C'}{A'C'}$

d) Summarize in your own words what parts of the similar triangles you were comparing that produced these ratios.

e) Go back to EF-45 and summarize in your own words what parts of the similar triangles you were comparing that produced the ratios you obtained there.

f) Briefly summarize the differences between parts (d) and (e). Add your team's best sentence to your tool kit.

168   UNIT 5

EF-47. You will now use ratios to find the missing side lengths of similar triangles.

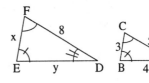

Triangles ABC and DEF are similar, with side lengths as indicated at right.

a) Use this information to write ratios for $\frac{EF}{BC}$ and $\frac{DF}{AC}$.

b) Are these ratios equal? Explain your reasoning.

c) Use these equivalent ratios to write an equation, called a **proportion**, so you can solve for x, the length of EF. Solve your equation.

d) Follow the process of parts (a) through (c) to write an equation to solve for y, the length of ED. Then solve your equation for y.

e) Add the following definition to your tool kit:

> A **PROPORTION** is an equation with two equivalent **ratios**. For example, $\frac{3}{6} = \frac{1}{2}$ is a proportion, as is $\frac{x}{3} = \frac{8}{5}$ in the similar triangles in problem EF-47.

EF-48. Make a Guess and Check table to solve this problem. Write your solution as a sentence.

Maya and Mike are 60 miles apart and traveling towards each other. Maya travels at 20 miles per hour on her bike while Mike travels 30 miles per hour in his car. In how many hours will they meet?

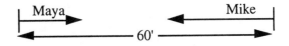

EF-49. Use the Distributive Property to rewrite the following:

a) $5(z + 4)$

b) $x(3 + y)$

c) $m^2 + 3m$

d) $12t^2 + 16t$

EF-50. A regular six-sided die is tossed.

a) What is the probability of getting a 4?

b) If the die is tossed 90 times, how many times should a 4 to come up? Write a ratio equation (proportion) to solve this problem.

EF-51. Which of the following equations would produce lines when graphed? Explain how you know.

    a)   y - 4 = 3x

    b)   y - $\sqrt{x}$ = 6

    c)   y = $3x^2$ - 6x + 1

    d)   3 + 2x = 8y

EF-52. Use the figure at right for your data.

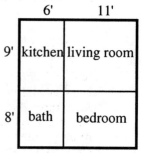

    a)   Find the total area of Susan's apartment.

    b)   What word best describes the shape of her apartment? Justify your answer.

    c)   Find the ratio of the area of the living room to the area of the bedroom. Reduce your fraction. What do you notice?

    d)   Find the ratio of the area of the bathroom to the area of the bedroom. Reduce your fraction. What do you notice?

EF-53. Express each of the following ratios as a percent. For example, $\frac{1}{5} = \frac{20}{100} = 20\%$.

    a) $\frac{1}{4}$

    b) $\frac{3}{10}$

    c) $\frac{17}{25}$

    d) $\frac{9}{5}$

RATIOS IN RIGHT TRIANGLES

 A SECOND LOOK AT SPIRALING

It's been awhile since we talked about how the book is put together. Before I go on, you and one other person on your team take a quick look at <u>one</u> of the two lists of problems from this unit and see what you find. The other two students on your team should look at the other list.

List A: EF-26, 43, 63, 89, 110, and 124

List B: EF-24, 25, 37, 49, 62, 77, and 125

Notice anything about these problems? I bet you did! Back in Unit 3 in PZL-14 I described how the authors took the skills and concepts of the course and spiraled them throughout the text. Notice that List A keeps your equation solving skills sharp, while List B makes sure you don't forget the important Distributive Property. Want more examples? Practice with percent continues in EF-16, 29, 53, and 126; writing and solving equations return in EF-17, 48, 64, 91 and 113; and graphing is the focus of EF-27, 51, 107, and 114.

Remember that spiraled problems are there for you to practice the important skills and concepts of Algebra 1. Some of the spiraled problems build on the basic idea they revisit. If you encounter a spiraled problem that stumps you, ask for help from your team or teacher. Often the best question is, "Where should I look in the book to find out how to do this problem?" In fact, before you ask this question, try to answer it yourself by using your tool kit, the index, or the table of contents.

EF-54. Add these definitions, including the diagrams, to your tool kit if you need them.

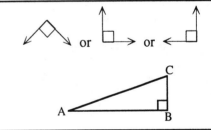

A **RIGHT ANGLE** is an angle that measures 90°. We indicate a right angle this way:

A **RIGHT TRIANGLE** is a triangle with exactly one right angle. For example, $\triangle ABC$ is a right triangle with its right angle at B.

EF-55. Look at right triangles MAT and TCH at right. How can you tell that they are similar? (Hint: review problem EF-44.)

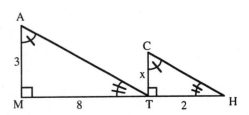

a) Find TC by writing and solving an equation.

b) Write another proportion that you could use to solve for TC.

Estimating Fish Populations: Numerical, Geometric and Algebraic Ratios

EF-56. The triangles in figure 1 are similar.

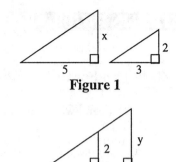
Figure 1

a) Write an equation to solve for the length x.

b) Compare your method of solution with your study team. Did anyone write a different proportion? If not, write one now. Verify that the solution is the same, then be sure that everyone has both equations written on their paper.

c) The triangles in figure 2 are also similar. Write an equation to solve for y.

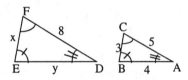
Figure 2

d) Compare the figures and your solutions for x and y. What is the basic difference between the pairs of triangles in the two figures?

EF-57. Kwame and Jiar arrived in class today to discover a team quiz on yesterday's big idea. When they shared strategies, they noticed their proportions were different. The quiz question was:

"Triangles ABC and DEF are similar triangles, with side lengths as indicated in the diagram. Find x and y."

Their proportions:

Kwame: $\frac{x}{8} = \frac{3}{5}$ Jiar: $\frac{x}{3} = \frac{8}{5}$

a) Decide who is correct. Explain why.

b) Solve for x.

c) Solve for y.

EF-58. The 4 x 400 m relay is a track event in which four runners each run 400 meters. Each 400 m portion is called a "leg." In a recent race, Maisha ran the second leg of a 4 x 400 m relay in 56 seconds. Her speed is graphed at right. Copy this graph onto your paper. Label your axes completely.

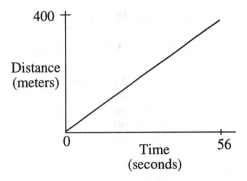

a) What is the ratio of the distance to time at the end of Maisha's leg?

b) Assume that Maisha ran at a constant speed. How long did it take Maisha to go 200 m?

c) What is the ratio of distance to time for the first 200m of the leg?

d) What is the ratio of distance to time for the first 100m of the leg?

e) Review the ratios you found. Write each ratio in simplest form and compare the results.

f) Draw similar right triangles on the graph at 56 seconds and 28 seconds.

g) What can you conclude about the ratio of the variables (in this case distance on the y-axis and time on the x-axis) on the line of a graph?

EF-59. Triangles ABC and DEF below are similar. Use a proportion to solve for x. Show all subproblems.

EF-60. The two right triangles at right are similar. △ABC is a reduction of △DEF.

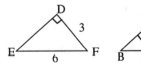

a) Find x, the length of AC.

b) What is the reduction ratio?

c) What is the ratio of the perimeter of △ABC to the perimeter of △DEF? Why?

d) What is the ratio of the area of △ABC to the area of △DEF is?

EF-61. A tree casts a 43 foot shadow. At the same time, a $4\frac{1}{2}$ foot boy casts a 10 foot shadow. Since it is the same time of day the sun is at the same angle, use the sketch at right and your understanding of ratios and similar triangles to complete a diagram and find the height of the tree.

EF-62. Dubious Dan is sure that $2(x + 4) = 2x + 4$. Doubtful Debbie is equally sure that $2(x + 4) = 2x + 8$.

   a) Which claim is correct?

   b) Use words or diagrams to give an explanation that demonstrates the correct equation.

EF-63. Solve the following equations:

   a) $5x + 12 = 3x - 8$

   b) $9(x - 1) = 45$

   c) $3(x - 2) - (5x - 3) = 2x + 10$
      Careful: review EF-25.

   d) $\frac{5x}{7} = 23$

EF-64. Write an equation and solve the following problem:

   Everything at Rose's Department Store is 20% off. If the sale price of basketball shoes are now $48.80, what was the original cost?

EF-65. Start with a point P at (-1, -5). Slide P four units up and then two units to the left. What are the coordinates of P's new position?

PERCENT AND PROPORTION

EF-66. We wish to graph the line $y = 2$. Since this equation does not force x to equal any specific or unique number, then y will always equal 2 no matter what x is.

Below is a table for the equation $y = 2$. Use it to graph your line.

x	-3	-2	-1	0	1	2	3
y	2	2	2	2	2	2	2

   a) Add the lines $y = -1$ and $y = 6$ to your graph.

   b) Add the line $x = 3$ to your graph.

   c) In your tool kit, make a sketch for $y = -3$ and $x = 5$. Be sure to label your lines and write a brief general observation about the equations of **horizontal** and **vertical** lines.

EF-67. A certain line contains the points (3, 1) and (6, 2). Find the coordinates of three other points on the line. Use graph paper to justify your answers.

EF-68. Solve each of the following equations.

a) $\dfrac{x}{3} = 7$

b) $\dfrac{x}{3} = \dfrac{4}{7}$

c) $\dfrac{x}{3} = \dfrac{x+1}{7}$

d) $\dfrac{3}{x} = \dfrac{7}{4}$

e) $\dfrac{3}{x+1} = \dfrac{8}{x}$

f) Compare your answers and strategies with your team members. Did you solve all the equations in the same way?

EF-69. △ABC is similar to △DEF below. Use ratios to solve for DE.

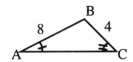
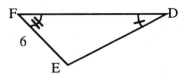

EF-70. Solve each of the following equations for x.

a) $\dfrac{x}{100} = \dfrac{43}{100}$

b) $\dfrac{x}{100} = \dfrac{7}{100}$

c) $\dfrac{x}{100} = \dfrac{6}{10}$

d) $\dfrac{x}{100} = \dfrac{9}{4}$

e) $\dfrac{x}{100} = \dfrac{5}{8}$

f) $\dfrac{x}{100} = \dfrac{2}{3}$

EF-71. Explain in one or two complete sentences why the values of x you found in the preceding problem represent percents. Explain how percents are ratios.

EF-72. In a city of three million people, 3,472 people were surveyed. Of those surveyed, 28 of them watched the last Lonely Alien movie.

a) What fraction of the people surveyed watched the movie?

b) What percentage of the people surveyed watched the movie?

c) If the survey represents the city's TV viewing habits, about how many people in the city watched the movie?

Estimating Fish Populations: Numerical, Geometric and Algebraic Ratios

EF-73. The local supermarket charges 98¢ tax on a $15.00 purchase. Write a ratio of the tax to the purchase price. What percent tax was charged?

EF-74. Write and solve an equation using ratios to find BY.

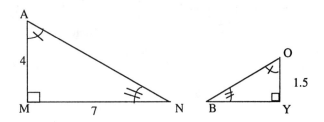

EF-75. Gimble's sells a radio for $15.00. The store's profit is $6.00. Write a ratio of the profit to the selling price. What percent of the selling price is the profit?

EF-76. Describe how to slide (move) a point from (3, 0) to (5, 2). Give the vertical directions first, followed by the horizontal directions.

EF-77. Use the Distributive Property to rewrite each of the following expressions.

   a) $2(x + 4)$ d) $x(x + 4)$

   b) $x(x - 2)$ e) $3x + 6$

   c) $4y - 8$ f) $m^2 + 5m$

EF-78.* Scientists use containers of many shapes for holding and measuring liquids. Imagine that we add water to each beaker. The graph below shows how the height of the water increases as water is added at a steady rate to beaker A.

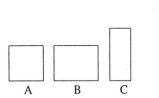 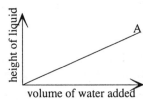

   a) Trace the beaker shapes and the graph for beaker A. Plot lines to show how beakers B and C fill up as water is added. Label each line.

   b) Explain why you drew the lines where you did.

EF-79. Graph $y = 2$ and $x = 4$ on the same set of axes. What are the coordinates of the point where they cross?

---

* Adapted from *The Language of Functions and Graphs*, Joint Matriculation Board and the Shell Centre for Mathematical Education, University of Nottingham, England.

EF-80. Janelle and Nikki are going to race. Janelle runs 5 meters per second while Nikki runs 4.5 meters per second. If the race is 100 meters long, by how many seconds does Janelle win?

EQUIVALENT RATIOS AND GRAPHS

EF-81. C.J.'s car gets 20 miles per gallon of gas.

a) Copy and complete the following table for C.J.

# miles traveled	# gallons of gas used
20	
100	
40	
10	
0	
m	

b) Use the table you made in part (a) to write an equation relating g, the number of gallons used, to m, the number of miles driven.

c) Does your equation in part (b) give the correct result for m = 20 miles? Show why or why not.

d) Graph your equation with **m**, the number of miles, on the horizontal axis, and **g**, the number of gallons, on the vertical axis. Label both axes. Scale the horizontal axis so one unit represents 10 miles. Scale the vertical axis so one unit represents one gallon.

e) Use your graph to estimate the value for g when m = 70.

f) Create similar triangles and write a proportion to help you find g when m = 70. Solve your equation to find out how much gas C.J. used in driving 70 miles. Compare your solution to your estimate in part (e).

g) Pick another point on your graph. Find the ratio of miles to gallons at this point.

Estimating Fish Populations: Numerical, Geometric and Algebraic Ratios

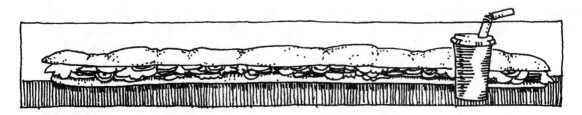

EF-82. Salami and More Deli sells a six-foot submarine sandwich for parties. It weighs 8 pounds. Assume the weight per foot is constant.

a) How much does a 0 foot long sandwich weigh?

b) Draw a graph showing the weight of the sandwich (vertical axis) compared to the length of the sandwich (horizontal axis). Label the axes with appropriate units.

c) Use your graph to estimate the weight of a one foot sandwich.

d) Write a proportion to find the weight of a one foot sandwich. Solve the equation.

e) Use your graph to estimate the length of a 12 pound sandwich.

f) Write a proportion to find the length of a 12 pound sandwich.

g) How close were your estimates in parts (c) and (e) to the actual values you found in parts (d) and (f), respectively?

h) What is the ratio of pounds to feet for any point on this graph?

For each of problems EF-83 and EF-84, write a proportion and solve it. If you need help writing your equation, try organizing the information in a table as shown in EF-81.

EF-83. Kim noticed that 100 vitamins cost $1.89. At this rate, how much should 350 vitamins cost?

EF-84. Joe came to bat 464 times in 131 games. At this rate, how many times at bat should he expect in a full season of 162 games?

EF-85. a) Humphrey and Shamu are swimming toward each other. Shamu swims twice as fast as Humphrey. If they start 12 miles apart, where will they meet?

b) If it takes them 2 hours to meet, how fast does Humphrey swim?

EF-86. An 18 foot board is divided into two parts as shown at right.

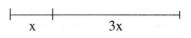

a) Solve for x.

b) Find the ratio of the first piece to the second.

EF-87. A photograph with width 5" and length 7" is enlarged so its new width is 17.5".

a) What is the length of the enlarged photograph?

b) What is the enlargement ratio?

EF-88. L.J.'s car has a gas tank which holds 19 gallons of fuel.

a) If he used eight gallons to drive 200 miles, does the car have enough gas to go another 250 miles? Show why or why not.

b) What assumptions did you make in your solution to part (a)?

EF-89. Solve each of the following equations.

a) $\dfrac{x}{3} = \dfrac{5}{7}$

b) $\dfrac{5}{y} = \dfrac{2}{y+3}$

c) $\dfrac{1}{x} = \dfrac{5}{x+1}$

d) Draw two similar triangles for which the equation in part (a) can be used to find a side length x.

EF-90. Two numbers are in a ratio of 1:5.

a) If the first number is 100, find the second number.

b) If the first number is x, find an expression to represent the second number.

EF-91. An apple tree produces three times as much fruit as a pear tree. During one season, two apple trees and three pear trees produced at total of 126 pieces of fruit. How many apples does an apple tree produce in one season?

Estimating Fish Populations: Numerical, Geometric and Algebraic Ratios

EF-92. Use a calculator to find each of the following products. Write each result in scientific notation.

a) $0.0738 \cdot (6.2 \cdot 10^{17})$

b) $360,000,000,000 \div (3 \cdot 10^{14})$

WRITING AND SOLVING EQUATIONS INVOLVING RATIOS

For problems EF-93 and EF-94, write an equation and solve it.

EF-93. Two numbers are in a ratio of 1:5. If their sum is 30, find the numbers.

EF-94. In Harry's backyard, his peach tree produces two pieces of fruit for every three pieces that grow on his apricot tree. If the two trees have a total of 265 pieces of fruit on them, how many pieces of fruit does each tree have?

EF-95. A line contains the points (3, 1) and (4, 3). Find the coordinates of three other points on the line. Graph paper is helpful with this problem.

EF-96. A number y is 14% of 300.

a) Is y larger or smaller than 300 ?

b) Write 14% as a ratio.

c) Use proportions to write an equation to determine the value of y. Solve your equation to find y.

EF-97. The number 54 is 16% of z.

a) Is z larger or smaller than 54 ?

b) Use ratios to write an equation that expresses this fact.

c) Solve your equation to find z.

EF-98. Use ratios to write an equation that can be used to solve each of the following problems. Then solve the equation.

a) What number is 25% of 40 ?

b) Twenty-five is what percent of 40 ?

c) Twenty-five is 40% of what number?

EF-99. A biscuit recipe uses $\frac{1}{2}$ teaspoon of baking powder for $\frac{3}{4}$ cup flour. How much baking powder is needed for three cups of flour? Set up a proportion and solve.

EF-100. Tina has 24 ounces of gas and one ounce of oil in her motorcycle. How much oil must be added to make the ratio of gas to oil 16:1 ?

EF-101. The length of a rectangle is six centimeters more than the width. Draw and label a diagram. If the ratio of the length to the width is five to three, find the rectangle's dimensions.

EF-102. Describe how to slide (move) a point from (0, -2) to (-3.5, 5). Give the vertical directions followed by the horizontal directions.

EF-103. Copy and complete the table. Find an equation to represent the rule used to determine y.

x	-5	-2	-1	0	1	2	3	4	x
y	7		-1				-9		

EF-104. Suppose you have a penny, a nickel and a dime and you toss them onto the ground.

a) Find the probability that the penny comes up 'heads'.

b) Find the probability that both the penny and the dime come up heads. (Hint: it helps to make a table of the eight possible outcomes.)

c) Using all three coins, find the probability that exactly two of the coins come up heads.

d) Write the probabilities you found in parts (a), (b), and (c) as percents.

EF-105. Use your formulas for circles in your tool kit to solve these pizza problems.

a) The pizza sauce covers an area of 201 square inches. How long is the pizza's diameter? (Assume the sauce covers the entire pizza.)

b) How long is its crust?

EF-106. Triangle UCB is similar to triangle UOP. Use a proportion to solve for x. (Hint: how long is OU?)

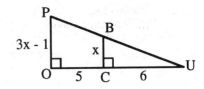

EF-107. Make a table of x- and y-values that make the equation $x + y = 5$ true. Then plot these points on a pair of xy-coordinate axes. Does your table contain all the points that make the equation true? Why or why not?

EF-108. ESTIMATING FISH POPULATIONS --A Simulation

Fish biologists need to keep track of fish populations in the waters they monitor. They want to know, for example, how many striped bass there are in San Francisco Bay. This number changes, however, throughout the year as fish move in and out of the bay to spawn. Therefore biologists need a way to gather current data fairly quickly and inexpensively.

Since it is impossible to count every animal, biologists use the "tag and recapture" process to help them estimate the size of a population. Tag and recapture involves collecting a sample of animals, tagging them, and releasing them back into the wild. Later, biologists collect a new sample of the animals and count the number in the sample distinguishing between first time captures and recaptures. They then use ratios to estimate the population size.

Each team has a lake (paper sack) full of fish (beans), and a net (a small cup). Be sure each team member records all the team's data, calculations and conclusions.

OVERVIEW

We will be trying to estimate the number of fish in the lake the way the department of Fish and Game does. Here is a preview of this simulation.

1. Collect an initial sample. Count and tag the specimens.
2. Return the tagged sample to the lake.
3. Collect an additional sample. Count the number of tagged and untagged fish.
4. Use the data (counts) to write proportions.
5. Solve the ratios to find the lake population.

When you finish these steps you will have three pieces of data to help you calculate the fourth:

1. "**total number of tagged fish in the lake**,"
2. "**number of tagged fish in the sample**," and
3. "**total number of fish in the sample**." Now solve for the fourth:
4. "**total number of fish in the lake**".

>>Problem continues on the next page.>>

a) Write an equation that incorporates these four data pieces into a proportion on the resource page.

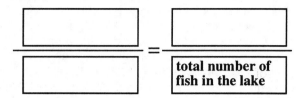

TAGGING THE INITIAL SAMPLE

b) Use your "net" to collect a sample (cupful) of fish (beans) from the lake. Count the number of fish you netted in your sample.

To "tag" the netted fish, either mark them with a marking pen or substitute a bean of a different color for each fish as directed by your teacher. This is the **only** sample you will tag in this simulation.

On the Resource Page record this number as the "**total number of tagged fish in the lake.**"

Put all the **tagged** fish back into the lake. Be careful not to let any of the fish jump out on to the floor. Gently shake the bag to thoroughly mix all the fish in the lake. Try not to bruise them.

COLLECTING A SAMPLE

c) Use your net to take out another cupful of fish. Count and record the "**total number of fish in sample.**"

d) Count and record the number of "**number of tagged fish in sample**."

ESTIMATING THE POPULATION

e) You now have three pieces of information:
the **total number of tagged fish in the lake**,
the **total number of fish in the sample**, and
the **number of tagged fish in the sample**.

Use this information to write a proportion and solve for the total number of fish in the lake.

f) Return your sample of fish to the lake. Gently mix the fish. Take another sample. Repeat the counting and recording procedures of parts (c) and (d). Use the new information to write a proportion as in part (e). Solve the equation you wrote for the total number of fish in the lake.

>>Problem continues on the next page.>>

Extension: Your solutions represent two estimates for the population of your lake. While it is important to get an accurate count of the fish population, each time you net a sample it costs the taxpayers $500 for your time and equipment. So far your samples have cost $1000. If you feel your estimate is accurate at this point, record it on the class chart with your cost. If you think you should try another sample for better accuracy, do the same steps as before. Draw as many samples as you feel you need, but remember each sample costs $500.

g) Record your team's data on the class chart. Then count the fish in your lake to find the actual population. Record the actual population on the class chart.

h) Was your estimate close? If not, what might have thrown it off? What do you think of this method for estimating fish populations?

EF-109. You have used subproblems to make complex problems more manageable. We will examine the subproblems involved in adding fractions with variables. Notice the x in the following problem. Treat this sum like others you have encountered.

$$\frac{2}{3} + \frac{x}{4}$$

a) What must you do first to add these fractions?

b) What is the common denominator?

c) The first subproblem is to find the common denominator and then rewrite each of the fractions with the same denominator. Show how you did this.

d) Now add the fractions.

EF-110. Use any method to solve for x. Justify your solution.

a) $\frac{3x}{7} = \frac{18}{49}$

b) $-4.5x - 13.8 = -6.3x + 4.9$

c) $5.3(2x - 9) = 10.6$

d) $\frac{x}{x} = 2$

EF-111. Use the figure at right to answer the questions below.

a) How much of the circular region is unshaded? Express your answer as a fraction, decimal, and as a percent.

b) If the area of the circle was 32 square centimeters, what is the area of the shaded portion of the circle? Describe how you arrived at your answer.

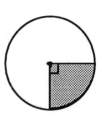

EF-112.

AT THE MOVIES

It is now 7:51 p.m. The movie that you have waited three weeks to see starts at 8:15 p.m. Standing in line for the movie, you count 146 people ahead of you. Nine people can buy their tickets in 70 seconds. Will you be able to buy your ticket before the movie starts? Show your calculations and predicted time.

EF-113. Solve the following problem and write an equation that represents it.

Admission to the football homecoming dance was $3 in advance and $4 at the door. There were 30 fewer tickets sold in advance than at the door and the ticket sales totaled $1590. How many of each kind of ticket were sold?

EF-114. Create a table and graph $y = 2 - x^2$ for $-4 \leq x \leq 4$.

EF-115. Plot and connect the points A(0, 0), B(7, 0), and C(5, 6).

a) What is the area of the region formed?

b) If C is at (3, 6), what is the area of triangle ABC?

c) What is the area of the triangle if C is at (x, 6)? Explain why.

UNIT REVIEW AND SUMMARY

EF-116. UNIT FIVE SUMMARY

With your study team, quickly list as many mathematical ideas or topics you remember from this unit. Do not flip back through the text or your notes! There will be time for that later. This exercise is to see what comes from memory. Be attentive to any suggestions offered by your teammates. Each member needs to write the list down.

a) With the list generated from the class, discuss with your study team the relative importance of each item. Which were the "big ideas" in the unit? Narrow the list to the most important four ideas.

b) For each topic, choose a problem from the Unit that demonstrates each main idea. Be selective. While there are many problems to choose from, certain problems are better examples than others. Perhaps you would like to choose a problem you are most proud of solving. Check that the problems you choose are directly related to the main ideas of your summary.

c) Solve each problem and show all work, especially the subproblems.

d) Finally, explain the method you used to arrive at your solution. Use complete sentences and be as descriptive as you can.

EF-117. Suppose a right triangle with sides of lengths of 5, 12, and 13 centimeters is similar to a right triangle whose shortest side is 15 centimeters long.

a) What is the perimeter of the larger triangle?

b) What is the ratio of the perimeter of the smaller triangle to the perimeter of the larger triangle?

c) How does the ratio in part (b) compare to the ratio of the lengths of corresponding sides of the triangles?

d) What is the ratio of the area of the smaller triangle to the area of the larger triangle?

EF-118. In this problem, you will construct a graph comparing the number of coins to their value. Plot the number of coins on the x-axis and their value on the y-axis. Create a table and use the data to graph the number of coins compared to their value in cents. Plot a line for pennies, nickels, dimes and quarters.

a) Write an equation for each coin (four in all). Let x = the number of coins and y = value of the coins.

b) What do you notice about the lines?

c) Which line is steepest?

d) Which line indicates the greatest value per coin?

e) Describe the relationship between the appearance of the line and the value of each coin.

EF-119. Mario's car needs 12 gallons of gas to go 320 miles.

a) How much gas is needed to go 0 miles?

b) Draw a graph that shows **g**, the number of gallons of gas used (vertical axis), compared to **m**, the number of miles driven (horizontal axis).

c) From your graph, estimate how much gas will be needed to go 70 miles.

d) Estimate how many miles can be driven with 10 gallons of gas.

e) Use proportions to write equations to find the exact answers in parts (c) and (d). Solve your equations.

EF-120. Find each of the following ratios.

a) One dime to one quarter.

b) Twelve feet to nine yards.

c) Sixty ounces of M&M's to two pounds of Gummi Bears.

EF-121. Enlarge this figure on dot paper by making each side of the new figure twice as long as its corresponding side in the original figure.

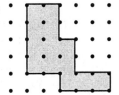

a) Write the enlargement ratio of the corresponding sides.

b) Write and reduce the ratios $\frac{S_{new}}{S_{original}}$, $\frac{P_{new}}{P_{original}}$ and $\frac{A_{new}}{A_{original}}$ for the figures.

EF-122. Find DE and DF.

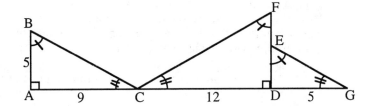

EF-123. Latisha is using a cake recipe which requires that the ratio of oil to water be 3:4. If the oil and water combined measure 2 cups, how many cups of oil did Latisha use?

EF-124. Solve each of the following equations for x.

a) $\frac{1}{2}(x+8) = 7$

c) $8.5 - 1.25x = 4x - 9.5$

b) $\frac{1}{x} = \frac{3}{x+5}$

EF-125. Use the Distributive Property to find each of the following products.

a) $5(x + 2)$

d) $5(2t^2 - 3t + 1)$

b) $4(3y - 2)$

e) $6(4x^{10} + 6x^3 + x)$

c) $2t(3t + 5)$

Estimating Fish Populations: Numerical, Geometric and Algebraic Ratios

EF-126. For each of the questions below, write an equation and solve it.

   a) What percent of 60 is 45 ?

   b) What is 45% of 60 ?

   c) Sixty percent of what number is 45 ?

EF-127. TOOL KIT CHECK-UP

Your tool kit contains reference tools for algebra. Return to your tool kit entries. You may need to revise or add entries.

Examine the list below. Each item should be included in your tool kit; add any that are missing now. You will find them in your text by skimming through the unit or by using the index in the back of the book.

- Similar Triangles
- Ratios of Similar Figures
- Proportion
- Vertical and Horizontal Lines

GM-9. LATISHA'S BIRTHDAY

Latisha's birthday was last week. Since she wanted lots of presents, she had a party and invited 50 of her closest friends.

Unbeknownst to her, the friends got together and decided 50 presents were too many. Instead, they got 50 boxes and numbered them from 1 to 50. They put presents in some of the boxes but not all of them.

When it was time for Latisha to open her gifts, the friends arranged all 50 boxes in a row. They informed Latisha that if she followed the instructions below she would discover which boxes held gifts.

- First, she was to go down the line and open every box.
- Then starting with box # 2, she was to close # 2, # 4, # 6, etc. going down the line.
- Starting with box # 3, she was to change every third box (she opened the box if it was closed and closed the box if it was open).
- Starting with box # 4, she change every fourth box.
- Starting with box # 5, she change every fifth box.

She went through the line of boxes 50 times continuing the pattern. When she was done, the only boxes left open had a pattern. Latisha squealed with delight when she recognized the pattern. She then searched these boxes to find her presents.

Your Task:

- Which boxes contained gifts? Describe the pattern Latisha found.
- Explain why you think this pattern occurred.
- Latisha wished she had invited 200 people to her party so there would have been 200 boxes. Using the same directions, which boxes would contain gifts this time? Explain how you know.
- What is the probability she would find a gift with a present if there were 50 presents and she randomly opened one box? What is the probability she would find a gift if there were 200 presents and she randomly opened one box? Why are these probabilities different?

Estimating Fish Populations: Numerical, Geometric and Algebraic Ratios

GM-10. **SPINNING MADNESS**

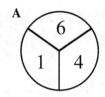

Raquel had two spinners. She spun both spinners at the same time and recorded which spinner landed on the highest number. Using the spinners at right, **A** usually won. Can you prove why?

She then found 3 blank spinners. She decided to use the numbers 1 through 9 putting a different number in each section. At first she set them up like this:

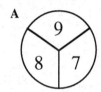

She quickly realized that **A** would always beat **B** and **A** would always beat **C**.

Your Task:

- Using the numbers 1 through 9 (putting a different number into each section), try to create three spinners so that **A** will usually win over **B**; **B** will usually win over **C**; and **C** will usually win over **A**. Explain why this is or is not possible.

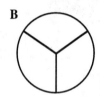

- This time, use the numbers 1 through 15 (putting a different number into each section) to create three spinners where **A** will usually win over **B**; **B** will usually win over **C**; and **C** will usually win over **A** Explain why this is or is not possible.

- Try it one more time, using the numbers 1 through 12. Explain why this is or is not possible.

- Summarize your findings.

Graphing and Systems of Linear Equations

Unit 6 Objectives
World Records: GRAPHING AND SYSTEMS OF LINEAR EQUATIONS

So far in this course we have focused a lot of our attention on solving word problems by writing, solving, and graphing equations. In this unit we will connect our skills for **solving word problems algebraically and by graphing**. We will see how these two approaches are related in solving linear equations like the one presented in the World Records Problem.

This unit requires many large, accurate graphs. Graph paper and a ruler should be used when graphing.

In this unit you will have the opportunity to:

- review all significant ideas from Units 0 - 5 while consolidating graphing and solving equations.

- solve problems that use linear equations to model two simultaneous situations by applying graphing, problem solving and algebraic skills.

- solve systems of two linear equations algebraically by substitution.

- use tiles to extend the Distributive Property to multiplying binomials.

We will pull together all of the skills you have gained in algebra so far this year. In this unit, the graphing and algebraic manipulation become two parts of one big idea: representing situations algebraically.

Read the following problem. **Do not try to solve it now**. During this unit, you will develop skills and understanding to help you solve this problem.

WR-0. **WORLD RECORDS**

In 1988, the women's world record for the 100-meter free-style in swimming was 54.73 seconds. The record time has been decreasing by approximately 0.33 seconds a year. The men's record was 49.36 seconds and has been decreasing at about 0.18 seconds a year. If both records continue to decrease at these same rates, in how many years will the men and women have the same record time?

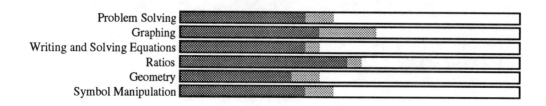

Unit 6
World Records: GRAPHING AND SYSTEMS OF LINEAR EQUATIONS

INTERPRETING GRAPHS, PART 1

WR-1. **MATCH-A-GRAPH, Part One**

Obtain the resource page from your teacher. **SAVE THIS PAGE; YOU WILL NEED IT LATER IN THIS UNIT.** Use what you know about graphing to match each of the following graphs with its equation below. As a team, decide how you are all going to work together to solve this problem.

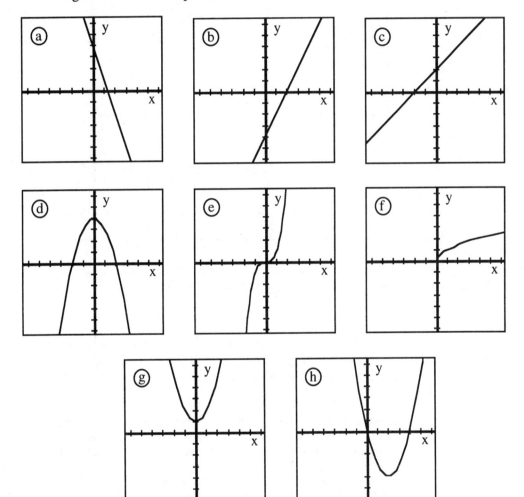

1) $y = 2x - 4$

2) $y = \sqrt{x}$

3) $y = x + 2$

4) $y = x^2 - 4x$

5) $y = 4 - 3x$

6) $y = x^3$

7) $y = x^2 + 1$

8) $y = 4 - x^2$

WR-2. Carefully describe what methods your team used to solve Match-A-Graph.

WR-3. WORLD RECORDS, Part One

Obtain a Resource Page from your teacher. **SAVE THIS PAGE; YOU WILL NEED IT LATER IN THIS UNIT.** The data points on the graph below represent winning times in the 100 meter free-style swimming race where world records were set from 1912 to 1988.

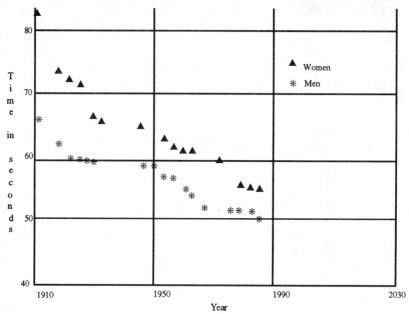

a) Explain what the different symbols on this graph represent.

b) Circle three points on the graph and explain what each point represents. Be as detailed as possible.

c) Use a ruler to draw two straight lines that show how the winning times have been decreasing through the years. Discuss with your team the best way to do this.

d) Which is decreasing faster, the women's or the men's winning times? Explain.

e) Why might one group's winning times decrease faster than the other?

f) In what year will men and women have the same winning time if the trend shown on the graph continues? Explain in one or two sentences how you determined your answer. What will that winning time be?

g) Could these trends continue at the same rate forever? Explain.

PZL-20 LEARNING PROFILES

Each lesson in this unit contains a description of a particular student learning type. These learning types, or profiles, list the most common sayings that each type of learner uses. Read them with your team each day and <u>briefly</u> consider whether this profile has qualities that are like you. Then consider how much of the profile describes members of your team or other students in your class. You might also want to extend comparisons to some of your other friends who are not in this class. As you read these profiles, keep track of which characteristics best describe YOU in your notebook.

Below is the first profile. These learning profiles are excerpted from *UP from Underachievement* by Diane Heacox, ©1991, with permission from Free Spirit Publishing Inc., Minneapolis, MN; (800) 735-7323. ALL RIGHTS RESERVED. CPM Educational Program thanks them for their permission to reproduce them for you.

THE STRUGGLING STUDENT

"I don't understand."
"It's too hard for me."
"I don't understand what the teacher wants me to do."
"I forgot. I need an example."
"I used to be smart."
"That's too much work to do--I'll never get it done in time."
"I thought that was what we were supposed to do."
"I JUST DON'T GET IT."

WR-4. Jan and Carol are competing in the 500m Freestyle swimming competition. In this event, swimmers make 10 "laps" in a 50-meter pool. Carol finished the race in 2 minutes, 40 seconds and swam $\frac{1}{4}$ m per second faster than Jan. What was Jan's time?

WR-5. When Ellen started with Regina's favorite number and tripled it, the result was twelve more than twice the favorite number. Define a variable, then write an equation and use it to find Regina's number.

WR-6. Solve each equation and check your solution.

a) $3x + 4 = 10$

b) $2x - 9 + 3x = 16$

c) $2(x + 3) = -4$

d) $-3(2x + 5) = 87$

e) $3x + 3 = 5x - 6$

f) $\frac{x}{8} = \frac{7}{3}$

g) $\frac{8}{x+2} = \frac{-5}{x}$

WR-7. Make a table and a graph for y = -2x + 3 for x = -4, -3, ... , 4, 5.

WR-8. Solve each equation below for y.

    a)   2x + y = 8   b)   2x - y = 8

WR-9. Find three consecutive even numbers so that when I double the largest and then add the smallest, I get 116. Write an equation and solve.

WRITING EQUATIONS TO SOLVE PROBLEMS

WR-10. MATCH-A-GRAPH, Part Two

Refer to your graphs and equations from Match-A-Graph, WR-1.

a) To see which <u>equations</u> in Match-A-Graph contain the point (0, 1), substitute x = 0 and y = 1 into all eight equations. Determine if the point makes the equations true or false.

b) Examine every <u>graph</u> in Match-A-Graph . Which graphs contain the point (0, 1)?

c) Using the point (0, 1), do you have enough information to match a graph and its equation? Explain.

d) How many graphs contain the point (2, 0)? Which equations will be true when you substitute (2, 0) into them? Find the matching equations for all the graphs that contain that point.

e) Find another point on each of the graphs you found in part (d). Use it to verify the equation you matched. Which point did you choose?

WR-11. Vern and Crystal are aides for Ms. Speedi. Vern can grade quizzes at the rate of three answers per minute. Four minutes after Vern starts to grade quizzes, Crystal joins him. She can grade at the rate of five answers per minute. Vern and Crystal continue grading at these rates until they are finished. The diagram below shows Vern's and Crystal's contributions to grading a set of quizzes. The whole segment represents the completed job.

                Vern   Crystal & Vern

a) Let x represent the <u>number of minutes</u> Vern spent grading quizzes. Write an expression using x for each of the following descriptions.

    (1) the number of minutes Crystal will be grading quizzes.

    (2) the number of answers Vern will check.

    (3) the number of answers Crystal will check .

    (4) the total number of answers that will be checked by both Vern and Crystal.

>>Problem continues on the next page.>>

b) After five minutes, how many answers has Vern checked? How many answers has Crystal checked?

c) After 10 minutes, how many answers has Vern checked? How many answers has Crystal checked?

d) Write an equation that states that Vern and Crystal have checked a total of 36 answers. Solve the equation to find out how long Vern has been grading when 36 answers have been checked. (Hint: use the pattern in parts (b) and (c) to help you write the equation.)

e) How long will Crystal work if she and Vern check 220 answers?

WR-12. Look for patterns in the figures below.

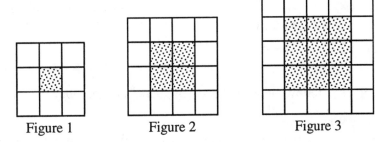

Figure 1   Figure 2   Figure 3

a) Make a table comparing the figure number to the total number of shaded tiles.

b) Describe the pattern formed by the shaded tiles. How many shaded tiles would be in the 15th figure? Explore figures 4 and 5 if you need more data to find a pattern.

c) Graph the first six figures. Let the x-axis represent figure number and the y-axis represent the number of shaded tiles. Write an equation for this pattern.

d) Make a new table comparing the figure number to the total number of white tiles. Describe the pattern formed by the white tiles. How many white tiles would be in the 15th figure?

e) Graph the first six figures for white tiles. Let the x-axis represent the figure number and the y-axis represent the number of white tiles. Write an equation for this pattern.

f) Compare the graphs of the white and shaded tiles. Why is one straight and the other curved?

g) **Extension:** Describe the pattern formed by the total number of tiles in each figure. What would the total number of tiles needed in the 20th figure? Write an equation for this pattern where x is the figure number. Explain how you arrived at your answer.

WR-13. The length of a rectangle is 4 less than twice the width. If the perimeter is 244, find the length, width, and area. Draw a diagram, write an equation, and solve it.

THE REBEL

"This is dumb."
"Why do we have to do this anyway?"
"This doesn't make any sense."
"This is a total waste of time."
"I have more important things to do."
"When are we ever going to use this stuff anyway?"
"My parents never needed algebra."
"I will not do that just because you want me to."
"WHY SHOULD I PLAY THE SCHOOL GAME?"

WR-14. Solve each of the following equations for y. Show all steps.

a) $2y - 8 = 5$

b) $6y + 7 = 8y - 13$

c) $2 - 7(y - 3) = -5$

d) $5 + 2y = y - 8$

e) $x + 2y = y - 8$

WR-15. Solve the following percent problems using a proportion.

a) What is 20% of 125?

b) 43 is what percent of 125?

c) 15 is 12% of what number?

WR-16. A box contains 24 red cubes and x blue cubes. The probability of pulling out one blue cube is $\frac{1}{3}$.

a) How many cubes are there altogether?

b) Write an expression from part (a) for the probability of pulling out a blue cube.

c) Write and solve an equation with two equal ratios to find the number of blue cubes.

d) Explain why your answer makes sense.

WR-17. Use the rectangle below to answer the following questions.

a) Write the area of the large rectangle as a sum and as a product of its dimensions AND as a sum of the four smaller interior rectangles.

b) Calculate the perimeter of the figure.

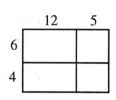

WR-18. Write an equation and solve the problem. Write your solution as a sentence.

In Spring, the daily high temperature in Boulder, Colorado rises, on average, $\frac{1}{3}$° per day. On Friday, May 2, the temperature reached 74°. Predict when the temperature will reach 90°.

WR-19. Julio runs $\frac{3}{10}$ mile in $1\frac{1}{2}$ minutes. If he keeps running at that rate, how long will it take him to run one mile? Eight miles? Show how you solved this.

INTERPRETING GRAPHS, PART 2

WR-20. Find your Resource Page from problem WR-3.

Gale and Leslie are engaged in a friendly 60-mile bike race that started at noon. The graph below represents their progress so far.

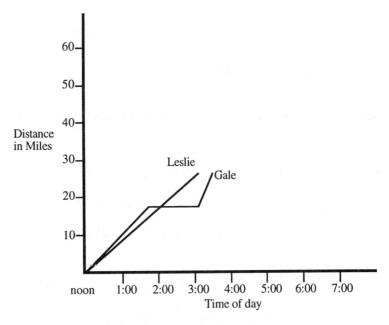

a) What does the intersection of the two lines represent?

b) At what time (approximately) did Leslie pass Gale?

c) About how far had Leslie traveled when she passed Gale?

>>Problem continues on the next page.>>

d) You can see that Gale will catch up to Leslie again if they both continue at their present rates. At approximately what time will this happen? About how far have they both traveled at that point?

e) What happened to Gale between 1:30 and 3:00?

f) Does Gale travel faster or slower after his rest? Explain.

g) If Leslie continues at a steady pace, about how long will it take her to complete the race?

WR-21. Jose, Racquel, and Lena are planning a bicycle riding vacation. They will ride approximately 75 miles everyday. Jose averages 10 mph on his bike while Racquel averages 15 mph. Lena averages 5 mph because she has a cast on her foot which slows her down. Lena plans to bike for only 3 hours and then will catch a ride in a van which travels at 30 mph.

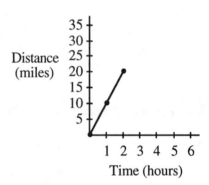

a) Copy this graph onto graph paper. Using a ruler, extend the distance axis to 75 miles and time to 10 hours. The line on the graph shows the first two hours of Jose's trip for one day.

b) Complete Jose's graph. On the same axes, graph Racquel's and Lena's trips. Use different markings (or colors) for each person's graph so that you can analyze this data. Be sure that Lena's graph reflects both the bike and van portions of her trip.

c) How long did each person's trip take?

d) Jose's and Racquel's graphs are straight lines. Which graph is steeper? Explain.

e) Compare Lena's graph to the other graphs. Is it steeper? Explain.

f) If someone rode in the van the entire time, describe what that graph would look like and how long the trip would take.

WR-22. Without drawing a graph, determine if the given points are on the graph of the given linear equation. Explain how you got your answer to part (a).

a) (-2, -5) for $y = 3x + 2$

b) (-1, -4), (2, 2), (1, -1), (3, 3) for $y = 3x - 4$

c) (-1, -1), (-2, -2), (3, -7), (4, -11) for $y = -2x - 3$

THE CONFORMIST

"I don't want to be a nerd."
"All of my friends did the same on the test."
"If I get done early, it just means more work."
"The best part of school is seeing my friends."
"If I want a date to homecoming..."
"I just don't want to work that hard."
"The regular math class is better than the honors class."
"But I won't have time for baseball practice."
"If I do too well, I'll get teased."
"DON'T NOTICE THAT I AM SMART."

WR-23. Use algebra to solve each of the following equations. Show your steps and check your work.

a) $3 + 5x = 28$

b) $3x + 15 = 3 - 3x$

c) $\frac{x}{3} + 6 = -45$

d) $\frac{x-2}{5} = \frac{10-x}{8}$

WR-24. Solve each of the following equations for y. Then make a table and graph.

a) $3y - 4x = 15$

b) $6x + 2y - 14 = 0$

WR-25. Valentine's Day is quickly approaching and Pablo in arranging flowers for his girlfriend. As a rule, he always uses twice as many roses as carnations and three more tulips than roses. If the arrangement he's designing will take 98 flowers, how many of each flower will he need?

WR-26. Here is a famous number pattern called **Fibonacci Numbers**:

$$1, 1, 2, 3, 5, 8, ...$$

a) Copy the pattern on your paper, then list the three numbers that would appear next.

b) Describe a rule for the pattern in one or two complete sentences.

WR-27. Write and solve an equation using ratios to answer each of the following questions.

    a) Latisha earned 45 out of 55 points on the last quiz. What percent did she receive?

    b) She wanted an A (90%). How many points would she have needed?

    c) Last week she got 60 points and that was 80% correct. How many points were on that quiz?

GRAPHING LINES USING TWO POINTS

WR-28. What is the least number of points necessary to determine a unique straight line?

WR-29. So far this year you have graphed equations by selecting many different values for x, then calculating the corresponding y-values. Each of the equations has been written in **y-form**; that is, y is alone on the left-hand side of the equation while x and numbers are on the right. Problems WR-29 through WR-31 will help you develop a faster method for graphing linear equations.

Here are two equations in **y-form**:

$$(1) \quad y = 2x + 3 \qquad (2) \quad y = 2x - 1$$

    a) Carefully graph each equation on the same set of axes by using any <u>two</u> inputs (x-values). Check with your team so that each person uses two different points to graph each line. Compare your graphs. Are your graphs the same? (Remember that lines continue infinitely in each direction.) Do you need more than two points to graph a line?

    b) Recall that the x-coordinate of any point on the y-axis is zero. Thus any point on the y-axis can be written in the form (0, y). Examine each of your graphs to find the **y-intercept** (the point where the line crosses the y-axis).

    c) Based on your observations, write a conjecture about how to find a line's y-intercept from the equation without graphing the line. If necessary, refer to your tool kit entry from Unit Five which defines the term "conjecture".

    d) In the equation $y = 2x + 3$, "3" is referred to as a **CONSTANT**. Compare the y-intercept for each line in part (a) with the constant in the corresponding equation. Compare what you now observe to your conjecture in part (c).

WR-30. Here are some linear equations in y-form. Use your observation in the previous problem to write the coordinates for the y-intercept of the graph of each equation.

    a) $y = 2x + 3$ d) $y = -3x + 2$

    b) $y = -x - 1$ e) $y = \frac{1}{2}x$

    c) $y = 2x - 4$

WR-31. "WHAT TWO POINTS SHOULD WE USE?"

a) What is the y-intercept of $y = 2x - 5$?

b) Now choose **any** x-value **other than zero** and find its corresponding y-value.

c) Plot the two points you found in parts (a) and (b) on a pair of coordinate axes and carefully use a ruler to draw a line through them. Label the line with its equation.

d) Pick another x-value, find its corresponding y-value, and check the graph to verify that the point is on your line.

TWO-POINT GRAPHING METHOD

You can graph linear equations by using the quick, efficient **two-point graphing method**.

1. Find the y-intercept and mark it on an xy-coordinate graph.
2. Choose any value for x, other than zero, and substitute this value into the equation to find y. Plot this second point on the graph
3. Use a ruler to carefully draw the line through the two points. Remember to extend the line beyond each point.
4. Pick a third x-value, find its corresponding y-value. Check the graph to verify that this third point is on your line.

WR-32. Choose three different equations from problem WR-30 and graph each one using the two-point method. Remember to use a third x-value to verify your graph.

THE VICTIM

"If you'd given more time, I would've gotten it."
"The assignment was just too hard for me."
"My family wasn't home last night, so I couldn't do my homework."
"I have gymnastics practice on Tuesdays."
"If you would quit pushing me, I might get it done."
"You expected too much."
"I never understood fractions."
"No one ever taught me to do that."
"The teacher doesn't like me."
"IT'S NOT MY FAULT."

WR-33. Write one expression to represent each of the following sums or differences. You may want to draw or visualize algebra tiles to help you rewrite these problems.

a) $(2x^2 + 3x + 5) + (x^2 + 2x + 8)$

b) $(3x^2 + 8x + 1) + (2x^2 + 8x + 4)$

c) $(3x^2 + 5x + 7) - (4x^2 + x + 1)$

d) $(x^2 + 9x + 8) - (x^2 + 4x + 8)$

e) $(7x^2 + x + 10) - (3x^2 + 12x + 12)$

f) Describe what happened to the x^2 terms in part (d) above.

WR-34. Solve each of the following equations for x. Show your steps.

a) $34 = 8 - 2x$

b) $8x - 72 = -8x - 40$

c) $3(x + 2) + 3 = -3$

d) $31 = 5 - 2(3x + 4) - x$

e) $5(5x + 4) - (x - 3) = 5$

f) $\frac{2x - 3}{5} = \frac{4x}{7}$

WR-35. The graph at right shows the distances covered by three cars over a time interval. List the cars in order by speed, from greatest speed to least speed, then explain your answer.

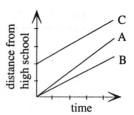

WR-36. The total area of the United States is about 3,623,420 square miles. This includes 79,537 square miles of water in the forms of rivers, lakes, and streams. If a satellite falls from its orbit and crashes into the United States, what is the probability that it will land in water? Write the probability as a ratio (fraction), as a decimal, and as a percent.

WR-37. If 50 empty soda cans weight $3\frac{1}{2}$ pounds, how much would 70 empty soda cans weigh?

WR-38. Graph the equation $y = x^2 - 3$ for x = -3, -2, -1, ..., 3, 4.

a) Is it appropriate to use the two-point method to graph this equation?

b) Is there a way to find the x- and y-intercepts without graphing? If so, do it. If not, explain why not.

SOLVING LINEAR SYSTEMS BY GRAPHING AND SUBSTITUTION

WR-39. Sometimes we need to find the value(s) that make two or more equations true for the same x- and y-values. This is called **solving a system of equations.**

   a) Use the two-point method to graph these two equations on the same set of axes:

   $y = -2x + 5$ $\qquad\qquad\qquad$ $y = x - 1$

   b) Locate the point where the two lines intersect and label it A.

   c) Write the coordinates (x, y) of point A.

   d) Substitute these x- and y-values into both equations to check your solution.

   e) Add the following definition to your tool kit.

> Point A is the **SOLUTION** to the **SYSTEM OF EQUATIONS**
>
> $y = -2x + 5$ and $y = x - 1$.
>
> Point A lies on <u>both</u> lines. This means that the x-value and y-value at point A make <u>both</u> equations true. This point is a common solution to the two equations.
>
> Both equations are said to be in **Y-FORM**, since they have been solved for y.

WR-40. Consider what happens when two items are each equal to the same third item.

   a) If Jasmine the same age as Mateo, and Jasmine is also the same age as Brooke, what can you determine?

   b) Oakland Tech has twice as many students as Fremont High School, and Oakland Tech also has twice as many students as San Lorenzo High School. What can you conclude about the populations at Fremont and San Lorenzo?

   c) So, if $y = -2x + 5$ and $y = x - 1$, what algebraic conclusion can you make? Write a clear explanation using complete sentences.

WR-41. We can sometimes find the point that makes two equations (called a "system of equations) true simultaneously (at the same time) by carefully graphing each equation as you did in WR-39. Most of the time, however, it is difficult to find an exact answer by graphing on paper; the graphs give us only close approximations of the x- and y-values we seek. We can use algebraic techniques to produce exact results.

The first algebraic method for solving a system of equations that we will study is called **substitution.** Since we are finding the one point that is a solution to both equations, we know that the "y" in each equation represents the same number. Therefore, we take the two equations, in **y-form**, and set them equal to each other.

Add this definition and example to your tool kit:

> **SUBSTITUTION** is an algebraic method used to find the point of intersection or **solution to a system of equations**.
>
> Start with two equations in y-form: $y = -2x + 5$ and $y = x - 1$
>
> Substitute equal parts (like WR-40): $-2x + 5 = x - 1$
>
> a) Solve the equation to find x.
>
> b) We now know the x-coordinate of the point of intersection. We also need to find the y-coordinate. To do this, substitute the solution for x in $y = -2x + 5$.
>
> c) Substitute the solution for x in $y = x - 1$. What do you notice about the result? Does it matter which equation we use to find y?
>
> d) Write the answers for x and y as an ordered pair (x, y).

WR-42. Use the algebraic substitution procedure you investigated in problems WR-40 and WR-41 to find the coordinates of each point of intersection:

a) $y = -x + 8$
$y = x - 2$

b) $y = -3x$
$y = -4x + 2$

c) $y = -x + 3$
$y = x + 3$

d) $y = -x + 5$
$y = \frac{1}{2}x + 2$

e) Check your solution by substituting the point of intersection in **each** equation, one at a time, for the systems in parts (b) and (d). Explain why your results verify that each solution is correct or incorrect.

WR-43. It is easier to graph equations that have been written in y-form, but not all equations are written in y-form, such as the one below:

$$4x + 2y = -10$$

You already know how to rewrite the equation $4x + 2y = -10$ in y-form because you know how to solve it for y.

Add this example to your tool kit:

To rewrite the equation $4x + 2y = -10$ in y-form, we need to solve it for y.

1) Subtract 4x from both sides:

2) Divide both sides by 2. Note: <u>both</u> terms on the right side of the equation must be divided by 2.

$$4x + 2y = -10$$
$$\underline{-4x \quad\quad = -4x}$$
$$2y = -4x - 10$$
$$\frac{2y}{2} = \frac{-4x - 10}{2}$$
$$y = -2x - 5$$

Thus, we can also write the above equation as $y = -2x - 5$ or as $y = -5 - 2x$. You will usually find the x term written first.

WR-44. Write the equation $3x + 2y = 2$ in y-form by solving it for y. Check your result with your study team.

WR-45. Write each of the following equations in y-form.

a) $3x + y = -4$

b) $y - 4x = 8$

c) $2x - y = 3$ (Notice -y.)

d) $6x + 2y = 10$

THE DISTRACTED LEARNER

"I worked until 10:00 last night."
"We had a track meet."
"Something came up at home."
"Things haven't been going real well lately."
"I lost my notebook."
"Do you think I could get an extension on the due date?"
"I JUST CAN'T HANDLE IT."

WR-46. There are two things to do in this problem. First, graph each of the following linear equations using the two-point y-form method. (You may first need to change some of the equations into y-form.) Then find and label the y-intercepts.

a) $y = -2x + 6$

b) $4x + y = 1$

c) $2x - y = 5$

d) $3x - 6y = -24$

WR-47. Do you think that two points are enough to determine the graph of a parabola? Explain.

WR-48. Solve each of the following equations for x. Show all steps.

a) $2x + 22 = 12$

b) $6x - 2(x + 8) = 24$

c) $2x - 3 = x + 4$

d) $2x + 15 = 2x - 15$

e) $\frac{5}{3} = \frac{x+2}{7}$

f) Explain your answer to part (d).

WR-49. The number of wild horses at the Lazy Z Dude Ranch could be found by counting them, but when Hank visits the ranch, he suggests that 23 fewer than five times the number of horses at the ranch is the same as 58 more than twice the number of horses on the property. If Hank is right, how many horses does the Lazy Z have? First define your variable.

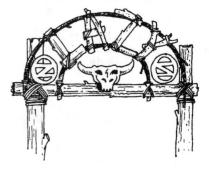

WR-50. Find x and y for each pair of similar triangles below.

a)
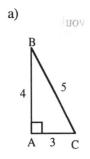 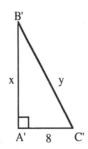

b)
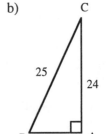 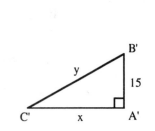

APPLYING LINEAR SYSTEMS OF EQUATIONS

WR-51. THE AMUSEMENT PARK

Juan and his friends are going to an amusement park and discover that they have two ticket options. One option is to buy an admission ticket for $5.00 and then pay 25¢ for each ride. The other option is to buy an admission ticket for $2.00 and then pay 75¢ a ride. What do you think Juan should do?

a) How much does Juan spend if he buys the $5.00 admission ticket and goes on eight rides? 12 rides? 20 rides? 0 rides? x rides?

b) How much does he spend if he buys the $2.00 admission ticket and goes on eight rides? 12 rides? 20 rides? 0 rides? x rides?

c) Let x represents the number of rides and y represent the total amount spent in dollars. Use the expressions for x rides in parts (a) and (b) to write two equations to represent the two admission and ticket options. Since y stands for the total Juan spends, start your equations with "y = ".

d) Graph both admissions options on the same axes. Scale the x-axis one unit per ride, for up to 20 rides. Scale the vertical axis one unit per dollar, for up to $20.00. Label each line as to which admission option it represents.

e) Use the graph from part (d) to find the point of intersection for the graphs of the two options. What does that point represent? How much will Juan spend at that point?

f) Use substitution to solve your equations from part (c). Refer to problem WR-41 if you need help. Compare the equation results with your solution by graphing in part (e).

g) Write an explanation to Juan as to what ticket option he should choose. Include enough information so that he can make an informed decision.

h) If you were going to this amusement park, what option would you choose? Explain.

WR-52. Two car rental agencies are competing for business. Deluxe Driving charges $35 a day and 15¢ a mile while Rent-It-Cheap charges $20 a day and 35¢ a mile. Which agency should you choose?

a) Let x represent the number of miles and y represent the total cost. Write an equation that represents the charges to rent a car at each agency. Refer back to problem WR-51 if you need help.

b) Use substitution to solve your two equations. State in words what that point means.

c) Graph both equations on the same axes. What is the point of intersection? Compare that answer with your solution to part (b).

d) When is it best to rent from Deluxe Driving? When is it best to rent from Rent-It-Cheap? When does it make no difference?

THE STRESSED LEARNER

"It's not quite ready yet."
"I could've done better if I'd had more time."
"I was sure I could do well with that assignment."
"If I keep it one more day, it will be so much better that I can make up the points I lose for turning it in late."
"What if I can't do it?"
"She's not a very good teacher. She gave me a decent grade on that paper I put together at the last minute."
"I'll work on it later."
"IT'S NOT GOOD ENOUGH."

WR-53. Triangles ABC, DEF, and GHI are similar right triangles. Find each missing side by matching corresponding sides, writing an equation, and then solving it.

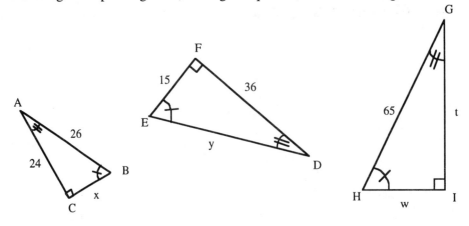

WR-54. Solve each of the following equations for x, if possible.

a) $4 + 2.3x = -5.2$

b) $2(4x - 7) = 8x + 14$

c) $6(x - 2) = 5(x - 11) - 21$

d) $3(x - 5) = \frac{1}{5}(10x - 25)$

e) $\frac{x}{x+4} = \frac{9}{2}$

WR-55. Graph each of the following equations on the same set of axes.

a) $y = 0.75x$

b) $y = -0.75x$

c) Describe the similarities and differences between these two graphs.

WR-56. Without drawing a graph, determine which, if any, of the following points are on the graph of $y = -3x - 5$: $(0, -5), (2, -11), (-3, -14)$.

WR-57. Find the area of the shaded region of this rectangle:

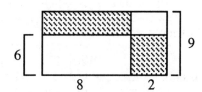

WR-58. There is a $34 tax on an $800 motorcycle.

a) How much tax would there be on a $1000 motorcycle? Write an equation and solve.

b) What is the tax rate?

MORE APPLICATIONS OF LINEAR SYSTEMS

WR-59. Latanya and George both want to buy new bicycles. The bicycles cost $300. Latanya opened a savings account with $50. She just got a job and is determined to save $30 a week. George started a savings account with $75. He is able to save $25 a week.

a) Write two equations that represent their savings plans. What do your variables represent?

b) Solve these equations using substitution. Explain what your solution means.

c) Graph each situation. Discuss with your team the best way to scale your axes. Label the lines.

d) What does the point of intersection on your graph represent?

e) How long will it take Latanya to save enough money to buy the bike?

f) How long will it take George to save enough money to buy the bike?

g) Do you think this is a realistic problem? Is there anything that could possibly interfere with George's and Latanya's plans to buy bikes? Explain.

WR-60. CHANGING POPULATIONS

Highland has a population of 12,200. Its population has been increasing at a rate of 300 people per year. Lowville has a population of 21,000 which is declining by 250 people per year. Assuming that the rates do not change, in how many years will the populations be equal?

a) Write an equation that represents each city's population trend. What do your variables represent?

b) Use substitution to solve the problem. Show your work.

c) Discuss with your team how to scale the axes for this problem. Then graph each situation. What are the coordinates of the point where the graphs meet? How large will the populations be when they are equal?

WR-61. You solved the previous problem using two methods: graphing and substitution. What is the benefit of using both methods? Which method do you like best at this point? Why?

THE SINGLE-SIDED ACHIEVER

"That class isn't important to me."
"That subject bores me."
"I only like certain classes."
"This class is different."
"I just don't like math."
"This class doesn't teach me anything I want to know."
"IT DOESN'T INTEREST ME."

WR-62. FUND RAISING

It is the end of the semester and the clubs at school are recording their profits. The Science Club started out with $20 and has increased its balance by an average of $10 per week. The Computer Club saved $5 per week, but started out with $50 at the beginning of the semester. The Math Club was just formed this year and managed to raise $15 per week.

a) Create an equation for each club. Let x represent the number of weeks, and y represent the balance of the accounts.

b) Graph all three lines on one set of axes.

c) During the semester, there are a few times when two clubs have the same balance. Can you find any?

d) Analyze which club is the wealthiest. Explain your answer in detail.

WR-63.

Fifty eighth grade and 100 ninth grade students wanted to attend a concert, but the bus would only hold 50 people. The decision was made to take 20 eighth graders and 30 ninth graders. Names were drawn at random for each grade level.

a) If you are an eighth grader who wants to go to the concert, what is the probability you will be selected?

b) What is the probability of being selected if you are a ninth grader?

c) Was the decision fair? Explain.

WR-64. Solve each of the following equations.

a) $14 = 8 + 5q$

b) $\frac{12}{x} = 4$

c) $m + 2(m + 1) + 3(m + 2) = 10m - 8$

d) $5x - (2x - 4) = 32$

WR-65. Change each equation to y-form. Write the coordinates of the y-intercept for each.

a) $y - 18 = 4x$

b) $3x + 2y = 6$

c) $3x - 2y = 6$

WR-66. Write the area as a sum and as a product.

a)

b)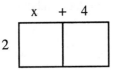

c)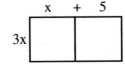

d) Find the perimeters of the three rectangles above.

WR-67. Copy and solve these Diamond Problems:

  a)   b)   c)   d)

WR-68. 29 is 80% of some number. Without solving an equation, Kee Wai reasoned that the answer was around 60. Is his estimate too high or too low? How do you know?

WR-69. Find the x- and y-intercept of the graph of $y = -2x + 9$. Write your answers in the form of (x, y).

MULTIPLYING BINOMIALS

 WR-70. In this semester we have developed several big ideas. One of these ideas has been the Distributive Property and multiplication of variables. In the next two days we will expand our understanding of multiplication to include pairs of sums.

Use your Algebra Tiles to construct each of the following rectangles with the specified dimensions. Draw the complete picture of each rectangle on your paper, then write its area as a product and as a sum. Label everything as shown in the example below.

Example: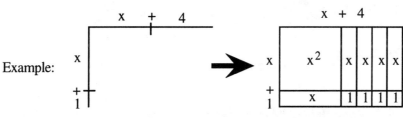

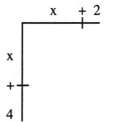

$(x + 1)(x + 4) = x^2 + 5x + 4$
     product   sum

a)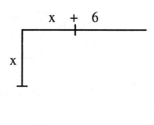

b)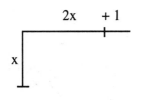

c)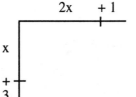

d)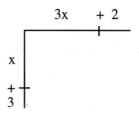

e)

f)

World Records: Graphing and Systems of Linear Equations

WR-71. We can make our work drawing tiled rectangles easier by not filling in the whole picture. That is, we can show a **generic rectangle** by using an outline instead of drawing in all the dividing lines for the rectangular tiles and unit squares. For example, we can represent the rectangle whose dimensions are x + 1 by x + 2 with the generic rectangle shown below:

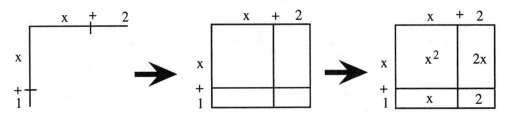

area as a product   **area as a sum**
$(x + 1)(x + 2) = x^2 + 2x + 1x + 2 = x^2 + 3x + 2$

In the last step we found the area of each of the parts of the generic rectangle (the four interior rectangles) by multiplying each length and width, and then recorded the area of each part as a sum.

Complete each of the following generic rectangles without drawing in all the dividing lines for the rectangular tiles and unit squares. Then find and record the area of the large rectangle as the sum of its parts. Write an equation for each completed generic rectangle in the form:

area as a product = area as a sum.

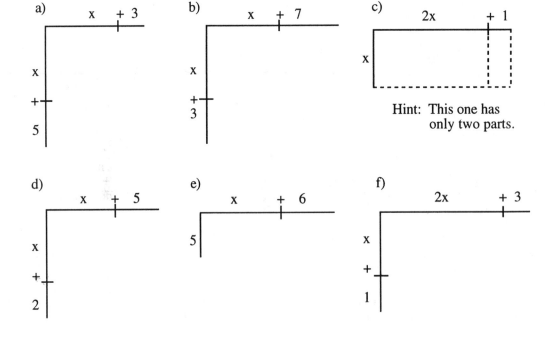

Hint: This one has only two parts.

WR-72. Carefully read this information about binomials. Then add a description of binomials and the example of multiplying binomials to your tool kit.

These are examples of **BINOMIALS**:

$$x + 2 \qquad 7 - 5x \qquad 2x - 7 \qquad (3x^2 - 17)$$

These are **NOT** binomials:

$$2x \qquad 3x^2 \qquad -5xy - 2x + 9$$

Write a description of a binomial that would be clear to a new student.

We can use generic rectangles to find various products. We call this process **MULTIPLYING BINOMIALS**. For example, multiply $(2x + 5)(x + 3)$:

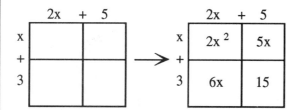

$(2x + 5)(x + 3) = 2x^2 + 11x + 15$
  area as a product   area as a sum

Note that a generic rectangle helps us organize the problem. It does not have to be drawn accurately or to scale. In the example above, you can see that the region representing 5x is probably not as big as it should be compared to the region for 6x.

WR-73. Find each of the following products by drawing and labeling a generic rectangle. Hint: you only need two parts for your generic rectangles in some of the problems.

a) $(x + 5)(x + 4)$

b) $(x + 2)(x + 8)$

c) $x(x + 10)$

d) $2x(x + 3)$

e) $(x + 30)(x + 20)$

f) $(2x + 3)(3x + 4)$

THE BORED STUDENT

"I learned all of this last year."
"How many times do we have to go over this?"
"Why can't the other kids understand this so we can go on to something else?"
"When do we get to the hard stuff?"
"I get so tired of the same thing every day."
"When can I learn what I want to learn?"
"I'd like to do it another way."
"THERE'S NOTHING NEW OR EXCITING TO LEARN."

WR-74. Find the point of intersection of these two lines. Use any method.

$$y = 2x - 3 \quad \text{and} \quad y = 4x + 1$$

WR-75. Is it possible to change a $20 bill into an equal number of nickels, dimes, and quarters? If so, how many of each coin will be needed? Write an equation and solve.

WR-76. In the example given for WR-70, an expression for the rectangle's perimeter is:

$$(x + 4) + (x + 1) + (x + 4) + (x + 1) = 4x + 10.$$

Write the expressions for the perimeters of the rectangles in parts (a), (c), and (e) in problem WR-70.

WR-77. Use the data in the tables below to write a rule for x that results in y. Write that rule as an equation. Use the two point method to quickly graph each equations.

x	y
8	23
2	5
-3	-10
10	29
x	

x	y
6	32
-2	-8
1	7
10	52
x	

WR-78. Reha saved $12.60 by buying his new sneakers on sale at 30% off. How much did the sneakers cost originally? (Be sure to write an equation.)

WR-79. The probability of rolling a sum of seven with two dice is $\frac{1}{6}$.

a) What is the percentage that corresponds to $\frac{1}{6}$?

b) If you rolled the dice 523 times, about how many times would you expect the sum to be 7?

WR-80. a) If 3 pounds of coffee cost $16, how much should $\frac{1}{2}$ pound cost?

b) If 12 rolls of toilet paper cost $2.99, what should 18 rolls cost?

c) A dozen donuts cost $3.50. How much should three donuts cost?

BINOMIAL SQUARES

WR-81. Use generic rectangles to multiply these binomials. Match the product with those given on the right. Even though you cannot draw the tiles, do the negatives effect the basic <u>process</u> of multiplying the dimensions?

    a) $(x + 6)(x - 8)$ 1) $x^2 - 13x + 12$

    b) $(x - 2)(x + 6)$ 2) $x^2 - 14x + 48$

    c) $(x - 12)(x - 1)$ 3) $x^2 - 2x - 48$

    d) $(x - 8)(x - 6)$ 4) $x^2 + 4x - 12$

WR-82. Algebra tiles can be used to build a series of squares. After you build each square, write the area as a product, sketch the tiles, and write the areas as a sum. Organize your work in a table as shown below.

    a) Complete this chart for the first 6 squares. The first square is $x + 1$ by $x + 1$, the second square is $x + 2$ by $x + 2$, and so on.

Area as a Product	Sketch	Area as a Sum
$(x + 1)^2$	(x+1 by x+1 square)	$x^2 + 2x + 1$

    b) In the rectangles you drew, explain where you see the squares.

    c) Describe all the patterns you found in the table.

    d) Using your patterns, and without drawing a sketch, fill in the chart's next two products and areas.

    e) Without drawing a sketch, what would the 10th product and area be?

    f) Add two more perfect square products and areas to your table without drawing a sketch.

WR-83. Use your patterns from the previous problem to determine whether the expression is a perfect square. If it is, write it as $(\ )^2$. If not, explain why not.

    a) $x^2 + 30x + 225$ b) $x^2 + 30x + 196$

World Records: Graphing and Systems of Linear Equations

WR-84. Copy the generic rectangle on your paper and multiply $(x + 9)(x^2 - 3x + 5)$. Write your answer as a sum.

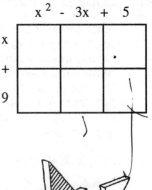

WR-85. UNIT 6 SUMMARY

Select 3 or 4 big ideas that you feel best represent Unit 6. Write a description of each main idea. For each topic, choose a problem that best demonstrates your understanding. Explain your solution as in the previous unit summaries. Show all your work.

WR-86. Ana and Gloria are swimming toward each other in different lanes from opposite ends of a 50 m swimming pool.

 a) If Ana swims half as fast as Gloria, where in the pool will they meet?

 b) If Ana swims three times as fast as Gloria, where in the pool will they meet?

 c) If Ana swims twice as fast as Gloria, and they meet after 1 minute, find Gloria's speed.

THE COMPLACENT LEARNER

"I'm doing as well as I want to."
"I'm satisfied. I don't know why you aren't."
"Quit pressuring me."
"You're never satisfied with what I do."
"I do okay on tests. Why should I do homework?"
"It's important to you, not me."
"Sure, I could've done better."
"I'M DOING JUST FINE."

WR-87. COURSE REFLECTION, Part One

Since this is the end of the first semester, it is an opportunity for you to reflect back on how much you have learned. We have now completed several summaries. Find your summaries for Units 3, 4, and 5 and look for improvement and growth in your understanding of algebra over time. Examine the degree of difficulty of the topics in those units. Write a paragraph stating your observations.

WR-88. COURSE REFLECTION, Part Two

Now we will shift our focus to your learning style. Write a paragraph examining how you approach mathematics. The questions below are merely suggestions to get you started.

What topics were difficult for you to learn at first? What learning are you most proud of? Is algebra what you thought it would be? Have the study teams helped your understanding? What talent are you most proud of? How have you helped others learn? Have your study habits changed? Have your feelings about math changed? What topics are still difficult for you? What are your goals for the next semester?

WR-89. Evaluate $h = -16t^2 + 192t + 30$ for

a) $t = \frac{1}{2}$, then 3, and finally $-\frac{1}{2}$.

b) If you were going to graph this equation, what shape would you expect to see? Why?

WR-90. Use the correct Order of Operations to simplify.

a) $\dfrac{-4 + \sqrt{24 - (-20)}}{2}$ b) $\dfrac{3 - \sqrt{5}}{6}$

WR-91. Consider the possible x-values that would make the following equations true. Are these equations always, sometimes, or never true?

a) $3x + 9 = 3x - 9$ c) $3x + 9 = x + 9$

b) $3x + 9 = 3(x + 9)$ d) $3x + 9 = 3(x + 3)$

WR-92. Light travels approximately 299,793 kilometers per second. Write your answers in scientific notation as you calculate the following. How far does light travel in:

a) one second?   b) one minute?

c) one hour?

d) Earth is about 150,000,000 kilometers from the sun. About how long does it take for the sun's light to reach the Earth? Explain your reasoning. Show your work.

World Records: Graphing and Systems of Linear Equations

WR-93. Jung's car travels 32 miles per gallon of gas. For each question below, use equivalent ratios to write an equation, then solve it.

a) How far will Jung's car go on 8 gallons of gas?

b) If Jung drives 118 miles, how much gas will be used?

WR-94. This graph contains the lines for $y = x + 2$ and $y = 2x - 1$.

a) Using the graph, what is the solution of this system?

b) Solve the system algebraically to confirm your answer to part (a).

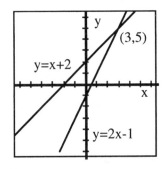

UNIT REVIEW

WR-95. WORLD RECORDS, Part Two

In 1988, the women's world record for the 100-meter free-style in swimming was 54.73 seconds. Since then, the record time has been decreasing by approximately 0.33 seconds a year. The men's record was 49.36 seconds and has been decreasing at about 0.18 seconds a year. If both records continue to decrease at these same rates, in how many years will the men and women have the same record time?

Drawing the graph of these two situations would be difficult because the numbers are difficult to graph. In this case we want to use the power, efficiency and convenience of algebra to solve the problem.

Let x represent the number of years from 1988.
Let y represent the time in seconds.

a) Write an equation for the women's record and the men's record. If you are stuck, look back at what you did in the Amusement Park problem, WR-51.

b) Solve your system of equations by substitution. Explain what your answer means in words.

c) In what year will the women's and men's records be the same? Compare this answer to the answer you found in World Records, Part One, problem WR-3. Which answer do you think is most accurate?

WR-96. Copy each expression and find its product.

a) $(x + 13)(x + 3)$

b) $(x - 8)(x + 10)$

c) $(x - 9)(x - 8)$

d) $(x + 5)(x - 9)$

e) $(2x + 1)(x - 4)$

f) $(x + 4)^2$

g) $4(5x - 2)$

h) $7x(3x + 5)$

i) $(x - 4)^2$

j) $(x + 11)^2$

k) $(3x + 7)^2$

WR-97. Can each of these equations be graphed with the two-point y-form method? Why or why not? (You do not need to graph the equations, but do explain your reasoning.)

a) $y = 2x + 1$

b) $y = 2x$

c) $y = x^2$

WR-98. Use algebra to solve each of the following equations for x. Check your answers with your study team.

a) $3 = 9 - 5x$

b) $\dfrac{10}{x - 6} = \dfrac{15}{x - 3}$

c) $4 - 6(x + 2) = 10$

d) $-3x - 2 = 2x - 8$

e) $2(x + 2) = 3(x - 5)$

f) $\dfrac{5 - x}{7} = 3$

THE COLLABORATOR

"When I don't understand a problem, I ask my group."
"Sometimes no one understands at first, but we work it out together."
"Explaining to some one else helps me learn."
"When we work together we get most of our assignments done during class."
"For a lot of the work four brains are better than one."
"Sometimes I come up with an idea no one else thought of!"

WR-99. Complete each part below.

a) Solve $2x - 5 = -4x - 2$.

b) Graph $y = 2x - 5$ and $y = -4x - 2$ on the same set of axes.

c) How can you use the graph in part (b) to check the solution from part (a)?

World Records: Graphing and Systems of Linear Equations

WR-100. Be sure to draw diagrams for the figures in the problems below.

    a) Find the measures of the sides of a square with an area of 75 sq. units.

    b) Sketch and label a rectangle with the same area.

    c) Draw a triangle with an area of 75 square units. Label the height and base lengths of each.

WR-101. Use the figure at right for the problems below.

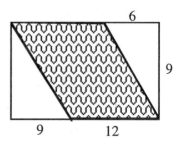

    a) Find the area of the shaded region. Show all of your subproblems.

    b) Write a ratio comparing the shaded region to the area of the entire rectangle.

WR-102. Density is a ratio of mass compared to volume: $D = \frac{m}{v}$. If the density of lead is $11.3 \frac{g}{ml}$, find the mass of 60 ml of lead. Write a proportion and solve.

WR-103. Speed is a ratio of distance compared to time. If Maisha runs 12 meters per second, how many seconds will it take her to run 30 meters? Write a proportion and solve.

WR-104. Use the figure at right for the problems below.

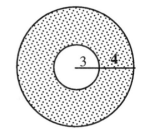

    a) Find the shaded area at right. Show your subproblems.

    b) Write the ratio of the shaded part of the circle to the whole circle.

    c) What is the probability of throwing a dart and hitting the unshaded region if every toss hits the dart board?

    d) Write this fraction from part (c) as a decimal and a percent.

WR-105. Triangle ABC is similar to triangle DEF.

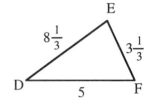

    a) What is the enlargement ratio?

    b) What is the perimeter ratio?

    c) What is the area ratio?

WR-106. TOOL KIT CLEAN-UP

Examine the elements in your extensive tool kit from Units 0 - 6. We will create a new, fresh, consolidated tool kit representing the first semester of algebra.

Include items you value the most in your tool kit. Spend time on this, as this is your tool kit for next semester!

a) Examine the list of tool kit entries from this unit. Check to be sure you have all of these entries. Add any you are missing.

b) Identify which concepts in your complete tool kit (Units 0 - 6) you understand well.

c) Identify the concepts you still need to work on to master.

d) Choose entries to create a Unit 0 - 3 tool kit that is shorter, clear, and useful. You may want to consolidate or shorten some entries. Use this shorter version with your tool kit entries for Units 4 - 6.

e) Have you used your tool kit in the last six units? Why or why not?

- Substitution
- Y-form
- Two-point Graphing Method
- Description of Binomials
- Multiplying Binomials
- System of Equations (Solution)

THE SAVVY STUDENT

"Once I realized I had to do the assignments every day, I started to understand."
"It's not so hard if you remember to use what you did yesterday"
"Having a Tool Kit really helps."
"The work is not always easy, but I can usually figure it out."
"When I can't do a homework problem, I just write down my question, and that counts as doing homework!"
"My secret to learning math is using what I already know and just THINKING."
"Memorizing is hard work; thinking is much easier."

CUMULATIVE REVIEW

WR-107. CORNFLAKES, ANYONE?

While interviewing for a managerial position at Foodway Market, Jamaica encountered the following question:

"If an 20 ounce box of Cornflakes is priced at $4.59, find the appropriate prices for 6, 12, and 36 ounce boxes of Cornflakes produced by the same company."

a) Luckily, Jamaica knew her algebra and got the job. Can you find the prices for the alternate sizes of Cornflakes? Describe a general method to determine the appropriate price for any size box of Cornflakes.

b) After surveying the cereal aisle, she recorded the following prices:

Weight of Box	Price
6 oz	$3.25
20 oz	$5.59
36 oz	$6.59

Her boss informed her that she was going to receive a new shipment of Cornflakes that included some new sizes -- and Jamaica would need to quickly determine the appropriate price for each new size.

Since Jamaica is new on the job, she needs your help! Draw a large graph on graph paper, plot the points for the various weights and prices of Cornflakes. Don't forget to include what the price would be for 0 ounces of Cornflakes. Determine the shape of the graph.

c) Write a note to Jamaica describing the trend of the prices.

d) The shipment has arrived. Use your graph to determine an appropriate price for each of the following new sizes:

Weight of Box	Price
12 oz	
26 oz	
48 oz	

WR-108. SPENDING SPREE

Ms. Speedi went on a whirlwind European vacation for five weeks during the summer! She wrote you the following postcards from exotic locations:

Hello from Prague! The castles are beautiful! Food is delicious, too! I can't believe it's already day 10 of my vacation and I have only $1248 left in my budget. I hope my money lasts... Miss you!

Ms. Speedi

¡Buenas Días from Barcelona! I've never seen such beauty! I hope all is well... it's now been 18 days of vacation and I haven't heard from you. I can't wait to show you all my pictures - that is, if I have enough money to process them! I've only got $864 left to spent on my vacation and have 17 more days to go! Miss you!

Ms. Speedi

a) Assume Ms. Speedi spent roughly the same amount of money each day of her trip. Graph this situation using appropriate scales along your x- and y-axes.

b) Roughly how much did Ms. Speedi spend each day?

c) How much money did Ms. Speed budget for her trip?

d) Will Ms. Speedi have enough money for her trip? Describe in complete sentences how you determined your solution.

World Records: Graphing and Systems of Linear Equations

WR-109. BULLS-EYE!

The math club is going all out for this year's Winter Festival. They are constructing a gigantic dart board to use as a fund-raiser. However, now that this dart board is built, they want to paint the five different rings all a different color. They need to know the area of each ring, so they can buy the right amount of paint.

The dart board is made up of concentric circles, as shown. The inner-most ring has a radius of 1 foot. After that, each circle's radius is 1 foot longer than the next smaller circle.

a) Create a table containing the areas of each ring, starting with the center ring. It will be helpful if you leave the π in your answer.

b) If there was a 10th ring, use your table to find a pattern and predict its area. Describe the pattern you found.

WR-110. ALL THINGS BEING EQUAL

Ms. Speedi bought a top of the line computer one year ago for $2,500. Since then, she has sadly realized that the value of her computer has been declining by roughly $45 per month. Mr. Algorithm's computer, purchased six months ago is currently worth $1,650, and is depreciating by $25 per month.

a) Graph this scenario. Scale your axes carefully so that the data fits.

b) What was the original price of Mr. Algorithm's computer? Describe how you arrived at your solution.

c) Write an equation to represent the value of each computer <u>today</u> that can also be used to predict their value in any future month.

d) When, based on today's values of the computers, will they be worth the same amount? Estimate your solution first using the graph, then use your equations to solve for the exact amount.

e) According to this model, when will Ms. Speedi's computer be worthless? Use your graph to estimate, then set up an equation and solve.

COMPILING YOUR LEARNING PROFILE

If you are following a semester calendar at your school, you are probably finishing the first semester about now (or did so recently). I hope you have found the PZL problems helpful. I've told you just about everything you need to know about the design of this course to help you succeed. If you need some reminders about how to find things in the book, go back and reread PZL 6 (just before SQ-28 in Unit 1). Information about the role of your teacher is in PZL-4 (before GS-27 in Unit 0), PZL-8 (before SQ-65 in Unit 1), PZL-11 (before KF-11 in Unit 2), and PZL-12 (before KF-108 in Unit 2). Skill and concept development (spiraling) is discussed in PZL-14 (before BC-23 in Unit 3) and PZL-19 (before EF-54 in Unit 5).

Anyway, it's time for me to go. Work especially hard throughout the rest of the book to make connections between the various ideas in the course. Remember--the authors' goal is to have algebra make sense to you. I hope this is happening in this course.

MISSION POSSIBLE (something to do)

I hope you have been writing down the learning profile traits that are similar to yours like I asked you to do at the beginning of this unit. Now it's time to write your profile. Please do the following two tasks:

a) Compile your list of learning traits, then give your profile a title and write the characteristics in a format similar to the eleven profiles in this unit.

b) Write a paragraph or two that evaluates the effectiveness of your learning habits. Include some suggestions that might make you a more effective learner. In fact, write at least two goals to improve your learning methods that you would be willing to work on during the second semester.

GM-11. CONSECUTIVE SUMS

A consecutive sum is an adding sequence of consecutive whole numbers.

Examples:
2 + 3
8 + 9 + 10 + 11 + 12 + 13
7 + 8 + 9 + 10
$x + (x + 1) + (x + 2)$

Your Task:

- Write the first 35 counting numbers (1 through 35) with as many consecutive sums as possible. The number 15 can be written as 7 + 8 or 4 + 5 + 6 or 1 + 2 + 3 + 4 + 5. Make sure you organize your work in order to find patterns.

- Describe as many patterns as you find. Are there any numbers that have no consecutive sums? Are some numbers easier to find? Write as many generalizations as you can.

- Demonstrate why your patterns are always true (algebra can be useful here).

- Can you find all the consecutive sums possible for any given number? For example, can you write all the ways 91 or 64 or 200 can be written as consecutive sums? Explain completely.

GM-12. GOING IN CIRCLES

In the olden days before the "Lottery" or the "Big Spin", there lived a very compassionate queen. She was very, very kind and very, very rich. In fact, she decided to share her wealth with some lucky people.

Once every year she would invite some of her subjects to a banquet dinner and choose one lucky winner who would be granted one wish. The chairs were numbered consecutively starting with number 1 and set up in order around a large circular table as shown at right.

```
        11  12   1
    10               2
   9                   3
    8                4
        7   6    5
```

After dessert, the court jester entered and, starting with chair number 1, followed this rule: he eliminated <u>every</u> <u>other</u> <u>person</u> in a clockwise rotation until only one person was left.

That person would be the lucky winner of the year and could ask the Queen for one wish.

For example, if twelve people were invited and seated around the table, the jester would eliminate them in the following order:

$$2, 4, 6, 8, 10, 12, 3, 7, 11, 5, 1$$

This would leave person number 9 the winner.

Your Task:

- When 12 people are invited, show why the lucky winner will be person number 9.
- Find the lucky winner for at least 20 different-sized groups of people. Describe any patterns you find.
- Using the patterns you found, tell where you would sit in order to be the lucky winner if you were one of 270 people invited to dinner. Explain why you think your answer makes sense.

Unit Seven
The Big Race
Slopes and Rates of Change

Unit 7 Objectives
The Big Race: SLOPES AND RATES OF CHANGE

We will strengthen our understanding of lines and equations in this unit with a focus on slope. When someone asks "What is the bank's interest rate?" they are referring to a **rate of change, or slope**. Slope is not only an important concept, it is also a useful tool. It will help us graph lines faster and will be used to predict future events.

Our work with systems of equations will continue to focus on substitution and graphing. Our understanding of slopes, graphing, and systems of equations will help us determine the winner of the Big Race.

In this unit you will have the opportunity to:

- develop your understanding of the relationship between graphs and their equations.

- formalize the notion of the steepness and direction of a line (slope) in terms of rate of change.

- continue learning how to solve systems of linear equations, especially expanding your understanding of algebraic solutions by substitution.

- learn how to graph linear equations using the slope-intercept method ($y = mx + b$).

- work with distance, rate and time relationships in context.

This unit typically starts the second semester of the course. You will continue to have opportunities to practice skills and concepts you encountered during the first semester, especially graphing, ratios, and solving equations. You will be expected to keep an organized notebook and continue your tool kit for Units 7 - 13.

BR-0.   THE BIG RACE

One of the annual events at school during Spirit Week is a tricycle race held in the gym. Leslie, a member of your class, has won the race each of the last three years and is starting to brag about it. The rest of the class is annoyed by this attitude and wants to end this winning streak. The race will be held at the end of this unit. Since you will also compete, you need to size up the competition.

Problem Solving
Graphing
Writing and Solving Equations
Ratios
Geometry
Symbol Manipulation

Unit 7
The Big Race: SLOPES AND RATES OF CHANGE

LINEAR EQUATIONS

BR-1. You will receive information every day in this unit about different competitors who will try to beat Leslie in the Big Race. On the resource page provided by your teacher, carefully graph all the given information to help you decide who might win. **BE SURE TO BRING THIS RESOURCE PAGE WITH YOU TO CLASS EVERY DAY.** Without it, following the Big Race will be difficult.

Leslie rides at a constant rate of 2 meters per second. On the resource page provided by your teacher, neatly graph and write an equation in terms of x and y that shows the distance Leslie travels. Let x represent time in seconds and y represent the distance in meters. We do not know how long the race will be, so extend your graph appropriately.

Find Leslie's equation and label her line appropriately. Keep this resource page in a safe place. We will use it every day during this unit.

BR-2. Write a rule using x and y for a line or curve that uses the following coordinates. Look for patterns to help find the rule.

    a)   (4, 2), (8, 4), (-10, -5), (16, 8)

    b)   (1, 6), (4, 9), (-1, 4), (10, 15)

    c)   (3, 9), (4, 16), (7, 49), (-2, 4)

SCALING AXES

Before setting up a graph, it its important to analyze if both positive and negative numbers are needed. Then you must decide how to scale the x- and y-axes. On <u>each</u> axis, one unit of graph paper must represent the same increment, but each axis can have different scales.

BR-3.   SLEEPY TIME

An equation used to relate the age of a person to the number of hours of sleep required each day is shown at right, where H represents the number of hours of sleep required and A represents the age in years.

$$H = \frac{34 - A}{2}$$

a) Graph this equation with H on the vertical axis and A on the horizontal axis. What is the smallest and largest number of hours you will need on this axis? What is the youngest age? The oldest age? Mark your axes clearly to reflect these values. Discuss the range of H and A with your study team.

b) Plot the point on the graph that represents your age and your typical number of hours of sleep. Are you on the line? Label each of your study team members on the graph with their name to see if they are on or off the line.

c) Does this formula work for you? Does it work for babies? Does it work for older people? For what ages does the formula seem to work best? Is there a point on this graph that makes no sense? Explain.

BR-4.   On graph paper draw and label △ABC.

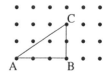

a) Write the ratio of $\frac{\text{the vertical side of } \triangle ABC}{\text{the horizontal side of } \triangle ABC}$.

b) Enlarge △ABC so that the new triangle, △DEF, has corresponding sides twice as long. Draw and label △DEF.

c) Write the ratio of $\frac{\text{the vertical side of } \triangle DEF}{\text{the horizontal side of } \triangle DEF}$. Simplify if possible.

d) What do you notice about the ratios for parts (a) and (c)?

e) Use similar figures to justify your findings about the ratios.

BR-5.   Suppose you multiply the sides of △ABC in BR-4 by four and call the new triangle △GHI. Draw and label △GHI. Write the ratio:

$$\frac{\text{vertical side of } \triangle GHI}{\text{horizontal side of } \triangle GHI}$$

Compare its ratio to the ratios in the previous problem.

BR-6. Transform (change) each of the following equations into y-form and then graph them using the two-point y-form method. Refer to your tool kit if necessary.

 a) y - 4x = -3
 b) 3x + 2y = 12
 c) 3y - 3x = 9
 d) 2(x - 3) + 3y = 0

BR-7. Write the coordinates of the y-intercept of each line in BR-6.

BR-8. Solve each equation below for x. Show all steps.

 a) 5 + 4(x + 1) = 5
 b) $\frac{2x}{3} - 5 = -13$
 c) 4(2x - 1) = -10(x - 5)
 d) 4(x + 5) - 3(x + 2) = 14
 e) $\frac{2x}{7} = \frac{4}{5}$

BR-9. Which of the following points are on the graph of the equation? Try to answer this question <u>without</u> a graph.

 a) (8, 4), (4, 0), (4, 3), (3, 4)   for $y = \frac{1}{4}x + 2$
 b) (0, -7), (2, -1), (4, 5), (6, 11)   for y = 3x - 7

BR-10. Look at the graphs below and answer the following questions.

 a) Compare and contrast lines l, m, n, & k. Be as detailed as possible.
 b) Which line appears steepest?
 c) Which lines appear to have the same steepness?
 d) How is line k different from the other lines?

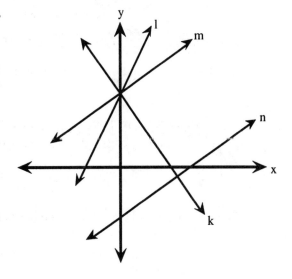

The Big Race: Slopes and Rates of Change   237

GRAPHING LINEAR EQUATIONS WITH INTERCEPTS

BR-11.

Dina wants to see if she could win the Big Race with a 3 meter head start. If she can ride her tricycle 1 meter per second, can she beat Leslie? Using her speed and her head start, add Dina to your graph on the BR-0 resource page. Label her graph with her name and equation. Be sure to extend her line as far as possible. Using different colors or markings will help to keep each competitor identified.

a) Compare both Leslie and Dina's lines.

b) How is Dina's head start represented on the graph?

BR-12. MONEY MATTERS, Part One

Six cousins were each given $20.00 in a new bank account for the Lunar New Year. On the same set of axes, graph their bank account balances. Use a different marking (or different color pen or pencil) for each person. Let x represent time in days and y represent money in the bank. You should look for the comparative steepness (or slope) of the lines as you graph the different financial scenarios. We will formally define slope on another day. For the purposes of this problem, slope is how steeply the line slants across the graph.

a) May saves two dollars every day. Graph her situation. Label the line with her equation and name.

b) Ling saves five dollars every day. Graph her situation. Label the line with her equation and name.

c) Tuan saves fifty cents each day. Graph his situation. Label the line with his equation and name. Hint: Do not forget the starting balance.

d) Compare May's, Ling's and Tuan's graphs and equations. How are they alike? Different?

e) Minh spends two dollars a day. Graph his situation. Label the line with his equation and name.

f) Tam spends three dollars every day. Graph her situation. Label the line with her equation and name.

g) Compare Minh's and Tam's graphs. How are they the same? How are they different?

h) Now compare May's, Ling's and Tuan's graphs to Minh's and Tam's. How are they different?

i) Kim is not going to spend or save any money. Graph her situation. What does this graph look like? What is her equation?

BR-13. Add this definition to your tool kit, if you have not done so already:

> ### X- and Y-INTERCEPTS
>
> Recall that the **X-INTERCEPT** of a line is the point where the graph crosses the x-axis; that is, where y = 0. To find the x-intercept substitute 0 for y and solve for x. Its coordinate is (x, 0).
>
> Similarly, recall that the **Y-INTERCEPT** of a line is the point where the graph crosses the y-axis, which happens when x = 0. Its coordinate is (0, y).

BR-14. In Unit Six, you used the y-form two point method to graph lines. Often the quickest and easiest two points to choose are the x- and y-intercepts. When you use the intercepts, the equation does not need to be in y-form. Find the x- and y-intercepts for the two lines below and use them to graph each line. Write the coordinates of the x- and y-intercepts on your graph.

   a) x - 2y = 4   b) 3x + 6y = 24

BR-15. The graph of the equation 2x - 3y = 7 is a line.

   a) Find the x- and y-intercepts and graph the line using these two points.

   b) If a point on this line has a x-coordinate of 10, what is its y-coordinate?

BR-16. MONEY MATTERS, Part Two

   a) May's, Ling's and Tuan's graphs have a **positive slope**. Minh's and Tam's graphs have a **negative slope.** Explain why this makes sense. What would Kim's graph represent? Explain.

   b) Who is spending money at the fastest rate? How does the graph show this? How does the equation show this?

   c) How does May's line differ from Minh's? How does the equation show this?

   d) Where would you expect the line for Kai, who <u>saves</u> $10 per day to be? Where would you expect his line to be if he <u>spent</u> $10 per day?

   e) Predict what would happen to Ling's line if she had started with $10 instead of $20. What would happen to Tam's line if she started with $30 instead of $20?

   f) Beside each graph, write the rate of savings or spending for each cousin. For example, May saves at a rate of $2 per day. This can be written as $2/day. Write these rates on the graph. Is there a relationship between the rates and the equations?

The Big Race: Slopes and Rates of Change

BR-17. In each part below, find the length of side BC if △ABC is similar to △DEF. Write your answer for the length as a fraction.

a) 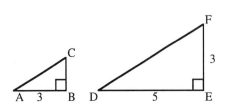   b)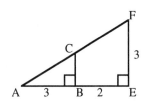

c) For the drawing in part (b), if the coordinates of point A were (0, 0), what would be the coordinates of points C and F?

BR-18. Use the graph at right to answer the following questions:

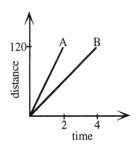

a) Which car is traveling at the greater rate?

b) Find the coordinates of point A and point B.

c) What is the average rate in miles per hour for car A? For car B?

BR-19. Write a rule using x and y for the following coordinates. Look for patterns.

a) (2, 4), (4, 8), (-5, -10), (8, 16)   b) (1, 4), (2, 3), (6, -1), (10, -5)

BR-20. Graph the point P: (4, -2).

a) Find the point 5 units up and 2 units to the left of P. Label it Q. Write the coordinates of Q.

b) Go back to P and now find the point 5 units up and 2 units to the right. Label it W. Write the coordinates of W.

c) Draw lines PQ and PW. How are those lines alike? How are they different?

d) **Extension**: Find the area of triangle PQW.

POSITIVE AND NEGATIVE SLOPE

BR-21. Look at Leslie's and Dina's graphs on your BR-0 Resource Page.

a) What happens to Leslie and Dina at the point of intersection?

b) Where is the point of intersection of the two lines?

c) Use substitution and your equations for Leslie and Dina to verify your point of intersection.

BR-22. Some of the lines below are quite steep; others are not. Below each line there is a ratio that describes how steep it is. This ratio is called the **slope** of the line. With your study team, discuss and compare the following graphs and ratios to discover how to determine the slope of a line.

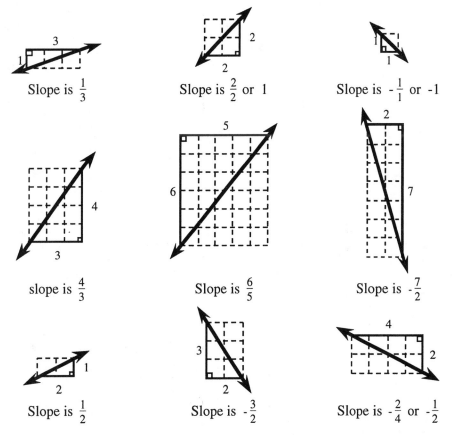

a) Use your observations of the preceding lines to find the slope for the following three lines.

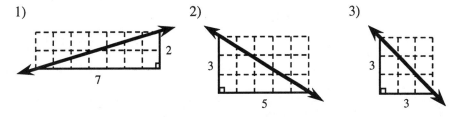

b) Describe how to find the slope of a line. If your team found more than one method, describe each method. Your descriptions should be clear enough for other teams to understand. Be ready to share your descriptions with the class.

c) How can you determine if the slope is positive or negative?

d) Does it matter if the triangle is above or below the line? Why or why not?

The Big Race: Slopes and Rates of Change

BR-23. On the Resource Page provided by your teacher, draw slope triangles and find the slope of the line segments below. Be sure to examine the scale carefully. Write the slope as a ratio. Compare your answers in your study team.

a)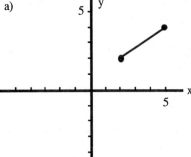

c)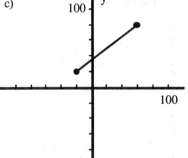

b)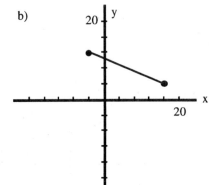

d)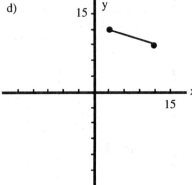

BR-24. Graph each to the following pairs of points on graph paper. Draw the slope triangle and determine the slope.

a) (1, 2) and (4, 5)

b) (3, 8) and (5, 4)

c) (-6, 8) and (-4, 5)

d) (7, 3) and (5, 4)

e) (-4, 5) and (5, 4)

f) (5, 0) and (0, 1)

g) Jim got "1" for the slope of the line through points (1, 2) and (4, -1). Explain to Jim the mistake he made and how to correctly find the slope.

BR-25. Include this information in your tool kit.

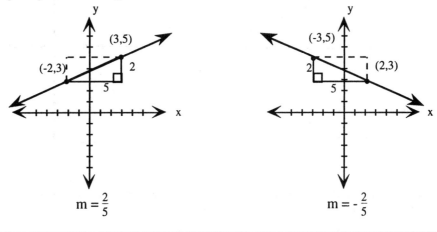

THE SLOPE OF A LINE

The **SLOPE** of a line is a measure of the rate at which something changes. It represents both "steepness" and direction.

$$\text{slope} = \frac{\text{vertical change}}{\text{horizontal change}} = \frac{\text{change in y-values}}{\text{change in x-values}}$$

Note that lines that go **upward** from left to right have **positive** slope, while lines that go **downward** from left to right have **negative** slope. The slope of a line is often denoted by the letter "m." Some texts refer to the vertical change as the "rise" and the horizontal change as the "run."

To calculate the slope of a line, graph two points on the line, draw the slope triangle (as shown in the examples), write the ratio, and lastly check to see if the slope is positive or negative.

$m = \frac{2}{5}$ $m = -\frac{2}{5}$

BR-26. Use the scale graph at right of the three points A (12, 334), B (288, 334), and C (288, 934) to answer the following questions:

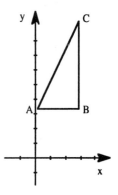

a) What kind of triangle is triangle ABC?

b) What is distance AB?

c) What is distance BC?

d) What is the slope of the line AC, that contains (12, 334) and (288, 934)?

e) What is the area of the triangle?

BR-27. Solve the following systems of equations to find both x and y.

a) $y = 3x - 5$
$y = 4x + 11$

b) $y = \frac{2}{3}x$
$y = x + 15$

BR-28. Refer to your BR-0 Resource Page.

a) What is the minimum distance the Big Race needs to be in order for Leslie to win? Explain your answer.

b) Under what condition would Dina win?

BR-29. Use the two-point method to graph $y = -3x + 4$.

a) What is the y-intercept?

b) What is the x-intercept?

c) Draw a slope triangle on your line and find its slope.

BR-30. Copy these generic rectangles on your paper. Multiply or factor to rewrite these expressions.

a)
	13x	-21
6		

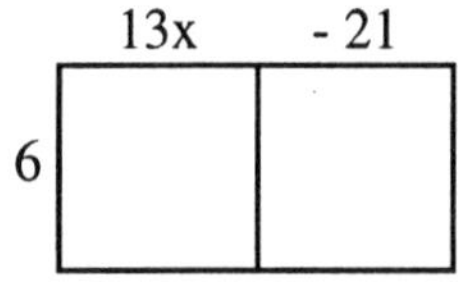

c)
	x	+3
x		
-5		

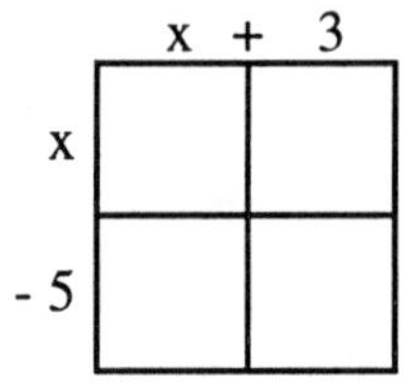

b)
| $16x^2$ | -24x | 4 |

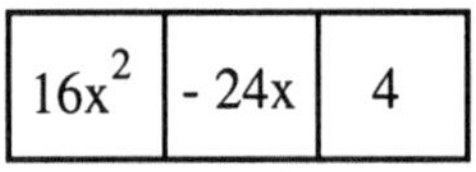

d)
	x	+4
3x		
-5		

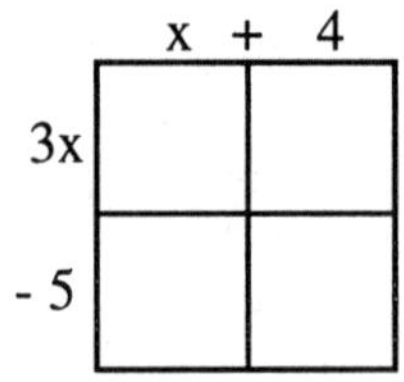

BR-31. Solve each of the following equations.

a) $2x + 8 = 3x - 4$

b) $8(x + 6) + 23 = 7$

c) $1.5(w + 2) = 3 + 2w$

d) $3(2x - 7) = 5x + 17 + x$

BR-32. The length of a certain rectangle is five times its width. Use this information to sketch and label a rectangle. Write an expression to represent the perimeter. Find the rectangle's dimensions if its perimeter is 30 centimeters.

SLOPE: RELATING THE EQUATION TO ITS GRAPH

BR-33. Study these slopes: $-\frac{2}{5}, \frac{-2}{5}, \frac{2}{-5},$ and $\frac{-2}{-5}$.

   a) Which of the slopes above are negative?

   b) Which of the slopes above are positive?

   c) Does the sign of the ratio depend on the position of the negative sign? Explain.

BR-34. Place a sheet of graph paper over these lines to estimate the slope of each line below. Remember to describe each slope with a positive or negative ratio.

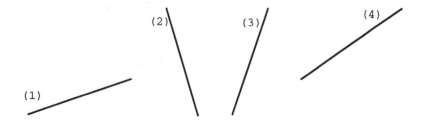

BR-35. Graph the line $y = \frac{1}{2}x$.

   a) Mentally verify that this line goes through the points O = (0, 0), A = (2, 1), B = (4, 2), C = (10, 5), D = (-6, -3), E = (-8, -4) and F = (2000, 1000). Explain how you verified point F.

   b) If you use points A and B to compute the slope and points A and D to compute the slope, how will the answers compare?

   c) Describe how to move (slide) from the y-intercept to point A.

BR-36. MATCH-A-GRAPH, Part Three

With your study teams, match the following graphs with their equations. Record the clues your team used to match each graph with its equation.

1) $y = \frac{1}{4}x + 4$

2) $y = 2x + 4$

3) $y = \frac{1}{2}x + 4$

4) $y = -\frac{2}{3}x + 4$

a)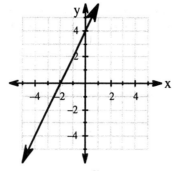

c)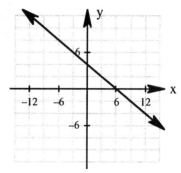

b)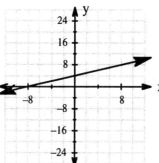

d)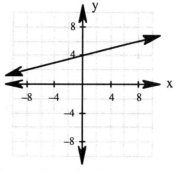

BR-37. Refer back to your Money Matters graph, BR-12. Determine the slope for each person's line. Determine the y-intercept for each line. How do these relate to the equations for each person? Be sure to consider whether the slope is positive or negative.

BR-38. Graph the line $y = -2x + 3$.

a) Mentally verify that this line goes through the points $O' = (0, 3)$, $A' = (2, -1)$, $B' = (3, -3)$, $C' = (-1, 5)$, $D' = (-2, 7)$ and $E' = (1000, -1997)$.

b) Compute the slope of this line.

c) What is the y-intercept of $y = -2x + 3$?

d) How do the answers to parts (b) and (c) compare with the original equation?

BR-39. Find the slope of the line $y = \frac{5}{3}x - 2$. How did you determine the slope?

BR-40. Ms. Speedi is planning on starting at the finish line and walking toward the starting line when the race starts. If she walks one meter per second and meets Dina 11 seconds after the start of the race, how long is the race?

BR-41. Sketch and label an example of a line with positive slope. Sketch and label an example of a line with negative slope. Graph paper may be helpful.

BR-42. Refer back to your graph of the race so far. Determine the slope of both Dina's and Leslie's lines. How do these slopes relate to the equations of the lines?

BR-43. Express each number below as a ratio of two integers. For example, $2.7 = \frac{27}{10}$.

a) 1.2

b) 0.003

c) 5

d) -4

e) 50%

f) 1250%

BR-44. In 35 minutes, Suki's car goes 25 miles.

a) If she continues at the same speed, how long will it take Suki to drive 90 miles?

b) How far will Suki go in 60 minutes?

c) Determine the car's average speed in miles per hour.

BR-45. Arnold is juggling four bowling balls and two seedless watermelons which amount to 66 pounds in all. Each watermelon weighs three pounds more than half of the weight of one bowling ball. How much does each watermelon weigh?

BR-46. Solve this system for the point of intersection.

$$x = 2y - 1$$
$$x = -5y + 10$$

The Big Race: Slopes and Rates of Change

SLOPE-INTERCEPT FORM OF LINEAR EQUATIONS

BR-47. Dean has decided he will enter the race. He estimates that he rides 3 meters every 4 seconds and wants a 2 meter head start.

a) On the Resource Page, graph a line representing Dean's distance from the starting line. Label his graph with his name.

b) How many meters does Dean ride each second?

c) Compare your answer to part (b) with the slope of the line.

BR-48. On a sheet of graph paper, draw a set of coordinate axes. Draw a line through the origin that has a slope of 1. Remember that the origin is the point (0, 0).

a) Write the coordinates of three points that are on the line.

b) Write an equation that describes these points.

BR-49. UNDERSTANDING THE EQUATION OF A LINE

Obtain a resource page from your teacher. With your study team, investigate each family of graphs. Compare the graphed solid lines to $y = x$, which is dotted.

1)

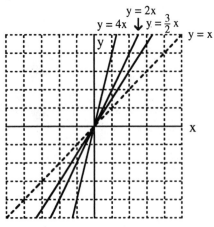

a) Copy and complete the table below and fill it in with the equation, slope, and y-intercept of each line. The first entry is done for you.

equation	slope	y-intercept
$y = x$	1	0

b) Examine the graph carefully and compare the slope of the lines to the slope of the line $y = x$. Describe your observations in complete sentences.

c) Examine the equations of the lines. What relationship does the slope have to the equation of the line?

>>Problem continues on the next page.>>

2)

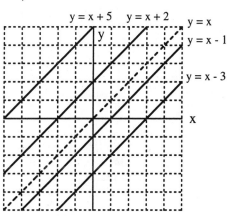

a) Copy and complete the table below and fill it in with the equation, slope, and y-intercept of each line. The first entry is done for you.

equation	slope	y-intercept
y = x	1	0

b) Examine the graph carefully and compare the slope of the lines to the slope of the line y = x. Describe your observations in complete sentences.

c) Compare the y-intercepts of the lines. What relationship does the y-intercept have to the equation of the line?

3)

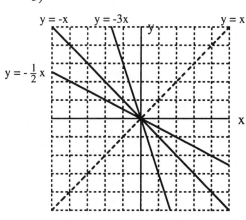

a) Copy and complete the table below and fill it in with the equation, slope, and y-intercept of each line. The first entry is done for you.

equation	slope	y-intercept
y = x	1	0

b) What is different about the lines in graph (3) at left and the lines in graphs (1) and (2) above?

c) Examine the equations of the lines. What makes these slopes different than the those in graphs (1) and (2) ?

>>Problem continues on the next page.>>

4)

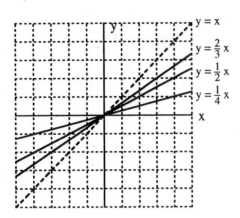

a) Copy and complete the table below and fill it in with the equation, slope, and y-intercept of each line. The first entry is done for you.

equation	slope	y-intercept
y = x	1	0

b) Examine the graph carefully and compare the slope of the lines to the slope of the line y = x. These lines are also different than those in graph (1) above. Describe your observations in complete sentences.

c) If the slope is less than one, describe the line.

BR-50. Without graphing, identify the slope and y-intercept of the following equations:

a) $y = 3x + 5$

b) $y = -6x + 3$

c) $y = -\frac{3}{4}x - 1$

d) $y = \frac{5}{-4}x$

e) $y = 7 + 4x$

f) $y = \frac{-x}{5} + 6$

BR-51. Explain to a student who was absent today how to find the slope and y-intercept of a line written in the y-form.

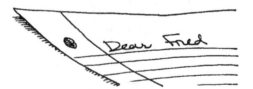

BR-52. Add this information to your tool kit.

SLOPE-INTERCEPT FORM

Any line can be described by an equation in the form of **y = mx + b**. The "m" represents the slope (it is the coefficient of x). The "b" represents the y-value of the y-intercept (where x = 0). Ordered pairs (x, y) that make the equation true are coordinates of points on that line.

BR-53. Interesting patterns occur when graphing several equations on one coordinate grid.

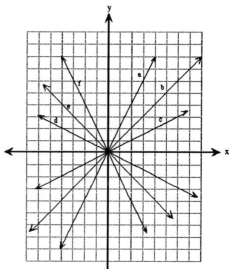

a) Find the slope of each of the lines at right.

b) Write the equation of each line.

c) What is the slope of the x-axis?

d) If the slope of a line were $\frac{-3}{2}$, between which two lines would the graph lie?

e) If the slope of a line were 100, between which two lines would the graph lie? (Hint: one of the lines may be one of the axes.)

f) If the slope of a line were $\frac{1}{6}$, between which two lines would the graph lie?

BR-54. Draw a line with the slope of $\frac{3}{4}$. Draw another line with the slope of $-\frac{3}{4}$.

BR-55. Use your explanation from BR-51 to state the slope and y-intercept of each of the following lines.

a) $y = 2x - 5$

b) $y = -3x + 10$

c) $y = \frac{3}{4}x + \frac{7}{4}$

d) $y = -\frac{5}{2} + 2x$

e) $4y = 3x + 6$
(Hint: first solve for y.)

f) $y + 2x = 4$
(Hint: first solve for y.)

g) Two pairs of lines above are parallel. Name them.

BR-56. Simplify the following. (Hint: $3x = \frac{3x}{1}$.)

a) $3x \cdot \frac{4}{x}$

b) $25 \cdot \frac{x}{5}$

c) $x \cdot \frac{1}{x}$

BR-57. Find the point of intersection for the two lines: $y = 6x - 9$ and $y = -\frac{1}{3}x - 2$.

GRAPHING USING SLOPE-INTERCEPT FORM

BR-58. Do you think it is always possible to draw a line given only one point and its slope?

BR-59. When first graphing a line in Unit Three, you set up a table and used many points to graph the line. In Unit Six, we developed the two-point method. Now, with your study team, you will explore a way to graph a line using one point and a slope.

a) Plot the point (2, 3).

b) Draw a line through that point with a slope of $\frac{1}{2}$. Discuss with your study team how to do this accurately. Is there more than one way to do this?

c) Find the equation of the line using the slope and y-intercept.

BR-60. Use the slope and the given point on the line to graph the following lines. Use a ruler and draw carefully so you can accurately find the y-intercept for:

a) a line through (8, 2), with a slope of $\frac{3}{4}$. Write the equation of the line.

b) a line through (8, 2), with a slope of $-\frac{3}{4}$. Write the equation of the line.

c) a line through (-4, 3), with a slope of -2. Write the equation of the line.

BR-61.  Elizabeth and Bob are considering joining the race. Bob usually rides 3 meters in 2 seconds and will get a 5 meter head start. Elizabeth rides 1 meter in 4 seconds and wants a 6 meter head start. Add lines for Elizabeth and Bob to the Big Race graph and label their lines with each equation and its name. Use a different color pen, pencil or highlighter if you can.

a) Who rides faster, Elizabeth or Bob? How does the graph show this?

b) When does Dean pass Elizabeth? How can you tell by examining the graph?

BR-62. Kyle made this sketch to find the slope of a line through (-15, 39) and (29, -2). Use his sketch to finish the problem. Remember to check whether the slope is negative or positive.

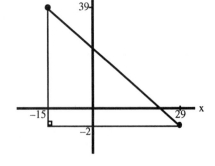

BR-63. Draw sketches similar to Kyle's to find the slope of the lines through the given points. We call this sketch a **slope triangle**.

a) (10, 2) and (2, 24)   c) (-6, 5) and (8, -3)

b) (-3, 5) and (2, 12)   d) (-6, -3) and (2, 10)

252 UNIT 7

BR-64. Refer to your BR-0 resource page for the Big Race.

a) Find the equation for Dean.

b) If the race were 8 meters long, who would be first, second and third? Explain your answer.

c) If the race were held today, how long does Dean's advantage over Leslie last? Use substitution to find the exact time that Leslie passes Dean.

BR-65. MORE OR LESS

Judy has $20 and is saving at the rate of $6 per week. Ida has $172 and is spending at the rate of $4 per week. After how many weeks will each have the same amount of money? Use your observations from Money Matters, problem BR-12, to answer the following questions.

a) Write an equation for Judy and Ida. Draw the graphs to estimate the solution. What will x represent? What will y represent? What scales will you use for each axis? Label your graphs completely. Carefully decide what scales to use.

b) Solve this problem using substitution.

c) Compare your answers from (a) and (b). Which answer is more accurate?

BR-66. Solve each of the following equations for x.

a) $\frac{x}{6} = \frac{7}{3}$

b) $\frac{6}{x} = \frac{4}{x+1}$

c) $3x + 2 = 7x - 8$

d) $6(x - 4) = 42$

BR-67. Write a shorter expression for each of the following polynomials by simplifying.

a) $3x^2 + 2x - 2 + 4x^2 + 6 - 5x$

b) $(4x^2 + 12x - 5) - (3x^2 + 6)$

c) $2(x^2 + 3) - 5(x^2 + 3)$

d) $(-5x^2 + 2) - (6x^2 + x)$

e) $3(x + 1) + 2(x + 1) + (x + 1)$

BR-68. Multiply. You may find the sketch of a generic rectangle useful.

a) $(x + 2)(x + 8)$

b) $(x - 2)(x - 8)$

c) $(x + 2)(x - 8)$

d) $(x - 2)(x + 8)$

e) $3x(2x + 4)$

f) $-5x(4x - 8)$

BR-69. Miguel bragged that he saved $12.60 by buying his new sneakers on sale at 30% off. How much did the sneakers cost originally?

The Big Race: Slopes and Rates of Change

SOLVING SYSTEMS OF LINEAR EQUATIONS BY SUBSTITUTION

BR-70. Use the recipe below, right and answer the questions below.

a) Copy the following sentence, but substitute the word "test" with "assignment".

There will be a test on Friday, and tests are worth 25% of your grade.

b) In the recipe at right, $\frac{1}{2}$ cup of apple sauce can be substituted for one egg and $\frac{1}{4}$ cup of butter. Rewrite the new recipe for Ms. Speedi's Brownies by replacing all the eggs and butter with applesauce.

Ms. Speedi's Brownies
4 squares of Chocolate
2 cups sugar
1 teaspoon vanilla
1 cup flour
3 eggs
$\frac{3}{4}$ cups butter

c) If $x = -2$, and $4x - 3y = -11$, then what is y? Explain what you do to solve for y and do it.

d) If $x = -3y + 1$ and $4x - 3y = -11$, discuss with your study team and write a complete sentence to explain why $4(-3y + 1) - 3y = -11$.

BR-71. Add this information to your tool kit.

MORE ON THE SUBSTITUTION METHOD

In Unit Six, we used Substitution to algebraically find the point of intersection for two linear equations. In those problems, the two equations were always in y-form. However, substitution can be used even if the equations are <u>not</u> in y-form.

Given: $x = -3y + 1$
        $4x - 3y = -11$

Use substitution to rewrite the two equations as one:

We can then write: $4(-3y + 1) - 3y = -11$ by replacing x with $(-3y + 1)$.

$x = \boxed{-3y + 1}$
$4(\quad) - 3y = -11$
$4(-3y + 1) - 3y = -11$

a) Solve the equation for y.

b) Substitute your answer from part (a) into $x = -3y + 1$. Write your answer for x and y as an ordered pair.

c) Substitute $y = 1$ into $4x - 3y = -11$ to verify that <u>either</u> original equation may be used to find the second coordinate.

d) Why is this method called substitution?

---

254   UNIT 7

BR-72. Use substitution to find the point of intersection (x, y) for each pair of linear equations below.

a) y = -3x
    4x + y = 2

b) 2x + 3y = -17
    y = x - 4

c) x = y - 3
    x + 3y = 5

d) Verify that your solution to (c) works in BOTH equations.

BR-73. The graphs for $y = -\frac{1}{2}x + 5$ and $y = \frac{3}{2}x + 1$ appear at right.

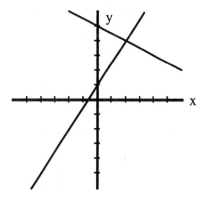

a) When we solve these equations using substitution, what solution should we get?

b) Solve using substitution to verify your answer. How is this solution represented on the graph?

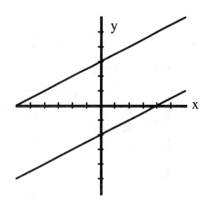

c) The graphs for $y = \frac{1}{2}x + 3$ and $y = \frac{1}{2}x - 2$ appear at right. When we solve using substitution, what solution should we get?

d) Use substitution and solve this system of equations. What happens? Why?

BR-74. The points (1, 2) and (7, 12) lie on a line. Find the coordinates of a third point on the line. How can you use the idea of slope?

BR-75. Eventually everybody passes Elizabeth in the Big Race.

  a) Write the order in which people passed her.

  b) Ignoring head starts, make a list of the competitors going from fastest to slowest.

  c) How do you know this from the graph?

  d) How do you know this from the equations?

  e) What is the slope of each line? How does this relate to their speed?

BR-76. Without making a table, graph each line. Start with the y-intercept, then use the slope.

  a) $y = \frac{2}{3}x - 2$

  b) $y = -\frac{2}{3}x + 2$

  c) $y = 3x$

  d) $y = 4 - 2x$

BR-77. Complete the following percentage problems using the method of your choice.

  a) Eighteen is what percent of 25 ?

  b) Eighteen percent of 25 is what?

  c) Twenty-five is what percent of 18 ?

BR-78. Find the area of a rectangle with one side 10,000 millimeters long and the other side 20 meters long. (Hint: the answer is not 200,000)

LINES WITH EQUAL SLOPES AND MORE SUBSTITUTION

BR-79. If Brian starts 3 seconds late and catches up with Dina 7 seconds after the race begins, add Brian's line to the graph.

  a) What is the x-intercept for Brian's line?

  b) Find Brian's speed.

BR-80. Simplify the following expressions. Check your results with your team. Making sense of the negative in front of the parentheses is important.

a) -1(4x - 6)

b) - (2x + 5)

c) 4x - (5 - 2x)

BR-81. Joe learned substitution yesterday and solved the following system of equations.

$$\left.\begin{array}{l}x = 4 - 2y \\ 3y - x = 6\end{array}\right\}$$

$3y - (4 - 2y) = 6$

$3y - 4 - 2y = 6$

$y - 4 = 6$

$y = 10$

$x = 4 - 2(10)$

$x = 4 - 20$

$x = -16$

a) Joe got an answer of y = 10 and x = -16. Check his answer.

b) Solve the problem and find the correct answer.

c) What did he do wrong?

BR-82. Solve the following systems. Remember to check your solution in both equations to make sure it is the point of intersection.

a) y = 2x - 3
    x - y = -4

b) y = x - 2
    2x - 3y = 14

BR-83. Use the figure at right for these questions.

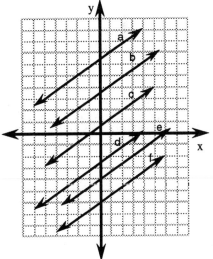

a) Estimate the y-intercept of each line.

b) What do all the lines have in common?

c) Between which two lines would the graph of $y = \frac{2}{3}x + 2$ lie?

d) Between which two lines would the graph of $y = \frac{2}{3}x - \frac{12}{7}$ lie?

e) Explain what happens to the graph of $y = \frac{2}{3}x$ when a positive number is added to the right side of the equation. What happens when a negative number is added there?

f) Write an equation for each of lines *a* through *f*.

The Big Race: Slopes and Rates of Change

BR-84. Solve the following systems. Remember to check your solution in both equations to make sure it is the point of intersection.

    a) $5x - 4y = 8$
            $y = x - 3$

    b) $4y - x = 6$
            $x = 12 - 2y$

BR-85. Use generic rectangles to multiply.

    a) $(x + 12)(x - 11)$

    b) $(2x - 1)(x + 5)$

    c) $2x(6x - 7)$

    d) $(4x - 11)(5x + 7)$

BR-86. Graph the lines: $y = -\frac{1}{3}x + 1$ and $y - 2x = 8$

    a)   Where do they intersect?

    b)   Name the y-intercepts for each line.

    c)   Find the area of the triangle formed by the two lines and the y-axis.

BR-87. Elizabeth is trying to choose her Internet Service Provider (ISP). She has three different companies competing for her business, and each has sent her some information, summarized below:

    WEBCom $4.95 per month, plus $1.95 per hour
    Ameri-Net $19.95 per month, unlimited hours

    a)   Analyze this information carefully. Create an equation to represent each company. Explain to Elizabeth which service you advise her to use.

    b)   Another company, CPMOnline, will soon be in business and is planning to offer 5 free hours each month, but then charge $2.49 for each hour thereafter. Add this company to your graph and find when CPMOnline is the best deal.

BR-88. Multiply or factor to rewrite the following expressions.

    a) $4x(9x - 5)$

    b) $9(3m^2 - 11m + 16)$

    c) $12m^2 - 8m$

    d) $15x^2 - 60x + 75$

BR-89. Draw a graph similar to the Big Race to solve this problem.

Mia started pedaling along the bike trail at 10 miles per hour. Lucy left the same place one hour later going 15 miles per hour. How long does it take before Lucy catches up with Mia?

BR-90. THE BIG RACE, Part One

Today is finally the Big Race. Obtain a fresh BR-0 Resource Page for this problem. You and the rest of your study team will compete against Leslie and Elizabeth at today's rally in the gym. Determine who will win the race.

Rules:
- Your study team must work cooperatively to answer all the questions on the cards.
- Each member of the team will select rider A, B, C, or D.
- You may not show your card to your team. You may only communicate the information contained on the card.
- Elizabeth's and Leslie's cards will be shared by the entire team.

Remember your guidelines for study teams. Be sure everyone on your team discusses the entire problem and its solution.

BR-91. THE BIG RACE, Part Two

a) Use a table to organize your information. Find the slope for each rider. Find the y-intercept for each rider. Write the equation for each rider.

b) Who won the race? What was the speed of this rider?

c) In what order did the participants finish the race? List their names (or letters) with the time it took them to finish.

d) After 8 seconds, which tricyclist had traveled the shortest distance from the starting line? Who had traveled the farthest distance?

e) Locate and label three times when one rider passed another rider. Is there a point where more than two tricyclists are tied?

f) If the race were only 20 meters long, does the order of the winners change? How?

g) After 16 seconds, how far had each rider traveled from the starting line?

h) Extension: How long does the race have to be for "C" to be able to win?

i) Extension: If the race is 22 meters long, how much of a head start does Elizabeth need to beat "D"?

The Big Race: Slopes and Rates of Change

BR-92. Before the Big Race, the track needs to be resurfaced. The price of the work depends on the area to be resurfaced. Use the dimensions of the track given at right to find the area of the track. (Assume the ends of the track are circular.)

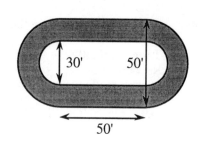

BR-93. Betty's Diner makes 15% profit on its lunches and 22% profit on its dinners. If the diner took in $2,700 on Tuesday and made $524 profit, how much was spent at lunch? Let x represent the sales at lunch and y represent the sales at dinner, write two equations, then solve.

BR-94. Use generic rectangles to multiply:

a) $(12x - 2)^2$

b) $(5a + 4)(2a - 7)$

c) $(4x - 3)(x - 2)$

d) $(y - 11)(y + 11)$

BR-95. Solve for x:

a) $2 - 12x = 14$

b) $9(x - 2) = 81$

c) $3.1x - 58 = 20.69x - 71.2$

d) $5(x + 1) - 2(x + 1) = 3$

BR-96. Two passenger trains started toward each other at the same time from towns 288 miles apart and met in three hours. The rate of one train was six miles per hour slower than that of the other. Find the rate of each train.

UNIT SUMMARY AND REVIEW

BR-97. UNIT SEVEN SUMMARY

With your study team, list 3 or 4 big ideas that you feel best represent Unit 7. Write a description of each main idea. For each topic, choose a problem that best demonstrates your understanding. Explain your solution as in previous unit summaries. Show all your work.

BR-98. Write the equations in the form y = mx + b for the following graphs:

a)
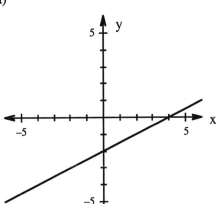

c)
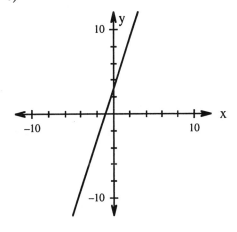

b)
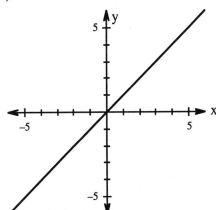

d)
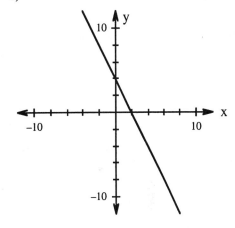

BR-99. GETTING IN SHAPE

Frank weighs 160 pounds and is on a diet to gain two pounds a week so that he can make the football team. John weighs 208 pounds and is on a diet to lose three pounds a week so that he can be on the wrestling team in a lower weight class. If they can meet these goals with their diets, when will Frank and John weigh the same, and how much will they weigh?

a) Solve the problem.

b) Clearly explain your method.

BR-100. Use generic rectangles to multiply:

a) $(x + 3)(x + 2)$ c) $(x + 7)(x - 5)$

b) $(x - 9)(x - 4)$ d) $(2x + 5)(x - 8)$

BR-101. Points $(-9, 3)$ and $(1, 1)$ are two points on a line.

a) Find the coordinates of a third point on the line.

b) Find the slope of the line.

BR-102. Use substitution to solve the following systems for x and y.

a) $3x + 7y = 71$
$y = 4x + 50$

b) $y = 4.5 - x$
$y = -2x + 6$

BR-103. Graph the lines and find the x-intercepts.

a) $y = -2x + 4$

b) $y = \frac{2}{3}x - 6$

BR-104. TOOL KIT CHECK-UP

Your tool kit contains reference tools for algebra. Return to your tool kit entries. You may need to revise or add entries.

Be sure that your tool kit contains entries for all of the items listed below. Add any topics that are missing to your tool kit NOW, as well as any other items that will help you in your study of algebra.

- X- and Y-intercepts
- Slope of a Line
- Substitution Method
- Slope-Intercept Form

NOTE: *A question similar to GM-13 is asked at job interviews at a major computer software firm.*

GM-13. CROSSING OVER

Elizabeth, Brian, Dean, and Leslie want to cross a bridge. They all begin on the same side and have only 17 minutes to get everyone across to the other side.

To complicate matters, it is night and there is only one flashlight. A maximum of two people can cross at one time. Any party that crosses, either 1 or 2 people, <u>must</u> have the flashlight with them. The flashlight must be walked back and forth; it cannot be thrown.

Each student walks at a different speed. A pair must walk together at the rate of the slower student's pace.

Elizabeth: 1 minute to cross
Brian: 2 minutes to cross.
Dean: 5 minutes to cross.
Leslie: 10 minutes to cross.

For example, if Elizabeth and Leslie walk across first, 10 minutes have elapsed when they get to the other side of the bridge. If Leslie returns across the bridge with the flashlight, a total of 20 minutes has passed, and you have failed the mission.

Your Task:
- How can they get everyone across in 17 minutes?
- Do you think this is a good question for a computer firm to ask future employees? Explain.

The Big Race: Slopes and Rates of Change

GM-14. **MAKE A GRAPH**

Anyone can become an artist by following graphing directions. For example, graph the following design:

$y = 2x + 8$	$0 \leq x \leq 2$
$y = 12$	$2 \leq x \leq 4$
$y = -x + 16$	$4 \leq x \leq 6$
$x = 6$	$8 \leq y \leq 10$
$y = \frac{4}{3}x$	$0 \leq x \leq 6$
$y = -\frac{4}{3}x$	$-6 \leq x \leq 0$
$x = -6$	$8 \leq y \leq 10$
$y = x + 16$	$-6 \leq x \leq -4$
$y = 12$	$-4 \leq x \leq -2$
$y = -2x + 8$	$-2 \leq x \leq 0$

Your Task: Make a design on graph paper which can be drawn by someone following your equation instructions. Think about ways you can get curves in your design. Be creative, but accurate. Include your design on graph paper along with a listing of equations and domain (input) values like those shown above.

The Amusement Park
Factoring Quadratics

Unit 8 Objectives
The Amusement Park: FACTORING QUADRATICS

Previously, much of your work focused on different methods of solving linear equations and graphing lines. In Units 8 through 12, you will learn various methods for solving **quadratic equations** such as

$$x^2 + 5x + 6 = 0.$$

You have used subproblems to find the area of rectangles as both a product of its dimensions and as a sum of the area of smaller rectangles. You have multiplied binomials using the Algebra Tiles to write the sum of the areas of the small rectangles. In this unit you will be given the **sum of the parts of rectangles** and asked to find the dimensions, so you can write the area as a **product of factors**. Algebra Tiles have provided a handy device for modeling multiplication, and now will be used in factoring expressions. These factoring skills will provide you with yet another method to solve equations.

In this unit you will have the opportunity to:

- learn how to "undo" multiplying binomials by factoring trinomials.

- explore how to factor special products, such as differences of squares and perfect square trinomials.

- combine your factoring skills, including common terms, special binomial products, and trinomials.

- solve factorable quadratic equations algebraically using the Zero Product Property.

- study quadratic equations in relation to their graphs.

This unit builds on the development of sums and products of polynomials based on a geometric model. It continues to give you practice with solving various kinds of equations, writing equations from word problems, and finding x- an y-intercepts graphically and algebraically.

Read the following problem. **Do not try to solve it now.**

AP-0. The figure at right shows the master plan for a proposed Amusement Park coming to our city. Park designers expect the Park size to change before it is built, so they made it square with a length of x. Rectangular rows of parking will be adjacent to two sides of the park, with a picnic area in a square corner. Based on data provided later in the unit, help the planners design the land space so that the three areas fit within the square design.

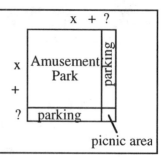

Unit 8
The Amusement Park: FACTORING QUADRATICS

MULTIPLYING BINOMIALS

AP-1.   Draw a rectangle to represent the product of $(x + 7)(x + 4)$.

    a) What are the dimensions of this rectangle?

    b) Write the area of this rectangle as a sum of the parts.

    c) Write the area of this rectangle as an equation of the form (length)(width) = area.

AP-2.   Obtain a resource page from your teacher. Determine if a rectangle is possible for each set of tiles described in the chart below. If so, draw a sketch; if not, answer "No." Write the area of each tile in the center of each piece. Write the dimensions along the edges.

When recording your rectangles, be sure that the big squares are arranged in the upper left corner and the small squares are located in the lower right corner. The rectangles fill in the rest of the figure, below and to the right of the big square, to form a composite rectangle.

Number of x^2's	Number of x's	Number of 1's	Is it possible?	Rectangle	Area as a sum and a product
1	3	2	Yes	(sketch: x^2, x, x, x, 1, 1 with dimensions $x+2$ and $x+1$)	$x^2 + 2x + x + 2$ sum = $x^2 + 3x + 2$ product = $(x + 1)(x + 2)$
1	5	3	No		
1	4	4			
1	6	5			
1	3	9			
1	4	3			
1	7	10			

AP-3. Write an algebraic equation for the area of each of the following rectangles as shown in the example below.

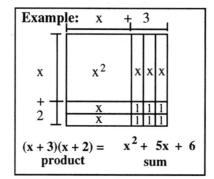

a)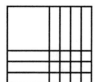

c)

e)

b)

d)

f)

AP-4. Read the following information and add the vocabulary to your tool kit. Then identify the following polynomial expressions in parts (a) through (h) as either a sum or a product.

> ### POLYNOMIALS AS SUMS AND PRODUCTS
>
> An expression of the form $3x^2$ or $5x$ or 7 is called a **monomial**, from the Greek words *monos* meaning *single,* and *nomos* meaning *part*.
>
> The sum (or difference) of two monomials is called a **binomial**. For example, $x + 2$, $x^2 - 4$, and $3 - 7x$ are binomials.
>
> More generally, the sum of two or more monomials is called a **polynomial** (from the Greek words *poly* for *many,* and *nomos* for *part*). Some polynomials, such as the **trinomial** $x^2 + 5x + 6$, can be written as a product of factors:
>
> $$x^2 + 5x + 6 = (x + 2)(x + 3)$$
>
> Here we have written the **SUM** $x^2 + 5x + 6$ as a **PRODUCT** of the two **factors**, $(x + 2)$ and $(x + 3)$. We say we have **FACTORED** the polynomial $x^2 + 5x + 6$ as "the quantity $x + 2$ times the quantity $x + 3$." The polynomials that have x^2 and no higher exponents are called **QUADRATICS.**

a) $2x + 1$

b) $x^2 + 7x + 12$

c) $(x + 1)(x + 4)$

d) $3x + 9$

e) $(x + 5)(x + 2)$

f) $x(2x + 5)$

g) $5x^3 + 8x^2 + 10x$

h) $2x(x^2 - 3x + 5)$

i) Which of the above expressions are quadratics? Hint: there are four.

AP-5. Combine like terms.

a) $(-6w + 3q - w) + (8q - 4 - 2w)$

b) $(3x + 4 + 2b) - (5b + 6x - 7)$

c) $(f^2 + 3f - 5) - (f^2 - f - 2)$

d) $(p - 2r - 2c) + (5c + 9p - r)$

e) When combining like terms, which operations were used?

f) Explain how you identify <u>like</u> terms. When are you able to combine algebraic terms?

AP-6. In the figure at right, points A, C, and E lie on the same line.

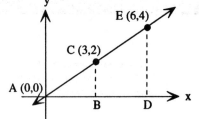

a) Demonstrate that △ADE is an enlargement of △ABC. (Hint: what is the enlargement ratio for corresponding sides?)

b) What is the ratio of the areas of △ADE and △ABC ?

c) Describe how to move (slide) from point A to point C.

d) Describe how to move (slide) from point A to point E.

e) What is the slope of this line?

f) What is the y-intercept?

g) Write the equation of the line in the form y = mx + b.

AP-7. Multiply or factor the following expressions. Hint: a sketch of a generic rectangle may help.

a) 3x(2x - 5)

b) $4x^2 - 8x$

c) -6a(3a - 4)

d) $14a^2 + 21b$

e) 4x(5x + 4)

f) $24c^2 + 16c$

AP-8. Find the point of intersection for the following sets of lines.

a) 6 + 2x = y
2x - 4y = 12

b) 4x - y = -5
y = 9x + 1

AP-9. Copy and solve each of the following Diamond Problems. Check for accuracy with your study team. You will need these for a later problem.

a) 10 / 7   b) 8 / 6   c) 15 / 8   d) 9 / 6

e) 20 / 9   f) -10 / 3   g) 12 / 8   h) -24 / 10

i) -24 / 23   j) -6 / -5   k) -12 / -1   l) 30 / -11

FACTORING

AP-10. Summarize the following information in your tool kit. Then answer the questions that follow.

FACTORING QUADRATICS

Yesterday, you solved problems in the form of (length)(width) = area. Today we will being working backwards from the area and find the dimensions. This is called **FACTORING QUADRATICS.**

Using this fact, you can show that $x^2 + 5x + 6 = (x + 3)(x + 2)$ because

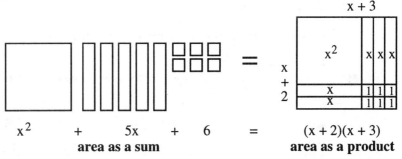

x^2 + 5x + 6 = (x + 2)(x + 3)
area as a sum   **area as a product**

Use your tiles and arrange each of the areas below into a rectangle as shown in AP-2, AP-3, and the example above. Make a drawing to represent each equation. Label each part to show why the following equations are true. Write the area equation below each of your drawings.

a) $x^2 + 7x + 6 = (x + 6)(x + 1)$

b) $x^2 + 4x + 4 = (x + 2)(x + 2)$

c) $x^2 + 3x + 2 = (x + 2)(x + 1)$

d) $2x^2 + 5x + 3 = (2x + 3)(x + 1)$

AP-11. Find the dimensions of each of the following generic rectangles. The parts are not necessarily drawn to scale. Use Guess and Check to write the area of each as both a sum and a product as in the example.

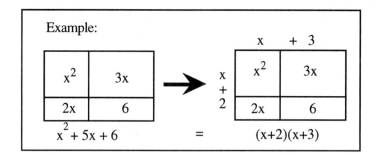

a)

x^2	$5x$
$3x$	15

b)

x^2	$4x$
$3x$	12

c)

x^2	$6x$
$3x$	18

d)

$2x^2$	$10x$

e)

x^2	$5x$
$2x$	10

f)

x^2	$4xy$
$4xy$	$16y^2$

AP-12. Draw generic rectangles to multiply the pairs of binomials below. Label the dimensions and area parts using the example from AP-11. Apply what you know about multiplying positive and negative numbers.

a) $(2x + 6)(3x - 5)$ c) $(2x + y)(4x - 3y)$

b) $(2x + 1)(4x - 3)$

AP-13. Graph a line through the point (3, -5) with a slope of $-\frac{3}{2}$.
a) Estimate its y-intercept.
b) Find the equation of the line.
c) Find the equation of a parallel line through (2, 1).

AP-14. Use a generic rectangle to multiply these binomials. Label the dimensions and write the area as the sum of the parts.
a) $(x + 10)(x + 3)$ c) $(2x - 11)(3x - 5)$
b) $(x + 7)(x - 9)$

AP-15. A damaged section of an apartment patio can be covered with 40 square tiles which are nine inches on a side. The owner decides to replace the section with square tiles which are three inches on a side. How many tiles will be needed? Drawing a picture may help you.

AP-16. Solve for x. Leave answers as integers or fractions--not decimals or mixed fractions.

a) $x + 3 = 0$
b) $x - 3 = 0$
c) $2x + 3 = 0$
d) $2x - 3 = 0$
e) $12x - 5 = 0$
f) $12x + 5 = 0$
g) $ax + b = 0$
h) $ax - b = 0$

AP-17. Recall from your previous classes that we have factored a number when it is written as the product of two or more integers. Since $6 \cdot 8 = 48$, 6 and 8 are factors of 48. All the factors of 48 are 1, 2, 3, 4, 6, 8, 12, 16, 24, 48. What is it called when an integer greater than zero has exactly two factors?

a) List the factors of 36.

b) What is the greatest common factor of 36 and 48 ?

c) Find the greatest common factor of 64 and 72.

The Amusement Park: Factoring Quadratics

FACTORING USING SUMS AND PRODUCTS

AP-18. USING ALGEBRA TILES TO FACTOR

What if we knew the area of a rectangle and we wanted to find the dimensions? We would have to work backwards. Start with the area represented by $x^2 + 6x + 8$. Normally, we would not be sure whether the expression represents the area of a rectangle. One way to find out is to use Algebra Tiles to try to form a rectangle.

You may find it easier to record the rectangle without drawing all the tiles. You may draw a generic rectangle instead. Write the dimensions along the edges and the area in each of the smaller parts as shown below.

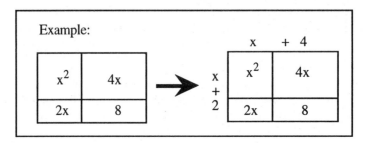

We can see that the rectangle with area $x^2 + 6x + 8$ has dimensions $(x + 2)$ and $(x + 4)$.

Use Algebra Tiles to build rectangles with each of the following areas. Draw the complete picture or a generic rectangle and write the dimensions algebraically as in the example above. Be sure you have written both the product and the sum.

a) $x^2 + 6x + 8$

b) $x^2 + 5x + 4$

c) $x^2 + 7x + 6$

d) $x^2 + 7x + 12$

e) $2x^2 + 8x$

f) $2x^2 + 5x + 3$

AP-19. USING DIAMOND PROBLEMS TO FACTOR

Using Guess and Check is not the only way to find the dimensions of a rectangle when we know its area. Patterns will help us find another method. Start with $x^2 + 8x + 12$. Draw a generic rectangle and fill in the parts we know as shown at right.

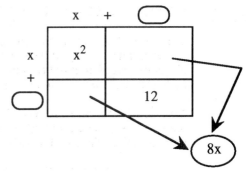

We know the sum of the areas of the two unlabeled parts must be 8x, but we do not know how to split the 8x between the two parts. The 8x could be split into **sums** of $7x + 1x$, or $6x + 2x$, or $3x + 5x$, or $4x + 4x$. However, we also know that the numbers that go in the two ovals must have a **product** of 12.

a) Use the information above to write and solve a Diamond Problem to help us decide how the 8x should be split.

b) Complete the generic rectangle and label the dimensions.

AP-20. Use either the Guess and Check method of problem AP-18 or the Diamonds in AP-19 to find the **factors** of each of the following quadratics. The factors are the binomials which represent the length and width of the rectangle.

Draw a picture for each problem and write an algebraic equation of the form:

area = (length)(width).

a) $x^2 + 7x + 10$ (Hint: it may be easier to think of the factors of 10 first and then find the pair whose sum is 7.)

b) $x^2 + 6x + 8$

d) $x^2 + 6x + 9$

c) $x^2 + 8x + 15$

e) $x^2 + 9x + 20$

The Amusement Park: Factoring Quadratics

AP-21. SUMS AND PRODUCTS

a) Use the worksheet your teacher provides you to record integer pairs whose sums and products are given below. Some of the pairs do not have integer solutions.

Sum	Product	Integer Pair
6	8	2, 4
7	10	
13	12	
8	12	
6	12	
6	9	
15	56	
8	15	
9	20	
2	-35	
1	-72	
3	-10	
10	-24	
0	-16	
23	-24	
0	-32	

Sum	Product	Integer Pair
-5	6	
-7	6	
-11	30	
-8	16	
-10	16	
0	16	
-10	24	
-9	18	
-5	-6	
-2	-35	
-3	-28	
-1	-12	
-10	-25	
0	-25	
-1	-42	
-12	-13	

b) Judy says the entries in the sum-product table in part (a) are just Diamond Problems in disguise. Explain what she means.

c) We can use most lines in the sum and product table above to write a quadratic polynomial and its factors. For example, $x^2 - 10x + 24 = (x - 6)(x - 4)$. Write three more sum and product factoring examples from the table.

d) Use the table to write a quadratic polynomial that has <u>no</u> integer factors.

AP-22. Solve each system of equations for x and y. Check your solution.

a) $y = 3x - 8$
    $y = 2x + 7$

b) $y = 3x - 8$
    $2x + 3y = 12$

AP-23. Graph points A (1, 1) and B (3, 4). Show your solution on graph paper.

a) What is the slope of the line containing A and B?

b) Point C is on the line which goes through points A and B. If the x-coordinate of C is 4, what is the y-coordinate? Draw two similar triangles to help you.

c) Write the equation of the line from the graph.

d) Verify that points A, B, and C make your equation from part (c) true. Revise if needed.

AP-24. Make a table with at least six entries using $-3 \leq x \leq 5$. Draw a large graph of the equation $y = x^2 - 4x$. You will use this graph again for a later problem so label it clearly. Is it a line or a curve?

AP-25. Draw a generic rectangle to factor these polynomials. Write an equation in the form: area = (length)(width).

a) $x^2 + 15x + 56$

c) $x^2 + 7x + 12$

b) $x^2 + 13x + 12$

AP-26. You have four cards: an ace of spades, an ace of hearts, a two of hearts, and a three of hearts.

a) List all possible combinations of two of the cards.

b) You draw two cards. What is the probability that both are aces?

c) You draw two cards. What is the probability they are both hearts?

d) If you draw only one card, what is the probability it is an ace?

e) If you draw two cards, what is the probability they are both spades?

AP-27. Write an equation and solve. Write your solution in a sentence.

A stick 86 centimeters long is cut into five pieces. The three long pieces are all the same length and the two shorter pieces are both the same length. Each of the shorter pieces is two centimeters shorter than one of the longer pieces. Find the length of a longer piece.

GREATEST COMMON FACTOR

AP-28. Use your experience with sums and products of integers to help you factor each of the following quadratics.

   a) $x^2 + 9x + 8$
   b) $x^2 + 6x + 8$
   c) $x^2 - 6x + 8$
   d) $x^2 - 2x - 8$
   e) $x^2 + 2x - 8$
   f) $x^2 + x - 20$
   g) $x^2 - x - 20$

AP-29. Use a Generic Rectangle to multiply.

   a) $3(x^2 - 3x + 2)$
   b) $4b(2b^2 + 3b - 1)$
   c) $-2(x^3 - 5x^2 + 6)$
   d) $3x(x^2 - 3x + 2)$

AP-30. Sheila was asked to factor $2x^2 + 10x$ with her Algebra Tiles. By moving the tiles around she produced the following configurations with these tiles. With your study team, decide which of the configurations below represent a factored form of $2x^2 + 10x$.

1)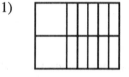

2)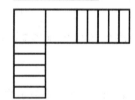

3)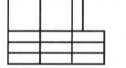

4)

   a) On your paper, draw a copy of the tile configuration(s) that is (are) in correct factored form. How did your study team determine that these were correct?

   b) For each configuration in part (a), write the dimensions of the composite rectangle. Then express the area as a product and a sum.

   c) Of the composite rectangles, which one is closest to a square? This composite rectangles is considered **factored completely**. With your study team, decide which factored form in part (b) is factored completely.

AP-31. We have seen cases in which only two types of tiles are given. Read the example below and add an example of the Greatest Common Factor to your tool kit. Then use a generic rectangle to find the factors of each of the polynomials below. In other words, find the dimensions of each rectangle with the given area.

GREATEST COMMON FACTOR

Example:

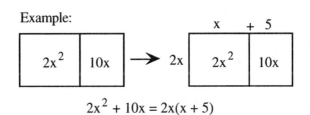

$$2x^2 + 10x = 2x(x + 5)$$

For $2x^2 + 10x$, "2x" is called the **GREATEST COMMON FACTOR**. Although the diagram could have dimensions $2(x^2 + 5x)$, $x(2x + 10)$, or $2x(x + 5)$, we usually choose $2x(x + 5)$ because the $2x$ is the largest factor that is common to both $2x^2$ and $10x$. Unless directed otherwise, when told to factor, you should always find the greatest common factor, then examine the parentheses to see if any further factoring is possible.

a) $x^2 + 7x$

b) $3x^2 + 6x$

c) $3x + 6$

d) How does a generic rectangle with only two sections relate to the Distributive Property?

e) Give an example of when you have used common factors before.

AP-32. Factor by using the greatest common factor as in AP-31. Notice that this is an extension of the Distributive Property and the generic rectangles will need 3 sections.

a) $8x^2 - 2x + 4$

b) $9x^2 + 15x - 24$

c) $25x^3 + 100x^2 + 50x$

d) $10r^3 + 30r^2 - 20r$

AP-33.* From the twelve graphs below, choose one that best fits each of the situations described in parts (a) through (d). Copy the graph that fits each description and label the axes clearly with the labels shown in the parentheses. If you cannot find a graph you want, sketch your own and explain it fully.

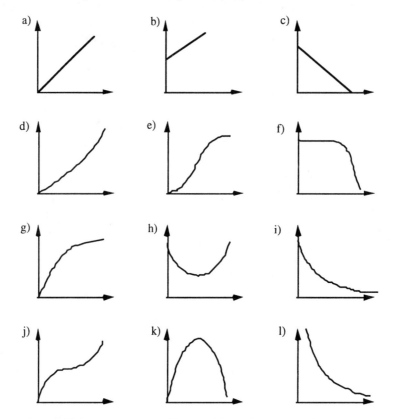

a) Sting held Lex Lugar over his head for a few unsteady moments, and then, with a violent crash, he dropped him. (height of Lex / time)

b) When I started to swim competitively, I initially made very rapid progress, but I have found that the better you get, the more difficult it is to improve further. (proficiency / amount of practice)

c) If schoolwork is too easy, you do not learn anything from doing it. On the other hand, if it is so difficult you cannot understand it, again you do not learn. That is why it is so important to plan work at the right level of difficulty. (amount learned / difficulty of work)

d) When biking, I try to start off slowly, build up to a comfortable speed, and then gradually slow as I near the end of my training. (speed / time)

e) Make up a story of your own to accompany one of the remaining graphs. In class you will give your story to a partner and see whether your partner chooses the same graph you wrote the story about.

---

* Adapted from *The Language of Functions and Graphs*, Joint Matriculation Board and the Shell Centre for Mathematical Education, University of Nottingham, England.

AP-34. Recall what you know about square numbers to help you answer the questions below.

a) What two numbers can you square to get a result of 25?

b) Compare the question in part (a) with the equation at right. What do you notice? $x^2 = 25$

c) Solve each of the following quadratic equations for x. Each equation has TWO solutions. Round decimal answers to the nearest 0.01.

1) $x^2 + 5 = 30$ 3) $x^2 + 5 = 25$
2) $x^2 + 25 = 650$ 4) $x^2 = 120$

AP-35. Find the equation of a line through the points (1, -5) and (-2, 1) at right.

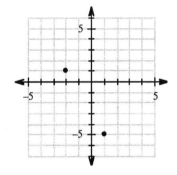

AP-36. Factor the following expressions. Be sure to look for the greatest common factor.

a) $x^2 + 10x + 16$ d) $2x^2 - 22x$
b) $x^2 + 15x + 56$ e) $5x^2 + 30x$
c) $x^2 + 11x + 30$ f) $3x^3 + 6x^2$

AP-37. The germination rate for zinnia seeds is 78%. This means that, on average, 78% of the seeds will sprout and grow. If Jim wants 60 plants for his yard, how many seeds should he plant?

AP-38. On the same set of axes, graph the lines y = 3, x = -2, y = 0, and x = 2. Find the area of the interior region defined by the four lines.

AP-39. Solve each of the following equations by Guess and Check.

a) $3x = 0$ d) $(x + 1)(x - 4) = 0$
b) $-\frac{1}{2}x = 0$ e) $(x - 2)(x + 5) = 0$
c) $x(x - 3) = 0$ f) $(x + 3)(2x - 3) = 0$

The Amusement Park: Factoring Quadratics

THE ZERO PRODUCT PROPERTY

AP-40. We have been working with quadratic expressions such as $x^2 + 4x - 5$. We will now explore the graphs of quadratic expressions and see how factoring relates to the graph.

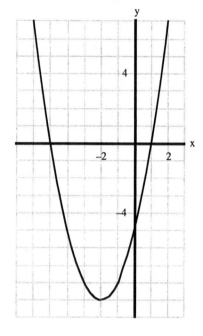

a) Highlight the line $y = 0$ on the resource page provided by your teacher. Where does the line $y = 0$ intersect the parabola $y = x^2 + 4x - 5$?

b) Factor $x^2 + 4x - 5$.

c) Use part (b) and Guess and Check to solve $0 = (x - 1)(x + 5)$.

d) What do you observe about the factored form of the equation and the solution in part (c)?

AP-41. ZERO PRODUCT PROPERTY

You have factored quite a few quadratic expressions similar to $x^2 + 2x - 3$. You have also graphed several parabolas in "y-form" where a quadratic expression appears on the right side of the equation, such as $y = x^2 + 2x - 3$. The following activities will help you to put these two ideas together with the **Zero Product Property**.

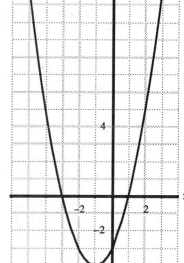

a) Factor $x^2 + 2x - 3$ and compare the constant in each binomial to the x-intercepts from the graph on the resource page.

b) Explore the following products carefully:

   (1) Find the value of the product (5)(8)(0)(4).

   (2) Find the value of (5)(0)(8)(4). How does this value compare to the value in part (1) above?

   (3) Find the product of (6)(x)(0)(7).

   (4) If (6)(x)(4)(7) = 0, what must be the value of x?

   (5) If (a)(b)(c) = 0, what can you say about the possible values of a, b, and c?

c) In part (b)(5) above, you concluded that if (a)(b)(c) = 0, then either a = 0 OR b = 0 OR c = 0. Now consider the equation $0 = (x - 1)(x + 3)$. Using this pattern,

   if $0 = (x - 1)(x + 3)$, then $x - 1 = 0$ OR $x + 3 = 0$.

   Solving each equation results in x = 1 OR x = -3.

d) Compare these results to the x-intercepts you found from the graph in part (a) above. Write a sentence or two about the relationship between the results of solving $0 = x^2 + 2x - 3$ using algebra and finding where the graph of the equation $y = x^2 + 2x - 3$ crosses the x-axis.

AP-42. ROOTS OF A PARABOLA

Test your conjecture from the previous problem, AP-41, part (d), in the problems below.

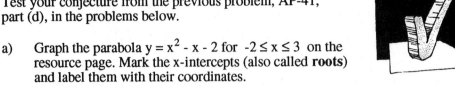

a) Graph the parabola $y = x^2 - x - 2$ for $-2 \leq x \leq 3$ on the resource page. Mark the x-intercepts (also called **roots**) and label them with their coordinates.

b) Substitute $y = 0$ into the equation $y = x^2 - x - 2$ to obtain an equation with one variable. Then find two solutions to the new equation by first factoring the expression $x^2 - x - 2$.

c) Explain how solving the equation $0 = x^2 - x - 2$ gives you enough information to name the x-intercepts of the parabola $y = x^2 - x - 2$ without having to draw the graph.

d) Why do people call the method you followed in parts (b) and (c) the "Zero Product Property"?

e) In order to use the Zero Product Property, what must the product be?

f) Add this definition to your tool kit.

> The x-intercepts of a parabola are also referred to as the **ROOTS** of the quadratic equation.

AP-43. Without graphing, use the Zero Product Property developed in the previous problems to find where each graph crosses the x-axis. (Hint: what is the equation of the x-axis?)

a) $y = (x - 3)(x - 4)$

b) $y = x^2 + 5x - 24$

c) $y = x^2 + 10x + 16$

AP-44. Add this information to your tool kit.

ZERO PRODUCT PROPERTY

The **ZERO PRODUCT PROPERTY** states that when the product of two or more factors is zero, one of these factors must equal zero. We use this property to solve quadratic equations that are factorable using integers. The general statement of the property is:

If $a \cdot b \cdot c = 0$, then $a = 0$ OR $b = 0$ OR $c = 0$.

For example, given the equation:	$x^2 + 11x = -24$
1) Bring all the terms to one side so the equation is set equal to zero.	$x^2 + 11x + 24 = 0$
2) Factor the quadratic expression.	$(x + 8)(x + 3) = 0$
3) Set each factor equal to zero and solve the equation.	$(x + 8) = 0$ or $(x + 3) = 0$ $x = -8$ or $x = -3$
4) Check both solutions in the original equation:	$(-8)^2 + 11(-8) = -24$, and $(-3)^2 + 11(-3) = -24$

AP-45. Use your experience with sums and products of integers to help you factor each of the following quadratics if possible.

a) $x^2 + 3x - 10$
b) $x^2 - 2x - 35$
c) $x^2 + 3x + 10$
d) $x^2 - 10x + 24$

AP-46. Solve each of the following quadratic equations. Check each solution.

a) $0 = (x + 3)(x - 6)$
b) $x^2 + 4x - 32 = 0$
c) $0 = x^2 + 7x + 6$
d) Find the coordinates of the x-intercepts of the parabola $y = x^2 + 4x - 32$ without graphing. (Hint: use what you found in part (b).)

The Amusement Park: Factoring Quadratics

AP-47. Solve each of the equations for the unknown variable. Use any method you choose.

   a) $2x + 6 = 3x - 4$

   b) $9s - 1 = 4$

   c) $2(w + 4) - 6(w - 7) = 50$

   d) $32 = \frac{1}{7}r + r$

   e) $(x - 7)(2x + 6) = 0$

   f) $3(9 - 2d) + 2(2d + 8) = 6 - 2d$

   g) $\frac{5}{x + 2} = \frac{3}{x}$

AP-48. Marci correctly graphed an equation and got a vertical line.

   a)   If the line contained the point (-2, 3), hat equation did Marci graph?

   b)   On the same axes, Marci's team member graphed the line $y = 4x - 1$. At what point did this line hit Marci's line? (Be sure to explain how you arrived at your answer.)

AP-49. Add each pair of fractions below.

   a)   Add $\frac{4}{5} + \frac{1}{7}$. What did you do first?   b)   Add $\frac{9}{x} + \frac{3}{5}$. What did you do first?

FACTORING DIFFERENCES OF SQUARES

AP-50. This investigation will tie together several concepts you have used and developed in this course: subproblems, area and factoring. We are going to look at a special factoring pattern known as the **difference of squares**.

   a)   Represent the difference of two numerical squares with graph paper by following these steps.

   1)   Cut a 20 by 20 square out of graph paper. Compute the area of the square and write its area using exponents.

   2)   Remove a 5 by 5 square by cutting it from the lower right corner of the 20 by 20 square. What is the area of this removed square? Write its area using exponents.

   3)   Write the measurements of the new shape along its edges. Find the remaining area. Is this expression a Difference of Squares?

   4)   We have been factoring quadratics by forming rectangles and finding their dimensions. Since this new shape is not a rectangle, we need to alter it by cutting off one of the extended pieces and attaching it in such a way that we form a new rectangle. What are the dimensions of this rectangle? Write its area as a product.

   5)   Describe in complete sentences how you can predict the dimensions of the new rectangle formed by any difference of squares.

>>Problem continues on the next page.>>

b) We will now represent the difference of squares with Algebra Tiles using variables. Repeat the process with unknown lengths by following these steps.

1) Trace and cut out an x^2. Label the sides of your square with x. Write the area of the square on your paper as x^2.

2) Place a small square tile in the lower right corner of the x^2 square you cut out. Trace and cut out the small square from the lower right corner as shown in the diagram at right. Label the dimensions and area of this new shape. Write an expression for the remaining area.

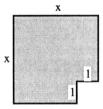

3) We have been factoring quadratics by forming rectangles and finding their dimensions. Form a rectangle with the remaining piece the same way as you did before. What are the dimensions of this new rectangle? Write its area as a product.

c) Extend these ideas to other examples by answering these questions.

1) What would be the dimensions of the rectangle formed by the difference of $x^2 - 2^2$? Write this as both a product and a difference.

2) What would be the dimensions of the rectangle formed by the difference of $x^2 - 16$? Write this as both a product and a difference.

3) Factor $4x^2 - 9$. Refer to the diagram at right. Write the expression as both a product and a difference.

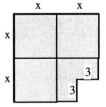

4) Factor $x^2 - y^2$. If you need to, cut a square as before to determine the dimensions. Write the expression as both a product and a difference.

d) Add this information to your tool kit.

DIFFERENCE OF SQUARES

We call this special case of factoring the **DIFFERENCE OF SQUARES**. The general pattern is written as

$$x^2 - y^2 = (x + y)(x - y).$$

Some students prefer to think of this special case as simply $x^2 + 0xy - y^2$

For example: $x^2 + 0x - 16 = (x + 4)(x - 4)$

---

e) We have shown geometrically that $x^2 - 1^2 = (x + 1)(x - 1)$.

1) Check that your factors or dimensions work by constructing a generic rectangle.

2) Graph the equation $y = x^2 - 1^2$. Find the x-intercepts. How does this relate to the factors?

AP-51. Factor each of the following quadratics. You may either use the pattern developed for the difference of squares in AP-50 or you can think of $x^2 - 49$ as $x^2 + 0x - 49$ and construct a generic rectangle.

a) $x^2 - 49$
b) $x^2 - 16$
c) $x^2 - 144$

d) Write another quadratic which is the difference of squares, then factor it. Have a study team member verify that it works.

e) Extension: Factor $16x^2 - 9$.

f) Extension: Suppose we write $16x^2 - 9 = 0$. Use part (e) and the Zero Product Property to solve for the x-intercepts.

AP-52. Factor each of the following quadratics, if possible.

a) $x^2 - 6x + 5$
b) $x^2 - x - 6$
c) $x^2 - x - 42$
d) $6x + x^2 - 16$
e) $x^2 - 100x + 2500$
f) $6x^2 - 12x$
g) $x^2 + 9x + 7$
h) $x^2 - 3x - 88$
i) $x^2 - 25$
j) $20x^2 - 100x$

AP-53. The following Zero Product Property problems were found but one of the problems was erased. Can you determine what problem (c) stated?

a) $(x - 3)(x + 2) = 0$
    $x = 3$ or $x = -2$

b) $(2x - 5)(x + 4) = 0$
    $x = \frac{5}{2}$ or $x = -4$

c) $(\ )(\ ) = 0$
    $x = -9$ or $x = 6$

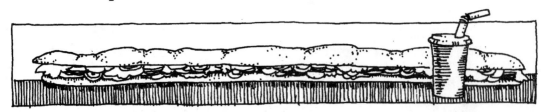

AP-54. Deli sandwiches can be bought at Giuseppi's in a variety of sizes. The smallest is the 6 inch long *Sub*, which weighs $\frac{3}{4}$ pounds and costs $2.35. The largest is a big version of the *Sub* called the *Party Giant*, which is 21 feet long.

a) How much does the *Party Giant* weigh?

b) Giuseppi charges by the weight of the sandwich. How much does the *Party Giant* cost?

AP-55. Solve each of the following equations or system of equations.

a) $3(x + 7) = 3x + 21$

b) $x^2 - 4x - 12 = 0$

c) $2x(x - 5) = 0$

d) $y = 3x - 1$
$x + 2y = 5$

AP-56. Multiply. Describe any pattern you observe. Remember in part (a) that $3 = \frac{3}{1}$.

a) $3 \cdot \frac{1}{3}$

b) $3x \cdot \frac{1}{3}$

c) $3x \cdot \frac{1}{x}$

d) $6x \cdot \frac{1}{3}$

e) $12x \cdot \frac{3}{4x}$

f) $32x \cdot \frac{3}{8}$

PERFECT SQUARE TRINOMIALS

AP-57. From the pattern found in the previous lesson, factor the following **difference of squares.**

a) $x^2 - 9$

b) $x^2 - 25$

c) $9x^2 - 25$

d) $x^2 - b^2$

AP-58. The following quadratics contain a new pattern. With your study team, factor them and discuss what special patterns appear.

a) $x^2 + 6x + 9$

b) $x^2 + 10x + 25$

c) $x^2 + 20x + 100$

d) $x^2 - 12x + 36$

e) $x^2 - 18x + 81$

f) $x^2 - 24x + 144$

g) These are called "perfect square trinomials." Why?

h) Each member of your study team should create a different example of a perfect square trinomial that fits this pattern. Factor it. Check each other's work by multiplying.

i) The polynomial $9x^2 + 30x + 25$ is also a perfect square trinomial. Find its factors. Sketch a generic rectangle to verify your factors.

AP-59. With your study team, compare the various methods and patterns you have used to factor polynomials. Write a brief summary of those methods, include an example and your opinion of which you prefer.

The Amusement Park: Factoring Quadratics

AP-60. Pay close attention to the similarities and differences between the two equations below.

  a) Solve the equation $x^2 + x - 6 = 0$ by factoring and the Zero Product Property.

  b) Where does the graph of $y = x^2 + x - 6$ cross the x-axis?

  c) Why are these two problems related?

AP-61. Solve each of the following equations. You may need to factor first.

  a) $(x + 2)(x - 3) = 0$

  b) $x^2 - 3x - 4 = 0$

  c) $(x + 22)^2 = 0$

  d) $x^2 + 3x - 10 = 0$

  e) $x^2 - 16x + 64 = 0$

  f) $(x + 3)(2x - 3) = 0$

AP-62. Graph $y = x^2 + 6x + 9$. Discuss with your study team how the factors from AP-58 part (a) tell you about a special property of this parabola.

AP-63. Factor the following.

  a) $x^2 - 9$

  b) $x^2 - 6x + 9$

  c) $x^2 + 9x + 20$

  d) $x^2 + 26x + 169$

  e) $9x^3 + 27x^2 + 108x$

  f) $x^2 - 9x + 20$

  g) $x^2 + 40x + 400$

  h) $6x^2 - 9x$

AP-64. Solve for the variable in each of the following equations.

  a) $0x = 5$

  b) $7x = 7x$

  c) $0c = 0$ What numbers could c be? Why?

AP-65. Show each subproblem in your solution. It may help to refer to your work from problem AP-49.

  a) Add $\frac{3}{4} + \frac{1}{3}$.

  b) Using part (a) as a model of subproblems, add $\frac{3}{x} + \frac{1}{3}$.

AP-66. Graph these equations using the slope and the y-intercept.

  a) $y = 4x - 2$

  b) $y = 3 - \frac{1}{5}x$

  c) $2x + y = 7$

  d) $2x + 3y = 7$

AP-67. Find the x- and y-intercepts of the following lines.

a) y = 3x - 6

b) 4x - y = 8

AP-68. Write a ratio and solve the following problems.

a) What is 12% of 210 ?

b) 28 is what percent of 210 ?

c) Chelsea answered 54 problems correct on a test. Brandee informed her that her score was 45%. How many points were possible on the test?

AP-69. A rectangular sign is twice as long as it is wide. Its area is 450 square centimeters. What are the dimensions of the sign?

AP-70. Some expressions can be factored more than once. Add this example to your tool kit. Then factor the polynomials following the tool kit box.

FACTORING COMPLETELY

Example: Factor $3x^3 - 6x^2 - 45x$ as completely as possible.

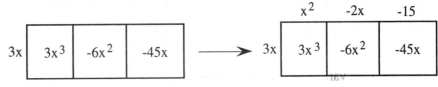

We can factor $3x^3 - 6x^2 - 45x$ as $(3x)(x^2 - 2x - 15)$.

However, $x^2 - 2x - 15$ factors to $(x + 3)(x - 5)$.

Thus, the **complete factoring** of $3x^3 - 6x^2 - 45x$ is $3x(x + 3)(x - 5)$.

Notice that the greatest common factor, 3x, is removed first.

Discuss this example with your study team and record how to determine if a polynomial is **completely factored**.

Factor each of the following polynomials as completely as possible. Consider these kinds of problems as another example of subproblems. Always look for the greatest common factor first and write it as a product with the remaining polynomial. Then continue factoring the polynomial, if possible.

a) $5x^2 + 15x - 20$

c) $2x^2 - 50$

b) $x^2y - 3xy - 10y$

The Amusement Park: Factoring Quadratics

AP-71. So far this year you have been exposed to three basic groups of equations, each of which have methods of solution and different types of answers. Be sure that you can identify each type of problem, know the method to use to solve it, and the number and form of the answers.

Add examples of each to your tool kit.

SUMMARY OF TYPES OF EQUATIONS

ONE LINEAR EQUATION	SYSTEMS OF LINEAR EQUATIONS	QUADRATIC EQUATIONS
One variable, one answer.*	Two variables, two answers which make an ordered pair (x, y).*	One variable squared, zero, one, or two answers.*
Remove any parenthesis, move all variables to one side, constants to another, and divide, if necessary.	Use substitution to combine the two equation into one, solve for the first variable, solve for the second variable.	Rearrange the equation so that it is set equal to zero, factor, and use the Zero Product Property.
Example: $$\begin{aligned}2(2x-1)-6 &= -x+2\\ 4x-2-6 &= -x+2\\ 5x-8 &= 2\\ 5x &= 10\\ x &= 2\end{aligned}$$	Example: $$\begin{cases} x+2y = -8 \\ y = 2x+1 \end{cases}$$ $$\begin{aligned}x+2(2x+1) &= -8\\ x+4x+2 &= -8\\ 5x+2 &= -8\\ 5x &= -10\\ x &= -2\end{aligned}$$ $$\begin{aligned}y &= 2x+1\\ &= 2(-2)+1 = -3\\ &(-2,-3)\end{aligned}$$	Example: $$x^2+9x+10 = -4$$ $$x^2+9x+14 = 0$$ $$(x+7)(x+2) = 0$$ $$x+7=0 \text{ or } x+2=0$$ $$x=-7 \text{ or } x=-2$$
*Except for cases where there is no solution or the solution is all real numbers.	*Unless parallel or the same line.	*Zero answers (parabola above or below x-axis); One answer (parabola vertex on x-axis); or, Two answers (parabola intersects x-axis in two points).

AP-72. Solve each of the following equations or systems by any method you choose. Show all your work. If you do not know where to begin, review the different types of equations in problem AP-71.

a) $0 = 2x - 3$
b) $0 = (2x - 3)(x + 5)$
c) $2x - 5 = -x + 7$
d) $x^2 + 5x + 6 = 0$
e) $x^2 - 4x + 4 = 0$
f) $y = 2x - 3$
   $x + y = 15$
g) $x^2 + 5x + 9 = 3$
h) $3(3x - 1) = -x + 1$
i) $\frac{5 - 2x}{3} = \frac{x}{5}$
j) $\frac{2x}{3} = 4$
k) $x^2 - 6 = -2$
l) $x^2 = y$
   $10 = y$

AP-73. Find the x-intercepts for the graph of each of the following equations without graphing. Describe your method in complete sentences.

a) $y = x^2 + 10x + 21$
b) $y = 2x - 1$
c) $y = 2x^2 - 5x + 2$

AP-74. Combine like terms.

a) $(5x^2 + x - 1) + (3x^2 - 4x - 8)$
b) $(6x^2 - 8x + 7) - (2x^2 + 6x - 7)$
c) $(8x^2 - 8x + 10) + (2x^2 + 8x - 6)$
d) $(x^2 - x - 1) - (2x^2 + x - 1)$

AP-75. Solve each of the following equations.

a) $2(x - 3) + 2 = 4$
b) $12 - 6x = 108$
c) $2x + 5 = 0$
d) $3x - 11 = 0$
e) Solve the equation $Dx - C = 0$ for x.

AP-76. Sonja notices that the lengths of the three sides on a right triangle are consecutive even numbers. The triangle's perimeter is 24. Write an equation and solve it to find the length of the longest side.

AP-77. Factor each of the following quadratic polynomials as completely as possible. (It may help to look for a common factor first.)

a) $2x^2 - 6x$
b) $x^2 - 12x + 35$
c) $6x^2 - 30x + 18$
d) $4x^2 - 16x + 12$

The Amusement Park: Factoring Quadratics

AP-78. The area of the rectangle at right is 60 square units.

a) Write the area as both a product and a sum.

b) Use the area as a sum to write an equation where the area is 60 square units.

c) Find the value of m.

d) Can there be more than one value for m? Explain why or why not.

AP-79. THE AMUSEMENT PARK PROBLEM

The city planning commission is reviewing the master plan of the proposed Amusement Park coming to our city. Your job is to help the Amusement Park planners design the land space.

Based on their projected daily attendance, the planning commission requires 15 rows of parking. The rectangular rows will be of the same length as the Amusement Park. Depending on funding, the Park size may change so planners are assuming the park will be square and have a length of x. The parking will be adjacent to two sides of the park as shown below.

Our city requires all development plans to include "green space" or planted area for sitting and picnicking. See the plan below.

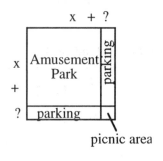

a) Your task is to list all the possible configurations of land use with the 15 rows of parking. Find the areas of the picnic space for each configuration. Use the techniques you have learned in this unit. There is more than one way to approach this problem, so show all your work.

b) Record the configuration with the minimum and maximum picnic area. Write an equation for each that includes the dimensions and the total area for the project. Verify your solutions before moving to part (c).

c) The Park is expected to be a success and the planners decide to expand the parking lot by adding 11 more rows. Assume that the new plan will add 11 additional rows of parking in such a way that the maximum original green space from part (b) will triple. Show all your work. Record your final solution as an equation describing the area of the total = product of the new dimensions.

d) If the total area for the expanded Park, parking and picnic area is 2208 square units, find x. Use the dimensions from part (c) to write an equation and solve for the side of the Park.

AP-80. Factor each of the following quadratics. Write as both a product and a sum.

a) $x^2 + 4x + 3$

b) $x^2 - 7x + 12$

c) $x^2 + 10x + 16$

d) $x^2 + 5x - 24$

e) $x^2 - 3x - 18$

f) $3x^2 + 9x - 30$

AP-81. Start at point A (-1, 2). Move (slide) to point B (3, 5). Find the coordinates of a point C so that A, B, and C are on the same line.

a) What is the slope of the line?

b) Graph the line.

c) Write the equation of the line from the graph.

AP-82. Scott wants to enlarge the rectangle at right by making each side 3 times as long. Write algebraic expressions for the perimeter and the area of the new enlarged rectangle. Be sure to draw a picture of the enlarged rectangle and show its dimensions:

AP-83. Write an equation and solve the following problem. It costs $25 per day plus $0.06 per mile to rent a car. What is the greatest number of miles that you can drive if you only have $40?

AP-84. Find each of the following sums.

a) $\frac{2}{3} + \frac{1}{5}$

b) $\frac{1}{x} + \frac{1}{3}$

c) $\frac{2}{x} + \frac{3}{5}$

d) $\frac{3}{x} + \frac{5}{2x}$

UNIT SUMMARY AND REVIEW

AP-85. UNIT 8 SUMMARY

Select 3 or 4 big ideas that you feel best represent Unit 8. Write a description of each main idea. For each topic, choose a problem that best demonstrates your understanding. Explain your solution as in the previous unit summaries. Show all your work.

AP-86. Penny drew this diagram of a frame for an x by x + 1 rectangular picture.

Write an algebraic expression to represent the total area of the frame. Show all subproblems.

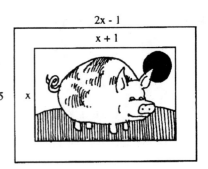

AP-87. Factor each of the following quadratics, if possible. Work together and treat this set as a puzzle. If the numbers you try do not work for one expression, they may work for another.

a) $x^2 - 2x - 24$
b) $x^2 + 11x + 24$
c) $x^2 - 10x + 24$
d) $x^2 + 5x - 24$
e) $x^2 - 10x - 24$
f) $x^2 + 8x + 24$
g) $x^2 - 23x - 24$
h) $x^2 + 25x + 24$

AP-88. Solve these equations:

a) $x^2 - 16 = 0$
b) $x^2 + 11x + 24 = 0$
c) $x^2 - 10x + 24 = 0$
d) $x^2 + 5x = 24$

AP-89. We can use the idea of a generic rectangle to find more complicated kinds of products. For each of the following products, draw a generic rectangle and label the dimensions. Then find the given product by finding the area of the rectangle. The first answer is given so you can check your method.

a) $(x + 3)(x^2 + 4x + 7)$
   $x^3 + 7x^2 + 19x + 21$

b) $(3x - 2)(4x^2 + 2x + 1)$

c) $(2x + 1)(x^2 - 3x + 4)$

d) $(x - 2)(x^2 + 2x + 4)$

AP-90. Solve for the missing area or side in each square. Do not use your calculator; use your number sense to estimate to a tenth.

a) 11 by 11, A =
b) 8 by 8, A =
c) 1.2 by 1.2, A =
d) s by s, A = 0.81
e) s by s, A = 5
f) s by s, A = 10

AP-91. Factor completely.

a) $x^2 - 17x + 42$

b) $2x^2 + 32x + 128$

c) $5x^3 - 5x^2 - 30x$

d) $3x^2 - 48$

AP-92. Solve for x and y:

a) $3x + 5 = 2y$
   $y = 6 - x$

b) $2(x + 5) = y + 6$
   $y + 6 = 12x$

AP-93. Graphs for equations (a) and (b) are drawn on your resource page. Graph part (c), then label each parabola clearly and identify the x- and y-intercepts for each one. If possible, use the factoring techniques you have learned to make your graphing easier.

a) $y = x^2 - 4$

b) $y = x^2 + 4$

c) $y = x^2 - 5x + 4$

d) Factor each of these equations, if possible.

e) What patterns do you notice relating the factors, equations and graphs of quadratics?

AP-94. TOOL KIT CHECK-UP

Your tool kit contains reference tools for algebra. Return to your tool kit entries. You may need to revise or add entries.

Be sure that your tool kit contains entries for all of the items listed below. Add any topics that are missing to your tool kit NOW, as well as any other items that will help you in your study of algebra.

- Sums and Products of Polynomials
- Greatest Common Factor
- Difference of Squares
- Summary of Types of Equation
- Roots of a Quadratic Equation
- Factoring Quadratics
- Zero Product Property
- Factoring Completely
- Factors

The Amusement Park: Factoring Quadratics

GM-15. **WEIGHING PUMPKINS**

Every year at Half Moon Bay, there is a pumpkin contest to see who has grown the largest pumpkin for that year.

Last year, one pumpkin grower (who was also a mathematician) brought 5 pumpkins to the contest. Instead of weighing them one at a time, he informed the judges, "When I weighed them two at a time, I got the following weights: 110, 112, 113, 114, 115, 116, 117, 118, 120, and 121 pounds."

Your Task: Find how much each pumpkin weighed.

GM-16. WITH OR WITHOUT FROSTING

Mr. Algebra baked a cake for the Midwest Mathematics Convention. He designed the cake in the shape of a big cube. As he was carrying the cake over to the frosting table, he slipped and sent the cake sailing into the vat of frosting.

Amazingly, the cake stayed in one piece, but all 6 sides were now frosted. He carefully got it out and put it on a platter.

The mathematicians were delighted when they saw the cube-cake with all sides frosted. One mathematician suggested the cake be cut into cube shaped pieces, all the same size. That way, some people could have a piece with no frosting, 1 side frosted, 2 sides frosted, or 3 sides frosted. Kawana, a very creative mathematician, said, "Cut the cake so that the number of pieces with no frosting is eight times more than the number of pieces with frosting on 3 sides. Then you will have the exact number of pieces of cake as there are mathematicians in this room."

Your Task:

- Using Kawana's clue, find out how many mathematicians were at the convention. Hint: build models of different size cakes.

- How many mathematicians would be at the convention if the number of pieces with 1 frosted side equaled the number of pieces with no frosting?

Unit Nine
The Birthday Party Piñata

Using Diagrams to Solve Equations

Unit 9 Objectives
The Birthday Party Piñata: **USING DIAGRAMS TO WRITE EQUATIONS**

In this unit, you will be increasing your problem solving and algebraic skills as you investigate problems involving **right triangles**. You will work with square roots, graphs and algebraic fractions.

In this unit you will have the opportunity to:

- use diagrams and models as an aid to writing equations for two and three dimensional situations.

- apply the Pythagorean Theorem to calculate the distance between two points.

- solve equations that involve fractions.

- explore square roots in terms of their meaning as numbers.

- use patterns to discover the methods for doing arithmetic calculations with square roots.

- solve equations that include square roots.

- write the equation of a line given the coordinates of two points on the line.

The primary focus of this unit is solving problems by drawing diagrams and writing equations. You will learn how to use a diagram to help to solve a problem. Some of the situations that arise will involve square roots. You will have a chance to understand square root as well as work with square roots in equations and compare their graphs to others you have studied in this course.

BP-0.   THE BIRTHDAY PARTY PIÑATA

For Katja's party, her friends are going to hang a piñata half way between two poles with a rope over the top of each pole. Both poles are 20 feet high, and they are 30 feet apart. The piñata must hang four feet above the ground. How much rope is needed to hang the piñata? With your study team, model and sketch this situation.

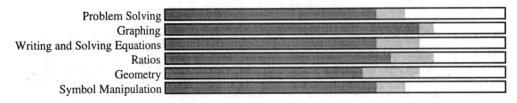

Unit 9

The Birthday Party Piñata: USING DIAGRAMS TO WRITE EQUATIONS

DEVELOPING THE PYTHAGOREAN THEOREM

BP-1. Copy the diagram and add the definitions to your tool kit.

> In a **right triangle**, the longest side is called the **HYPOTENUSE** while the other two sides are called **LEGS**. Notice that the legs meet to form the right angle.
>
>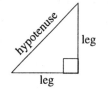

BP-2. What is the relationship between the legs and hypotenuse of a right triangle? How complicated is the relationship? To answer these questions, complete the table on the resource page provided by your teacher for the five given triangles. Verify your answers with your study team. Look for a relationship among the three columns on the right.

a) 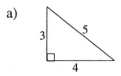   b) 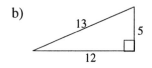   c)

d)   e)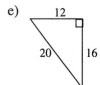

	length of leg #1	length of leg #2	length of hypotenuse	(length of leg #1)^2	(length of leg #2)^2	(length of hypotenuse)^2
a)						
b)						
c)						
d)						
e)						

BP-3. With your study team write an equation in words to describe the relationship you found for the legs and the hypotenuse of a right triangle. Be prepared to present your team's description to the class. The side lengths for the triangles in BP-2 were all integers. Do you think the pattern you have developed works for every right triangle, including those whose side lengths are fractional? Explain your answer.

BP-4. Solve for each unknown side using your "word equation" from BP-3.

a)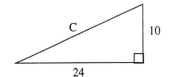

c)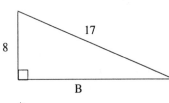

b)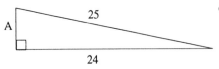

d)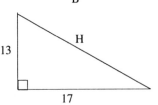

BP-5. Draw a diagram or generic rectangle to illustrate each of the following products. Write the products in simplest form by combining like terms.

a) $x(2x + 1)$

c) $(x + 4)(x^2 + 5x + 3)$

b) $(x + 2)(x + 3)$

BP-6. SQUARING A SQUARE ROOT

a) Pick a number between zero and twenty. Take the square root of your number, then square the result. What happens?

b) Pick a number from twenty to forty. Square your number, then take the square root of the result. What happens?

c) When picking a positive number, what do you notice about the process of squaring and square rooting, or square rooting and squaring?

d) Now try the process on any positive number you choose. Record your result.

e) Extension: This process does not always work. Can you find a number for which it does not appear to work? Explain why you think this happens.

BP-7. Use the Order of Operations to evaluate.

a) Does $3^2 + 4^2 = 7^2$?

b) Does $\sqrt{3^2 + 4^2} = 7$?

c) Using your observations from parts (a) and (b), does $\sqrt{a^2 + b^2} = a + b$?

BP-8.   Solve each of the following equations for x. Remember there are two solutions. Round to the nearest hundredth.

a) $x^2 = 152 - 122$

b) $x^2 + (26.4)^2 = 29^2$

Solve each of the following equations for b. Remember there are two solutions.

c) $122 - 52 = b^2$

d) $25^2 + b^2 = 35^2$

e)   Extension: $a^2 + b^2 = c^2$

BP-9.   Multiply. Describe any patterns you observe.

a) $8\left(\frac{1}{8}\right)$

b) $x\left(\frac{6}{x}\right)$

c) $3x\left(\frac{5}{3x}\right)$

d) $18xy\left(\frac{3}{9x}\right)$

e) $12\left(\frac{5x}{6}+\frac{y}{3}\right)$

Hint: distribute first.

BP-10.   a)   Without a calculator, add: $\frac{3}{11} + \frac{2}{7}$.

b)   Describe what you did in part (a) to find the sum of the two fractions.

c)   Now find the sum of this pair of fractions: $\frac{3}{x} + \frac{1}{7}$.

FINDING THE LENGTH OF A LINE SEGMENT

BP-11.   Answer these questions to help you determine the hypotenuse in any right triangle.

a)   Sketch a right triangle. Label the legs x and y and label the hypotenuse z.

b)   Write a sentence or two to explain to a friend how to decide which side is the hypotenuse and which sides are the legs.

c)   In your study team, discuss and write another way to tell the hypotenuse from the legs.

d)   If a right triangle had the side lengths 6, 10, and 8, could the hypotenuse be the side with length 8? Explain why or why not.

BP-12.   Add the following information and diagram to your tool kit:

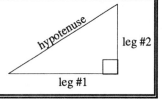

The **PYTHAGOREAN THEOREM** states the relationship between the legs and hypotenuse of any right triangle:

$$(\text{leg \#1})^2 + (\text{leg \#2})^2 = (\text{hypotenuse})^2$$

The Birthday Party Piñata: Using Diagrams to Write Equations

BP-13. LENGTH OF A LINE SEGMENT

Copy each of the following graphs onto graph paper unless you have the resource page. For each graph, draw a right triangle and determine the length of the given line segment. Write the length as both a square root and its decimal approximation to the nearest hundredth.

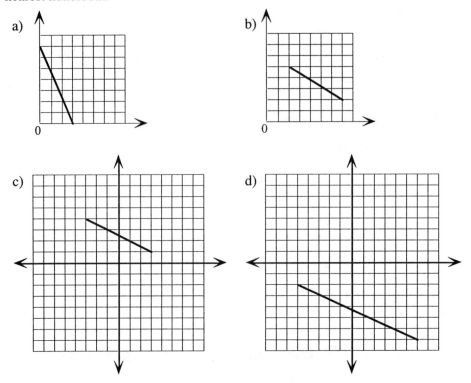

BP-14. Draw a set of coordinate axes on a piece of graph paper. Plot the point (10, 6). Label it A. Move (slide) the point seven units down, then eight units to the left and stop.

   a) Write the coordinates of the stopping point. Label it B.

   b) Find the distance between A and B.

   c) How could we find the lengths of the legs of the right triangle without graphing?

BP-15. GRAPHING SQUARE ROOTS

Obtain the graph resource page from your teacher. (You need to keep track of this page because you will complete a portion of it each day for the next several days).

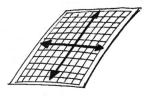

a) Graph $y = \sqrt{x}$. Record your points in the table on the resource page.

b) Try $x = -3$ in the rule. What is the result? Explain.

c) Describe the possible values for x that will yield solutions for the rule $y = \sqrt{x}$.

d) What y values are possible?

e) Describe your graph in words.

BP-16. Are 10, 24, and 26 the side lengths of a right triangle? Explain how you know.

BP-17. Use the substitution method to solve each pair of equations for x and y.

a) $y = 7 - 2x$
$y = 6x - 1$

b) $y = 7 - 2x$
$3x - y = 3$

c) What do the solutions to parts (a) and (b) represent on a graph?

BP-18. Evaluate each of the following expressions for $x = 3$, then $x = -2$.

a) $2x^2$

b) $-5x^2$

c) $(2x)^2$

d) $3x^3$

e) $(3x)^3$

f) $(-5x)^2$

BP-19. Find each of the following products. Generic rectangles may be helpful.

a) $(3x + 1)(x + 4)$

b) $(2x - 7)^2$

c) $(x - 2)(x^2 + 2x + 1)$

d) $(2x + 5)(7 - x)$

BP-20. Pat received 9 out of 10 on his quiz for the following work. He does not know what he did wrong. Study his work and find his mistake. Write Pat a note explaining to him his mistake and what his answer should have been.

Problem: Factor $x^2 - 8x + 12$

Pat's work:
$x^2 - 8x + 12$
$(x - 6)(x - 2)$
$x = 6$ and $x = 2$

	x	- 6
x	x^2	-6x
- 2	-2x	12

The Birthday Party Piñata: Using Diagrams to Write Equations

BP-21. You can solve for x the equation: $x^2 - 8x + 12 = 0$

but not for the expression: $x^2 - 8x + 12$.

a) Explain why using a complete sentence.

For each of the following, decide if there is enough information to solve for the variable. If there is, then do so. If not, merely factor and move on to the next part.

b) $b^2 - 4b + 4$

c) $b^2 - 5b + 4$

d) $b^2 + 10b + 21 = 0$

e) $b^2 - 6b + 5 = 0$

f) $b^2 - 17b + 16$

g) Extension: $2x^2 - 3x + 1$

BP-22. Equations (d) and (e) in problem BP-21 each had two solutions. What do these solutions represent on the graph of the equation? What shape is each graph?

BP-23. Find the area of the triangles you drew in BP-13 parts (a) and (b).

USING DIAGRAMS TO WRITE EQUATIONS

BP-24. The beautiful young princess of Polygonia is very distressed. Her father has chosen a husband for her: a very rich, but very old king from another land. However, the princess is in love with someone else. In order to ensure that the princess will not escape and elope with her prince charming, her father has locked her in the tower until the old king arrives for the marriage.

She could escape through the window, but it is 50 feet above the ground (rather a long distance to jump). An alligator-infested moat, which is 10 feet wide, surrounds the tower. Naturally her prince charming is planning to rescue her. His plan is to use an arrow to shoot a rope up to her window. She can then slide down the rope to the other side of the moat into his waiting arms, and off they will ride on his beautiful stallion.

However, even though the prince is charming, he is not too bright. Every night he has tried to rescue the princess with a rope that was too short. The old king arrives tomorrow and the prince needs your help. Remember: this is his last chance to rescue the princess!

With your study team, find the minimum amount of rope the prince needs. (Assume they need an extra 1.5 feet of rope at each end in order to tie the rope.) Write a complete explanation for the prince, including a diagram, that shows how to determine the minimum length of rope needed.

BP-25. On a baseball diamond, the bases are 90 feet apart. (Every baseball diamond is a square.) How far is it from home plate directly to second base? Sketch and label a diagram. Use it to write an equation, then solve the equation.

BP-26.  A 10 foot ladder is leaning against a tree. The foot of the ladder is 3.5 feet away from the base of the tree trunk.

a) Draw a diagram and label the ladder length and its distance from the tree.

b) How high on the tree does the ladder touch? Write an equation and solve it.

BP-27. Jehrico is studying the expression "$2(x + 1)(2x + 3)$" and is not sure where to start. He thought he should first multiply 2 and $(x + 1)$, and then multiply the result by $2x + 3$. But a teammate told him to first multiply $(x + 1)(2x + 3)$ and then multiply that result by 2. He wants your team's help to decide which way is correct.

In your study team, choose some members to multiply the 2 and $(x + 1)$ first, and then multiply the result by $2x + 3$. The other members should first multiply $(x + 1)(2x + 3)$, and then multiply that result by 2.

a) Write down both results. What do you notice?

Find each of the following products. Verify your answers with your study team.

b) $3(x - 1)(5x - 7)$ c) $5(x + 3)(2x - 5)$

BP-28. A careless construction worker drove a tractor into a telephone pole, cracking the pole. The top of the pole fell as if hinged at the crack. The tip of the pole hit the level ground 24 feet from its base. The stump of the pole stood seven feet above the ground. If an additional five feet of the pole extends into the ground to anchor it, how long should the replacement pole be? Draw a diagram. Write an equation. Show all subproblems as you solve.

The Birthday Party Piñata: Using Diagrams to Write Equations

BP-29. Make a table and draw the graph of $y = x^2 - 4$.

  a) Mark and label the points where the graph crosses the x-axis.

  b) Factor the expression $x^2 - 4$.

  c) Substitute $y = 0$ in the equation $y = x^2 - 4$ to get $0 = x^2 - 4$. Use the Zero Product Property to find the solutions of the equation $0 = x^2 - 4$ and explain how the solutions are related to the graph and the factors.

  d) Solve $x^2 = 4$.

  e) Solve $x^2 - 4 = 0$.

  f) Did you get the same results in parts (c), (d), and (e)? What does this tell you about their graphs? Explain.

BP-30. Write one polynomial to represent each of the following differences.

  a) $(8x^2 + 5x + 7) - (3x^2 + 2x + 2)$ d) $(16x^2 + 3x - 8) - (7x^2 + 8x - 10)$

  b) $(8x^2 + 5x + 7) - (3x^2 - 2x + 2)$ e) $(10x^2 - 13x + 1) - (3x^2 - 10x + 1)$

  c) $(5x^2 + 14x + 3) - (2x^2 - 9x + 5)$

BP-31. Factor each of the following polynomials. Part (d) is an extension.

  a) $y^2 - 3y - 10$ c) $x^2 + 27x + 50$

  b) $x^2 - 15x + 50$ d) $5x^2 + 6xy + y^2$

BP-32. Solve for x in each right triangle below.

  a)   b) 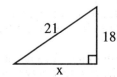   c)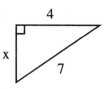

BP-33. Graph the points (2, -6) and (-3, 5).

  a) Find the distance between them.

  b) Find the slope of the line through these two points.

  c) Are the distance between the points and the slope of the line the same? Write a few sentences about what "distance" and "slope" means, and their relationship, if any.

BP-34. Zal claims the y-intercept of the graph of $2x - y = 4$ is 4. Lida says that it is -4. Which claim is correct? Explain how you know.

BP-35. Each of the following equations is in $y = mx + b$ form. For each equation, state the value of m and the value of b.

a) In an equation of the form $y = mx + b$, the **m** represents the _____ of the line and the **b** represents the_____. This is why the form $y = mx + b$ is called the **slope-intercept form** of a linear equation.

b) $y = 4x$

c) $y = -4x + 7$

d) $y = 4$

e) $y = x - \frac{3}{2}$

MAKING A MODEL: THE FLAT FAMILY'S ROOF

BP-36. We have been studying right triangles from a two dimensional viewpoint. However, often triangles occur in three-dimensional settings.

The roof on the Flat Family's house is one large flat rectangle. It is parallel to the ground. Their TV antenna is mounted in the center of the roof. Guy wires are attached to the antenna five feet below its highest point, and are attached to the roof at each corner and at the midpoint of each edge. These wires support the antenna in the wind.

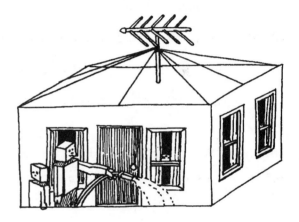

Each guy wire is the hypotenuse of a right triangle. How long does each wire need to be to keep the antenna upright?

a) Build a model of the Flat Family Roof. The model will serve as a reference while your study team is writing the subproblems necessary to find the lengths of the sides of the right triangles in a 3-dimensional setting.

1) For the roof, use a rectangular piece of cardboard. Draw the diagonals of the rectangle to locate the center.

2) Tape the string guy wires to the straw. Don't forget to leave a gap that represents the five feet between the strings and the top of the antenna. How many guy wires are there?

3) Attach the antenna (straw or stick) to the roof. Use tape, tie knots, or make slits to anchor your antenna and guy wires to the roof (cardboard).

b) Locate as many vertical right triangles as you can on your model. How many right triangles did you find?

c) Identify which right triangles are the same size and which are different sizes. How many different sized right triangles did you find?

d) Sketch each of the different sized right triangles on a different color of paper. Cut out the triangle. Tape the triangle in position on your model. Use this model to help you solve problem BP-37.

The Birthday Party Piñata: Using Diagrams to Write Equations

BP-37. The Flat Family's roof is 60 feet long and 32 feet wide. The TV antenna is 30 feet tall. The wires are attached five feet below the top of the antenna.

How long must each guy wire be? Show all subproblems in your solution.

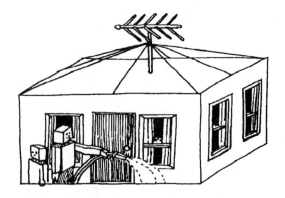

BP-38. Bryce leaned his five foot rake against the shed. The base of the rake was 2 feet from the shed. How high up the shed did the tip of the rake reach? Draw a right triangle diagram. Write an equation and solve.

BP-39. Find the distance between each pair of points.

a) (5, -8) and (-3, 1) b) (0, -3) and (0, 5)

BP-40. So far, in problems AP-49, AP-65, AP-84, and BP-10 we added fractions with variables using patterns we found in Unit Two. The problems below extend this notion of adding fractions to solving equations with fractions. Examine the work from those problems mentioned above and complete the following:

a) Add $\frac{x}{3} + \frac{x}{5}$. b) Solve. $\frac{5x + 3x}{15} = 2$

BP-41. Solve for x and y: $5x - 2 = y$
$3x - y = -2$

BP-42. Solve each equation by factoring:

a) $x^2 - 13x + 40 = 0$ c) $x^2 - 169 = 0$

b) $x^2 + 15x + 56 = 0$ d) $100 - x^2 = 0$

BP-43. Find the slope and y-intercept of each line, using any method you prefer.

a) $y = -\frac{2}{3}x - 1$ c) $2x + 3y = 48$

b) $y = 5 + \frac{8}{5}x$

BP-44. **GRAPHING SQUARE ROOTS**

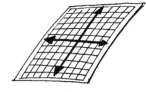

a) On your resource page from problem BP-15, graph $y = \sqrt{x} + 2$.

b) Try $x = -3$ in the rule. What is the result? Explain.

c) Describe the possible values for x that may be used for the rule $y = \sqrt{x} + 2$.

d) What y values are possible?

e) Compare your two square root graphs. How did adding 2 change the graph?

BP-45. Multiply.

a) $6x \cdot \dfrac{4}{x}$
b) $6x \cdot \dfrac{11}{6}$
c) $6x \cdot \dfrac{3}{2x}$

SOLVING EQUATIONS WITH ALGEBRAIC FRACTIONS

BP-46. Problems on preceding days of this unit have involved using diagrams to write equations you could solve. BP-46 will provide you with yet another equation-solving technique to rewrite more complicated equations into ones that you can more easily solve. Read the following example and add it to your tool kit:

FRACTION BUSTERS
Solving Equations Containing Algebraic Fractions

Solve: $\dfrac{x}{3} + \dfrac{x}{5} = 2$

The complicating issue in this problem is dealing with the fractions. We could add them by first writing them in terms of a common denominator, but there is an easier way.

There is no need to use the often time consuming process of adding the fractions if we can "eliminate" the denominators. To do this, we will need to find a common denominator of all fractions, and then we will <u>multiply</u> <u>both</u> <u>sides</u> <u>of</u> <u>the</u> <u>equation</u> by that common denominator. In this case the lowest common denominator is 15, so we <u>multiply</u> <u>both</u> <u>sides</u> of the equation by 15.

The result is an equivalent equation without fractions!

The number we use to eliminate the denominators is called a **FRACTION BUSTER**. Now the equation looks like many we have seen before and we solve it in the usual way.

Finally, remember to check your answer.

$\dfrac{x}{3} + \dfrac{x}{5} = 2$

The lowest common denominator of $\dfrac{x}{3}$ and $\dfrac{x}{5}$ is 15.

$15 \cdot (\dfrac{x}{3} + \dfrac{x}{5}) = 15 \cdot 2$

$15 \cdot \dfrac{x}{3} + 15 \cdot \dfrac{x}{5} = 15 \cdot 2$

$5x + 3x = 30$

$8x = 30$

$x = \dfrac{30}{8} = \dfrac{15}{4} = 3.75$

$\dfrac{3.75}{3} + \dfrac{3.75}{5} = 2$

$1.25 + 0.75 = 2$

>>Problem continues on the next page.>>

Copy the following problem on your paper. Fill in each of the lines labeled (a) through (e) to explain how the equation to its right was obtained from the equation above it.

Solve the equation: $\quad \dfrac{4}{x} + \dfrac{3}{2x} = \dfrac{11}{6}$

a) _____ $\quad 6x(\dfrac{4}{x} + \dfrac{3}{2x}) = 6x \cdot (\dfrac{11}{6})$

b) _____ $\quad 6x \cdot (\dfrac{4}{x}) + 6x \cdot (\dfrac{3}{2x}) = 6x \cdot (\dfrac{11}{6})$

c) _____ $\quad 24 + 9 = 11x$

d) _____ $\quad 33 = 11x$

e) _____ $\quad 3 = x$

BP-47. In each equation, find the lowest common denominator of all fractions first, as in problem BP-46, then use Fraction Busters, and solve.

a) $\quad \dfrac{x}{2} + \dfrac{x}{3} = \dfrac{1}{6}$ 　　　　　　c) $\quad \dfrac{4}{x} - 1 = 7$

b) $\quad x + \dfrac{x}{2} + \dfrac{x}{3} = 22$ 　　　　　d) $\quad \dfrac{1}{x} + \dfrac{1}{2x} = 3$

BP-48. THE BUCKLED RAILROAD TRACK

Railroad designers are always looking for ways to improve the safety and comfort of train travel.

Suppose technology made it possible to make and transport straight rails one mile long. If one winter two pairs of these rails were installed along a two mile portion of track in the desert, then each rail  would expand approximately one foot in length in the summer heat. Suppose that instead of buckling, these rails keep their (straight) shape. In this case, the ends where they meet could jut upward to form a triangle. How high above the ground would the ends of the rails be?

Let H be the height (in feet) of the tracks above the ground where the two rails come together.

a) How large do you think H is? Is it big enough for you to stick your arm between the ground and the tracks? Is it big enough walk through? Could you drive a car under the buckled tracks? Make a guess.

b) Draw a picture and calculate H. (Note: a mile is 5280 feet long.)

c) How does your calculated value for H compare with your guess in part (a)?

BP-49. A child's shoe box measures 4" x 6" x 3". What is the longest pencil you could fit into this box? An empty box may help you visualize the varies ways you could fit the pencil in the box. If possible, draw a diagram to show the pencil's position. Show your subproblems.

BP-50. Solve for x. Use Fraction Busters from BP-46. Leave answers in **fraction** form.

a) $\frac{x}{2} + \frac{x}{6} = 7$

b) $\frac{x}{2} + \frac{x}{3} - \frac{x}{4} = 12$

c) $\frac{x}{9} + \frac{2x}{5} = 3$

d) $\frac{5}{2x} + \frac{1}{6} = 8$

BP-51. Solve each of the following systems for x and y.

a) $y = 4x$
$x + y = -1$

b) $-2x + y = 3$
$x = 3y$

BP-52. Find the distance between each set of points.

a) (0, 5) and (5, 0)

b) (-4, 7) and (29, 76)

BP-53. MONEY MONEY MONEY

Janelle has $20 and is saving at the rate of $6 per week. Jeanne has $150 and is saving at the rate of $4 per week. After how many weeks will each person have the same amount of money?

a) Make a graph to solve the problem.

b) Write equations to solve the problem algebraically.

c) If you were presenting your solution to students who did not know the problem, would the graph help them visualize the problem? Explain your graph to that student.

BP-54. Solve each of the following equations by factoring.

a) $g^2 + 5g + 6 = 0$

b) $m^2 + 5m - 6 = 0$

c) $e^2 - 10e + 21 = 0$

d) $h^2 - 16h - 36 = 0$

BP-55. Graph using the slope and y-intercept.

a) $y = \frac{3}{4}x - 2$

b) $y = 3x + 4$

BP-56. A jar has an even number of pennies, nickels, and dimes. When Juan puts 24 more pennies in the jar the probability of pulling out a penny becomes $\frac{1}{2}$. How many pennies, nickels and dimes were in the jar to begin with?

The Birthday Party Piñata: Using Diagrams to Write Equations

SIMPLIFYING EXPRESSIONS WITH RADICALS

BP-57. Use your calculator to determine which of the following statements containing unlike square roots are true and which are false.

a) $\sqrt{2} + \sqrt{3} = \sqrt{5}$

b) $\dfrac{\sqrt{12}}{\sqrt{4}} = \sqrt{3}$

c) $(\sqrt{2})(\sqrt{3}) = \sqrt{6}$

d) $\dfrac{\sqrt{7}}{\sqrt{2}} = \sqrt{3.5}$

e) $\sqrt{5} - \sqrt{2} = \sqrt{3}$

f) $\sqrt{16} + \sqrt{3} = \sqrt{19}$

g) $\dfrac{\sqrt{6}}{\sqrt{3}} = \sqrt{2}$

h) $(\sqrt{5})(\sqrt{7}) = \sqrt{35}$

BP-58. Use your results from problem BP-57 above.

a) Which kinds of square root operations in BP-57 were true?

b) Which kinds of square root operations in BP-57 were false?

c) With your study team, write conjectures about using the different operations (addition, subtraction, multiplication and division) with unlike square roots. Put these in your tool kit. For each conjecture give one or more examples that support it.

BP-59. Use your conjectures to simplify these expressions as much as possible. Be sure not to leave perfect squares (such as 9 or 16) under a radical sign. Also, combine square root terms if possible. If the expression cannot be simplified, leave it as is. Check your answers with a calculator.

a) $(\sqrt{3})(\sqrt{7})$

b) $\dfrac{\sqrt{20}}{\sqrt{5}}$

c) $\sqrt{3} + \sqrt{7}$

d) $\sqrt{5} \cdot \sqrt{20}$

e) $\sqrt{7} - \sqrt{3}$

f) $\dfrac{\sqrt{21}}{\sqrt{3}}$

g) $\sqrt{20} + \sqrt{5}$

h) $\dfrac{\sqrt{144}}{\sqrt{3}\,\sqrt{4}}$

BP-60. GRAPHING SQUARE ROOTS

a) Graph $y = \sqrt{x-2}$ on your resource page.

b) What x values are possible in part (a)?

c) Compare this graph to BP-15 and BP-44.

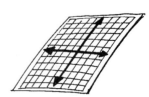

BP-61. A fly was sitting on the ground in the back left corner of your classroom. He flew to the ceiling in the opposite corner of the room. How far did he fly if he went in a straight line? Draw a diagram of your classroom, complete with the correct measurements. Write equations. Show your subproblems.

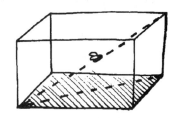

BP-62. Using graph paper, find the equation of the line through each pair of points below. Write your answer in **y = mx + b** form.

a) (3, 4) and (4, 6)

b) (-5, -8) and the origin

c) (1, -5) and (-1, 1)

d) (1, -5) and (-2, 0)

BP-63. Use Fraction Busters to solve for x.

a) $\frac{2}{3} - \frac{x}{5} = 10$

b) $\frac{1}{x} - 5 = \frac{4}{3}$

BP-64. Since $x + x + x = 3x$, Hannah thinks that $3\sqrt{5}$ might be the simplified way to write $\sqrt{5} + \sqrt{5} + \sqrt{5}$.

a) Is Hannah correct? Use your calculator. Explain how you tested Hannah's conjecture.

b) Since $x + y + z$ cannot be simplified, can Mary simplify $\sqrt{5} + \sqrt{6} + \sqrt{7}$? Explain.

BP-65. GRAPHING SQUARE ROOTS

a) Graph $y = \sqrt{x + 5}$ on your resource page.

b) What happens if you use $x = -6$ in part (a)?

c) What values are possible for x?

d) Compare this graph to the graphs from BP-15, BP-44, and BP-60. Describe the effect of adding or subtracting a number from x under the square root symbol.

BP-66. Find the distance between this pair of points: (-2, -8) and (-2, 56).

BP-67. In this problem, write two equations to describe the situation, and solve.

To fill Katja's piñata, her father used Choco-giggles and Fruity Scooters. Each Choco-giggle costs $0.25, and each Fruity Scooter costs $0.30. If he paid $73 for all 264 pieces of candy, how many of each type did he use?

BP-68. Right triangles ABC and △A'B'C' are similar, with right angles at B and B'. If AB = 12 cm, BC = 5 cm, and B'C' = 15 cm, draw a diagram and label it. Then find the length of A'C'.

ADDITIONAL WORK WITH SIMPLIFYING RADICALS

BP-69. Based on your conjectures about unlike square roots (the results of BP-58), quickly determine which of the following statements are true and which are false.

a) $\sqrt{25} + \sqrt{9} = \sqrt{34}$

b) $\sqrt{8} - \sqrt{3} = \sqrt{5}$

c) $(\sqrt{1.5})(\sqrt{1.5}) = \sqrt{2.25}$

d) $\sqrt{9} + \sqrt{8} = \sqrt{17}$

e) $\sqrt{100} - \sqrt{36} = \sqrt{64}$

f) $(\sqrt{5})(\sqrt{3}) = \sqrt{15}$

g) $\sqrt{4^2 + 5^2} = \sqrt{9^2}$

h) $\dfrac{\sqrt{90}}{\sqrt{3}} = \sqrt{30}$

i) $(\sqrt{3704})^2 = 3704$

j) $\sqrt{512^2} = 512$

k) $\sqrt{5^2 \cdot 2} = 5\sqrt{2}$

l) $\sqrt{3 \cdot 3 \cdot 7} = 3\sqrt{7}$

m) Check your answers using a calculator. Be sure you follow the correct Order of Operations.

n) For any of your answers that disagree with your check in part (m), review your work and compare it to your conjectures.

BP-70. Add this information to your tool kit.

SIMPLIFYING SQUARE ROOTS

Before calculators were universally available, people who wanted to use approximate decimal values for numbers like $\sqrt{45}$ had other techniques available to them:

1. Carry around long square root tables.

2. Use Guess and Check repeatedly to get desired accuracy.

3. "Simplify" the square roots. A square root is **SIMPLIFIED** when there are no more perfect square factors (square numbers such as 4, 25, and 81) under the radical sign.

Simplifying square roots was by far the fastest method. People factored the number as the product of integers hoping to find at least one perfect square number. They memorized approximations of the square roots of the integers from one to ten. Then they could figure out the decimal value by multiplying these memorized facts with the roots of the square numbers. Here is an example of this method.

Example: Simplify $\sqrt{45}$

Rewrite $\sqrt{45}$ in an equivalent factored form.
Factor 45 so that one of the factors is a perfect square.

$$\sqrt{45} = \sqrt{9 \cdot 5}$$
$$= \sqrt{9} \cdot \sqrt{5}$$

Simplify the square root of the perfect square.

$$= 3\sqrt{5}$$

On your calculator compute $3\sqrt{5}$ and $\sqrt{45}$. Show they are equal.

Here are two more examples:

$$\sqrt{27} = \sqrt{9}\sqrt{3} = 3\sqrt{3}$$

$$\sqrt{72} = \sqrt{36}\sqrt{2} = 6\sqrt{2}$$

Note: We chose to write $\sqrt{72}$ as $\sqrt{36} \cdot \sqrt{2}$, rather than $\sqrt{9} \cdot \sqrt{8}$ or $\sqrt{4} \cdot \sqrt{18}$, because 36 is the largest perfect square factor of 72. However, since

$$\sqrt{4} \cdot \sqrt{18} = 2\sqrt{9 \cdot 2} = 2\sqrt{9} \cdot \sqrt{2} = 2 \cdot 3\sqrt{2} = 6\sqrt{2},$$

we can still get the same answer if we do it using different subproblems.

We live in the age of technology. When we want a decimal approximation of a square root we use a calculator. An exact answer uses the $\sqrt{}$ symbol. We showed you the method of "simplifying" square roots because this simplified form may be useful in some situations.

The Birthday Party Piñata: Using Diagrams to Write Equations

BP-71. Consider each factored choice below, then write the square root in simplified form.

a) To simplify $\sqrt{50}$, would you factor it into $\sqrt{5} \cdot \sqrt{10}$ or $\sqrt{25} \cdot \sqrt{2}$? Why?

b) To simplify $\sqrt{60}$ would you factor it into $\sqrt{6} \cdot \sqrt{10}$ or $\sqrt{4} \cdot \sqrt{15}$? Why?

c) To simplify $\sqrt{32}$ would you factor it into $\sqrt{4} \cdot \sqrt{8}$ or $\sqrt{16} \cdot \sqrt{2}$? Why?

BP-72. Write each of the following square roots in simplest form.

a) $\sqrt{75}$

b) $\sqrt{18}$

c) $\sqrt{48}$

d) $\sqrt{24}$

e) $\sqrt{250}$

f) $\sqrt{1000}$

BP-73. Match the number in the left-hand column with its equivalent in the right-hand column. Whenever possible do this without using a calculator.

a) $\sqrt{96}$ $2\sqrt{15}$

b) $\sqrt{12}$ $3\sqrt{2}$

c) $\sqrt{60}$ $3\sqrt{5}$

d) $\sqrt{72}$ $3\sqrt{6}$

e) $\sqrt{24}$ $5\sqrt{2}$

f) $\sqrt{54}$ $2\sqrt{3}$

g) $\sqrt{50}$ $4\sqrt{6}$

h) $\sqrt{45}$ $2\sqrt{5}$

i) $\sqrt{20}$ $2\sqrt{6}$

j) $\sqrt{18}$ $6\sqrt{2}$

BP-74. Draw a sketch of two line segments which have the same slope but which are not part of the same line. What one word best describes the relationship of the two line segments?

BP-75. Find the equation of the lines graphed below:

a)
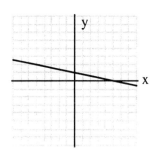

b)
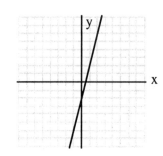

BP-76. **GRAPHING SQUARE ROOTS**

Use your calculator to graph $y = \sqrt{2x}$ on your resource page from BP-15.

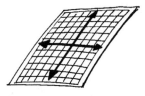

a) What happens if you use $x = -3$ in the square root?

b) What values are possible for x?

BP-77. Use your calculator to graph $y = \sqrt{x} + 2$ on your resource page.

a) What happens if you use $x = -1$ in part (a)?

b) What values are possible for x?

c) Compare this graph to that from BP-15. What is the effect of adding a number to a $\sqrt{x}$ graph?

d) How are the equation and the graph different from BP-44?

BP-78. Factor each of the following expressions. (Hint: Look for the greatest common factors.)

a) $4x + 8$

b) $4x^2 + 8x$

c) $4x^2 + 20x + 24$

BP-79. Write the greatest common factor of the terms in each of the following polynomials, then factor.

a) $2x^2 + 2x$

b) $27x^2 - 3$

c) $12x^3y - 3xy^3$

d) $5x^2y - 30xy + 45y$

BP-80. Solve each of the following equations.

a) $\frac{x}{3} = \frac{4}{5}$

b) $\frac{x}{x+1} = \frac{5}{7}$

c) $\frac{6}{15} = 2 - \frac{x}{5}$

d) $\frac{2}{3} + \frac{x}{5} = 6$

BP-81. Simplify the following:

a) $\sqrt{24}$

b) $\sqrt{120}$

c) $\sqrt{300}$

d) $3\sqrt{18}$

e) $6\sqrt{40}$

f) $-3\sqrt{12}$

The Birthday Party Piñata: Using Diagrams to Write Equations

BP-82. What is the longest stick that can be placed corner to corner inside a shoe box that has dimensions 24" by 30" by 18"?

FINDING THE EQUATION OF A LINE GIVEN TWO POINTS

BP-83. Add this information to your tool kit.

HOW TO FIND THE EQUATION OF A LINE WITHOUT GRAPHING

We can find the equation of the line passing through (3, 4) and (4, 6) by drawing a graph, using the slope triangle, and estimating where the graph crosses the y-axis. But what if the y-intercept were 396 or $1\frac{3}{7}$? In such cases it would be hard to use a graph to find an equation, so we will use the algebra we already know.

Example: Find the equation of the line passing through (3, 4) and (4, 6).

First, write the general equation of a line:

$y = mx + b$
where m is slope, and b is y-intercept

Second, calculate slope by drawing a generic slope triangle, m = 2 so the equation must be $y = 2x + b$.

$m = \frac{2}{1} = 2$

$y = 2x + b$

Third, find the y-intercept, "b," without a graph. Remember, we know that (3, 4) is on the line. This means that this point is a solution of the equation. Substitute (3, 4) for (x, y) and solve for b.

$4 = 2(3) + b$
$4 = 6 + b$
$-2 = b$

Finally, the equation of the line is:

$y = 2x - 2$

a) In the third step we used (x, y) = (3, 4). See what happens if we use the other given point, (x, y) = (4, 6). Explain your results in a sentence.

b) Check your equation by drawing a line through the two points (3, 4) and (4, 6) and marking the y-intercept. Do the slope and intercept numbers in the equation actually fit the line?

BP-84. Use the method described in BP-83 to find the equation of the line passing through each pair of points. In each case write the equation in the slope-intercept (y = mx + b) form. Show all your steps.

a) (4, 6) and (6, 7)

c) (-3, 2) and (4, 5)

b) (9, 8) and (3, 5)

BP-85. A certain line with slope $\frac{1}{2}$ goes through the point (6, 1).

  a) Find the equation of the line in the form of y = mx + b.

  b) Suppose the line y = 2x + b goes through the point (1, 4). Find the equation of the line.

BP-86. Explain to a student who was absent how you would find an equation of the line through two given points.

BP-87. Start with triangle ABC, where A is the point (-1, 5), B is (2, 7), and C is (3, 4). Move (slide) △ABC five units down and six units to the left. (You can do this by sliding each of the vertices A, B, and C.) Call the new triangle △A'B'C'. Find the coordinates of vertices A', B' and C', then determine the lengths of segments AA', BB', and CC'. What do you notice?

BP-88. Solve each of the following equations for x.

  a) $\frac{x-3}{5} = 12(x-1)$ b) $\frac{10}{x} + \frac{10}{2x} = 10$ c) $x^2 - 12x + 35 = 0$

BP-89. Solve for x. Show your work.

The Birthday Party Piñata: Using Diagrams to Write Equations

BP-90. Add an example of F.O.I.L. to your tool kit.

MULTIPLYING BINOMIALS: A HISTORICAL PERSPECTIVE

In multiplying binomials, such as $(3x - 2)(4x + 5)$, you might use a generic rectangle.

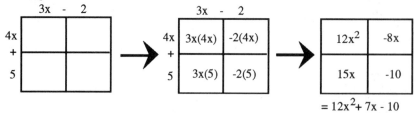

You might view multiplying binomials with generic rectangles as a form of **double distribution**. The $4x$ is distributed across the first row of the generic rectangle and then the 5 is distributed across the second row of the generic rectangle.

Another approach to multiplying binomials is to use the mnemonic 'F.O.I.L.' F.O.I.L. is an acronym for First, Outside, Inside, Last:

F.	multiply the FIRST terms of each binomial	$(3x)(4x) = 12x^2$
O.	multiply the OUTSIDE terms	$(3x)(5) = 15x$
I.	multiply the INSIDE terms	$(-2)(4x) = -8x$
L.	multiply the LAST terms of each binomial	$(-2)(5) = -10$

Finally, we combine like terms: $12x^2 + 15x - 8x - 10 = 12x^2 + 7x - 10$.

Notice how the generic rectangle relates to the F.O.I.L. method:

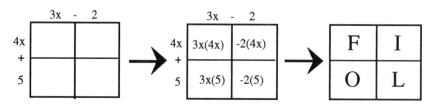

Use the approach to multiply binomials that makes the most sense to you.

BP-91. Multiply these binomials. Check your work.

a) $(7x - 4)(3x + 2)$

b) $(9x + 7)(4x - 3)$

c) $(2x - 5)(3x - 10)$

d) $(5x - 4)(3x - 2)$

e) $(x + 4)(x^2 + 5x + 7)$

f) Will F.O.I.L. work in problem (e)?

BP-92. John Bluecloud builds four square pens (each with equal area) side by side in a row. He uses 83.2 feet of fencing material.

a) What are the dimensions of each pen?

b) Extension: John figures he can make each pen larger if he places the squares in a square pattern instead of in a row. Is he right? Justify your answer.

BP-93.  The top of Susan's telephone table is a semicircle (a half-circle) and is shown at left. Its diameter, against the wall, is 2.5 feet. Find the area of its top surface.

BP-94. Does $\frac{x-4}{x+8} = -\frac{1}{2}$? Substitute a value for x to support your answer.

BP-95. THE BIRTHDAY PARTY PIÑATA

For Katja's party, her friends are going to hang a piñata half way between two poles with a rope over the top of each pole. Both poles are 20 feet high, and they are 30 feet apart. The top of the piñata must hang four feet above the ground.

a) How much rope is needed to hang the piñata? Draw a diagram and write an equation to solve this problem. Show all your subproblems.

b) Katja's mom forgot to bring the rope. Bob found some rope in the trunk of his car. His rope is only 40 feet long, however the poles can be moved closer together. How far apart should the poles be placed so that the piñata remains four feet above the ground?

c) Scott is much taller than the other children so it was decided that the piñata should be raised to six feet above the ground during his turn. Using Bob's 40 foot rope, how far apart should the poles be for Scott's turn?

The Birthday Party Piñata: Using Diagrams to Write Equations

BP-96. In this unit, you have been working with numbers in square root form. It is time to consolidate what you know. Compare your tool kit entries from BP-58 and subsequent explorations to the general laws below. Revise them as necessary. Be sure that for each law you have recorded the law and a numerical example in your tool kit.

LAWS FOR SQUARE ROOTS FOR POSITIVE NUMBERS

	Numerical Example	In General
1)	$\sqrt{3} \cdot \sqrt{7} = \sqrt{21}$	$\sqrt{x} \cdot \sqrt{y} = \sqrt{x \cdot y}$
2)	$\sqrt{20} = \sqrt{4}\sqrt{5} = 2\sqrt{5}$	$\sqrt{x \cdot y} = \sqrt{x} \cdot \sqrt{y}$
3)	$\dfrac{\sqrt{18}}{\sqrt{6}} = \sqrt{\dfrac{18}{6}} = \sqrt{3}$	$\dfrac{\sqrt{x}}{\sqrt{y}} = \sqrt{\dfrac{x}{y}}$
4)	$\sqrt{3^2} = (\sqrt{3})^2 = 3$	$\sqrt{x^2} = (\sqrt{x})^2 = x$
5)	$3\sqrt{7} + 8\sqrt{7} = 11\sqrt{7}$ $9\sqrt{11} - 1\sqrt{11} = 8\sqrt{11}$	$a\sqrt{x} + b\sqrt{x} = (a+b)\sqrt{x}$

BP-97. Does $3x^2 = (3x)^2$? Think about what is being squared in each expression. Draw a picture to justify your answer.

BP-98. Solve for x.

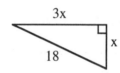

BP-99. Solve each system of equations for x and y:

a) $y = 0$
   $y = x^2 + 5x + 4$

b) $y = \dfrac{-2}{3}x + 4$
   $\dfrac{1}{3}x - y = 2$

BP-100. **GRAPHING SQUARE ROOTS**

For the graph $y = \sqrt{x - 4}$, what is its domain (the possible x-values)? Examine the other graphs on your resource page from problem BP-15. Predict its general shape and location. You do not need to graph it unless you want to verify your answer.

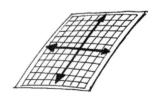

BP-101. Start with the generic point (x, y). Imagine moving (sliding) the point 4.5 units up and 6 units to the left.

a) What is the slope of the line through the two points?

b) What are the coordinates of the point after the move?

c) Draw a diagram. What is the length of the segment that joins the original point and its image after the move?

BP-102. A young redwood tree in Muir Woods casts a shadow 12 feet long. At the same time, a five-foot tall tourist casts a shadow two feet long. How tall is the tree? Draw a diagram, use ratios to write an equation, and then solve the equation.

BP-103. Solve each equation by factoring.

a) $x^2 + 18x - 19 = 0$

b) $x^2 - 4x = 5$

BP-104. Suppose the line $y = \frac{1}{2}x + b$ goes through the point (2, 3).

a) Find the equation of the line.

b) Another line with slope $\frac{2}{3}$ goes through the point (-3, 5). Find the equation of the line.

BP-105. Solve each of the following equations for x.

a) $x^2 - 140 = 1$

c) $1\frac{5}{6x} = \frac{x}{6}$

b) $x(x + 1)(x + 2) = 0$

d) $2x + 3y = 5$

The Birthday Party Piñata: Using Diagrams to Write Equations

UNIT SUMMARY AND REVIEW

BP-106. UNIT 9 SUMMARY

Select 3 or 4 big ideas that you feel best represent Unit 9. Write a description of each main idea. For each topic, choose a problem that best demonstrates your understanding. Explain your solution as in the previous unit summaries. Show all your work.

BP-107. GRAPHING SQUARE ROOTS

Using your Square Root Graph Resource Page from problem BP-15, write several observations about square root graphs and their equations. On your own graph paper, sketch the curve $y = \sqrt{x} + 1$ using your observations.

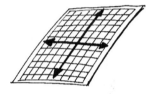

BP-108. Determine whether each of the following statements is true or false. Show why or why not.

a) $\sqrt{12} = \sqrt{4} \cdot \sqrt{3}$

b) $3\sqrt{3} = \sqrt{27}$

c) $\sqrt{48} = 4\sqrt{3}$

d) $\dfrac{\sqrt{10}}{2} = \sqrt{5}$

e) $\sqrt{2} + \sqrt{3} = \sqrt{5}$

f) $\sqrt{100} - \sqrt{64} = \sqrt{36}$

g) $\sqrt{3} \cdot \sqrt{3} = 3$

h) $\sqrt{7} \cdot \sqrt{7} = \sqrt{49}$

i) $\sqrt{5} \cdot \sqrt{5} = (\sqrt{5})^2$

BP-109. Find the length of the diagonal of a square whose sides have length 5 inches. Sketch a diagram and solve.

BP-110. Janis is going to fence off a rectangular garden. She will use an existing wall along the back, and she wants the length to be twice as long as the width. The total amount of fencing material she has is 84 meters long. What are the width and length of her garden? (Be sure to draw a diagram and write an equation.)

BP-111. The length of a certain rectangle is 6.3 feet longer than the width. The perimeter is 32.6 feet. Find the length and width.

BP-112. Find the equation of each line described below. Show all of your subproblems and find the equation of:

   a) the line through points (-1, 4) and (2, 1).

   b) the line through points (6, 3) and (5, 5).

   c) the line with slope $\frac{1}{3}$ through the point (0, 5).

   d) the line parallel to y = 2x - 5 through the point (1, 7).

BP-113. Solve each of the following equations for x.

   a) $\frac{x}{2} + \frac{x}{3} = 5$

   b) $\frac{5}{x} - 8 = 12$

   c) $x^2 + 6x - 7 = 0$

   d) $\frac{5+x}{7} + \frac{2x}{3} = 5$

   e) 2x + y = 29

BP-114. Solve each system for x and y:

   a) y = 9 - x
       y = 9 + x

   b) 3x - 2y = 20
       x = 5 - y

BP-115. A circle has an area of 100 square inches.

   a) What is its radius?

   b) If another circle has a circumference of 100 inches, what is its radius?

BP-116. Factor.

   a) $x^2 + 46x + 45$

   b) $x^2 - 18x + 45$

   c) $x^2 - 4x - 45$

   d) $x^2 + 12x - 45$

BP-117. Put these lists of numbers in order from smallest to largest.

   a) $3\sqrt{2}, 4, 7.5, 2\sqrt{3}$

   b) $5\sqrt{2}, 2\sqrt{5}, 6.9, 6.12$

BP-118. Find at least ten points that are 10 units away from the origin and have integer coordinates. (Hint: the point (6, 8) works.)

The Birthday Party Piñata: Using Diagrams to Write Equations

BP-119. Use a graph to find the coordinates of the midpoint of each pair of points.

a) (6, 2) and (10, 12)

b) (-4, 3) and (5, -6)

c) (1, 7) and (-7, -5)

d) (3, 9) and (7, -2)

e) Try this one without a graph: (2, 24) and (-124, 135)

BP-120. TOOL KIT CLEAN-UP

Tool kits often need to be reorganized to continue to be useful. Your tool kit spans entries from three different units: 7, 8, and 9, plus first semester.

a) Examine the list of tool kit entries from this unit. Check to be sure you have all of these entries. Add any you are missing.

b) Identify which concepts you feel you understand.

c) Which concepts are still not clear to you?

d) Choose entries to create a Unit 7 - 9 tool kit that is shorter, clear, and useful. You may want to consolidate or shorten some entries.

e) How have you used your tool kit in the last two weeks?

- Hypotenuse and Legs of a Right Triangle
- Pythagorean Theorem
- Fraction Busters
- Operations with Square Roots
- Find Equation of a Line Using Two Points
- Simplifying Square Roots
- F.O.I.L.
- Laws for Square Roots for Positive Integers

GM-17. HAPPY NUMBERS

Some numbers have special qualities that earn them a title, such as "Square Number" or "Prime Number." This problem will explore another type of number, called "Happy Numbers."

The number 23 is a Happy Number. To determine if a number is a Happy Number, square each of its digits and add.

$2^2 + 3^2 = 13$

Repeat this process.

$1^2 + 3^2 = 10$

When the final answer is 1, the original number is called a **"Happy Number."**

$1^2 + 0^2 = \mathbf{1}$

The number 34 is not Happy Number, as demonstrated below:

1) $3^2 + 4^2 = 25$
2) $2^2 + 5^2 = 29$
3) $2^2 + 9^2 = 85$
4) $8^2 + 5^2 = \mathbf{89}$
5) $8^2 + 9^2 = 145$
6) $1^2 + 4^2 + 5^2 = 42$
7) $4^2 + 2^2 = 20$
8) $2^2 + 0^2 = 4$
9) $0^2 + 4^2 = 16$
10) $1^2 + 6^2 = 37$
11) $3^2 + 7^2 = 58$
12) $5^2 + 8^2 = \mathbf{89}$

*Since 89 is repeated in this series, the "Happy Number" process is in a never-ending loop and, consequently, will never equal 1. Therefore, 34 is **not** a Happy Number.*

Your Task:

- There are 17 two digit "Happy Numbers." Find as many as you can. Describe your technique for finding happy numbers.

- Remember to keep all your work and ideas so you can refer back to them when writing up what you discovered. It will save you time and help you look for patterns if you keep an organized record of what you try.

- Find five 3 digit happy numbers.

- Find five 4 digit happy numbers.

GM-18. THE STREETS OF SAN FRANCISCO

Ms. Speedi lives at the corner of Chestnut and Mason and drives to school, which is located at the corner of Jackson and Grant, every morning. She usually drives down Mason, then turns left on Jackson. However, after going 12 blocks, she's late for school! See if you can find a shorter route.

The streets in downtown San Francisco are set up in a grid with Columbus Avenue running diagonally between them, as shown on the map at right. Columbus directly meets the intersection of Chestnut and Taylor, as well as the intersection of Washington and Montgomery.

One-way streets are shown with arrows. Kearny is unusual as it only allows traffic that heads toward Columbus.

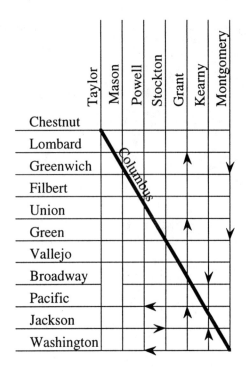

Columbus is a two-way street. You can turn on or off of Columbus from any street that intersects it.

Your Task:

- Help Ms. Speedi find the <u>shortest</u> route from home to school. Be sure to check out alternative routes!
- Find the shortest route for Ms. Speedi to take home after school. Is this route shorter or faster than her route to school?

Unit 10
Yearbook Sales
Exponents and Quadratic Equations

Unit 10 Objectives
Yearbook Sales: EXPONENTS AND QUADRATICS

In this unit you will further develop your factoring skills to include work with more complicated expressions. These factoring skills will enable you to simplify algebraic fractions. You will also work with **exponents** and learn another technique for solving **quadratic equations**.

In this unit you will have the opportunity to:

- extend your ability to factor trinomials to cases where the coefficient of x^2 is not 1.

- explore exponents to develop basic procedures for working with positive, negative and zero exponents.

- simplify elementary rational expressions in preparation for more complicated cases in Units 11 and 12.

- learn how to solve quadratic equations using the Quadratic Formula.

This unit builds on several fundamental ideas to solve more complicated problems. Particular attention is given to quadratic equations. The patterns discovered when working with exponents are extended to simplifying rational expressions.

Read the following problem but **do not try to solve it now**. In this unit, you will learn the skills needed to solve it.

YS-0. YEARBOOK SALES

Last year, the yearbook at Central High School cost $50 and only 500 books were sold. It cost so much because so few were sold and so few were sold because it cost so much! The student body officers want more people to be able to buy yearbooks this year. A student survey discovered that, on the average, for every $5 reduction in price, 100 more students will buy yearbooks. The yearbook company has also raised the minimum order per school to $27,500. What price should the school charge so that more students will buy the yearbook <u>and</u> the school can achieve the minimum sales amount?

Problem Solving
Graphing
Writing and Solving Equations
Ratios
Geometry
Symbol Manipulation

Unit 10
Yearbook Sales: **EXPONENTS AND QUADRATICS**

FACTORING QUADRATICS: ADDITIONAL CASES

YS-1. Add this information to your tool kit.

EXTENDING FACTORING

In earlier units we used Diamond Problems to help factor sums like $x^2 + 6x + 8$.

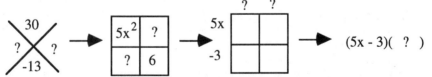

We can modify the diamond method slightly to factor problems that are a little different in that they no longer have a "1" in front of the x^2. For example, factor:

$2x^2 + 7x + 3$

Try this problem: $5x^2 - 13x + 6$.

YS-2. Factor each of the following quadratics using the modified diamond procedure.

a) $3x^2 + 7x + 2$

b) $3x^2 + x - 2$

c) $2x^2 - 3x - 5$

d) $x^2 - 4x - 45$

e) $5x^2 + 13x + 6$

YS-3. Remember that some quadratics have a common factor which requires that you factor twice. Factor each of the following quadratics completely.

EXAMPLE: $2x^2 + 10x + 12 = 2(x^2 + 5x + 6) = 2(x + 2)(x + 3)$

a) $3x^2 - 9x - 30$

b) $5x^2 - 20$

c) $4x^2 + 4$

d) $3x^2 + 11x + 6$

YS-4. Use the Zero Product Property to solve each quadratic equation.

a) $(x + 2)(x - 7) = 0$

b) $(2x + 1)(x - 5) = 0$

c) $x^2 + 3x - 10 = 0$

d) $y^2 - 7y + 10 = 0$

YS-5. Factor each of these quadratics.

a) $15x^2 - 37x + 18$

b) $42x^2 + 17x - 15$

YS-6. Examine the information provided in the graph at right. Find y. Show your work. Note: this graph is not to scale.

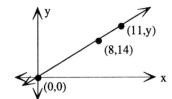

YS-7. Look at the graph in the previous problem.

a) What is the distance from (0, 0) to (8, 14)?

b) What is the slope of the line?

c) What is the equation of the line?

YS-8. Rewrite x^3 and $\dfrac{x^3}{x^2}$ as repeated products of x (without exponents.)

YS-9. For each triangle, solve for x. Write your answer in square root form, simplified square root form, and the decimal approximation.

a)

b)

YS-10. Mr. Nguyen is dividing $775 among his three daughters. If the oldest gets twice as much as the youngest, and the middle daughter gets $35 more than the youngest, how much does each girl get? Write an equation and solve.

YS-11. Multiply each of the following binomials. Use a generic rectangle, if necessary.

    a) $(x + 1)(x + 1)$

    b) $(x + 5)(x + 2)$

    c) $2x(x + 5)$

    d) $(2x + 1)(x + 5)$

    e) $(x + y)(x + y)$

YS-12. Factor each of the following quadratics, if possible.

    a) $x^2 + 3x - 10$

    b) $y^2 - 7y + 10$

    c) $x^2 + 3x + 10$

    d) $2y^2 + 11y - 21$

YS-13. Look for and then apply the different types of factoring that you have encountered this year as you factor each of the following. Decide with your study team how to best check your answer.

    a) $x^2 + 10x + 21$

    b) $3x^2 + 7x + 2$

    c) $2x^2 + 7x - 15$

    d) $x^2 - 64$

    e) $6y^2 - 2y - 48$

    f) $m^2 - 14m + 49$

YS-14. Solve each of the following equations or systems by any method you choose. Show all your work.

    a) $x^2 + 10x + 21 = 0$

    b) $\frac{x+1}{3} = \frac{x}{2}$

    c) $2x^2 + 7x - 15 = 0$

    d) $\frac{x}{4} + \frac{5x}{3} = 1 + x$

    e) $x - 2y - 1$
           $y - x = -5$

    f) $x^2 + 5x = 24$

    g) $2(x + 1) - 4(x - 2) = x + 3$

    h) $y = -x - 2$
           $y = \frac{1}{2}x + 4$

YS-15. Graph these two equations on the same set of axes and estimate where they intersect. Use $-2 \leq x \leq 2$ in increments of 0.5 (e.g., -2, -1.5, ..., 1.5, 2).

$$y = x^3 - x + 2 \quad \text{and} \quad y = x + 2$$

YS-16. The graph below shows 3 lines with their equations. Use subproblems and systems of equations to answer the following questions.

a) Where do the lines intersect?

b) Find the area of the triangular region formed by the three equations.

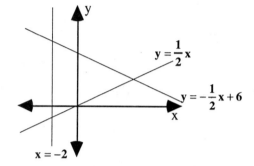

YS-17. Factor each of the following quadratics by looking for common factors.

a) $x(x + 2) + 3(x + 2)$

b) $2y(y - 3) + 5(y - 3)$

c) $(x + 4)x^2 + 3(x + 4)$

d) $(y + 1)^2 + (y + 1)$

YS-18. Use the slope and y-intercept to graph each of the following equations.

a) $y = -2x + 4$

b) $y = \frac{2}{3}x - 2$

YS-19. Use the points $(2, -1)$ and $(1, 2)$.

a) Draw a generic triangle to determine the slope. Is the slope negative or positive?

b) Find the equation of the line.

c) Write the equation of a line that is parallel to the equation in part (b).

d) Write the equation of a line that would intersect the line in part (b).

YS-20. Use the diagram at right to find the lengths of the triangle's legs and its area.

a) Using x, write an expression for the area of the triangle.

b) Find x to the nearest hundredth.

c) Find the area of the triangle.

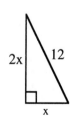

YS-21. Simplify each of the following fractions:

a) $\dfrac{5}{5}$

b) $\dfrac{-2.3}{-2.3}$

c) $\dfrac{y}{y}$

d) $\dfrac{x^3}{x^3}$

e) $\dfrac{6}{18x}$

f) $\dfrac{24x}{100}$

YS-22. Solve this system for (x, y): $y = 5x - 3$
$3y - 10x = -12$

YS-23. Complete the following.

a) Factor $3x^2 - 5x - 2$

b) Solve $3x^2 - 5x - 2 = 0$

c) What do your answers in part (b) tell you about the graph of $y = 3x^2 - 5x - 2$?

YS-24. Divide. Write your answer in scientific and standard notation.

a) $\dfrac{4.2 \cdot 10^5}{3.0 \cdot 10^2}$

b) $\dfrac{3.0 \cdot 10^6}{1.5 \cdot 10^4}$

PROPERTIES OF EXPONENTS

YS-25. Write each of these expressions as simply as possible using the method shown below.

> Knowing that $x^3 = x \cdot x \cdot x$ and $\dfrac{y}{y} = 1$, then:
> $x(x^3) = x \cdot (x \cdot x \cdot x) = x^4$ and
> $y^3 \div y = \dfrac{y \cdot y \cdot y}{y} = y^2$

a) $x^2 \cdot x$

b) $y^2 \cdot y^5$

c) $x^3 \cdot x^6$

d) $x^3 \div x^2$

Yearbook Sales: Exponents and Quadratics

YS-26. Most of the work you have done this semester has been with x or x^2. During the next few lessons you will use patterns of multiplication and division to work with other powers of x. Record the following definitions in your tool kit:

BASE, EXPONENT, AND VALUE

In the expression 2^5, 2 is the **base**, 5 is the **exponent**, and the **value** is 32.

2^5 means $2 \cdot 2 \cdot 2 \cdot 2 \cdot 2 = 32$
x^3 means $x \cdot x \cdot x$

YS-27. Use exponents to write each of the following expressions as simply as possible. Look for patterns as you do this with your study team. Write out the variables to show the meaning whenever necessary.

a) $(x^2)(x^5)$
b) $x^7 \cdot x^5$
c) $y^8 \cdot y^6$
d) $y^7 \div y^4$
e) $\dfrac{x^3}{x^1}$

f) $x^3 \cdot x^4$
g) $m^{13} \cdot m^{14}$
h) $x^{32} \cdot x^{59}$
i) $x^{31} \div x^{29}$
j) $\dfrac{x^3}{x^3}$

YS-28. Write out the meaning of each expression below and then simplify using exponents.

EXAMPLE: $(x^3)^2 = x^3 \cdot x^3 = (x \cdot x \cdot x)(x \cdot x \cdot x) = x^6$

a) $(x^4)^2$
b) $(y^2)^3$
c) $(x^5)^5$

d) $(x \cdot y)^2$
e) $(x^2 \cdot y^3)^3$
f) $(2x)^4$

YS-29. Using the patterns for exponents that you have found, write each expression below as simply as possible using exponents.

a) $x^3 \cdot x^4$
b) $(x^3)^4$
c) $x^4 \div x^3$

d) $(xy^2)^2$
e) $(2x^2)^3$
f) $(x^2y^2)^4$

YS-30. Use your calculator to write these exponential numbers as a decimal and as a fraction.

a) 10^{-1} b) 10^0 c) 5^{-1} d) 5^{-2}

e) What effect does a negative sign have when it appears in an exponent? Was this what you expected?

f) What effect does zero have when it appears as an exponent?

YS-31. Use the modified Diamond method from YS-1 to factor.

a) $6x^2 - 5x + 1$ b) $4x^2 + 4x + 1$

YS-32. Factor each of the following

a) $x^2 - 6x + 9$ d) $x^2 - 25$

b) $5x^2 + 4x - 1$ e) $9x^2 - 25$

c) $9x^2 - 18x$ f) $2x^2 - 7x + 3$

YS-33. Look for a relationship between the two equations in parts (a) and (b).

a) Solve the equation $3x^2 + 4x + 1 = 0$ by factoring.

b) What are the x-intercepts for the quadratic equation $y = 3x^2 + 4x + 1$?

c) How are (a) and (b) related?

YS-34. Solve for x: $\frac{4}{x} + \frac{3}{5} = \frac{7}{10}$. Look in your tool kit for Fraction Busters if you need help.

YS-35. The two triangles in the diagram below are similar. Find x.

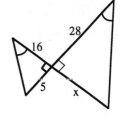

YS-36. Find the missing side lengths in each triangle. Write each length in simplified square root and decimal forms.

a)   b)   c)

Yearbook Sales: Exponents and Quadratics

YS-37. Write each expression as simply as possible using exponents.

a) $x^5 \cdot x^4$

b) $(x^2)^4$

c) $\dfrac{x^4}{x^1}$

d) $(2x)^3$

ZERO AND NEGATIVE EXPONENTS

YS-38. With your study team, summarize the patterns that you found in yesterday's exponent problems. Describe how to write the examples below as simply as possible.

a) $(x^2)(x^3)$ What patterns did you find? Use the words <u>base</u> and <u>exponent</u>.

b) $\dfrac{x^6}{x^3}$ What patterns did you find? Use the words <u>base</u> and <u>exponent</u>.

c) $(x^3)^4$ What patterns did you find? Use the words <u>base</u> and <u>exponent</u>.

YS-39. Use the patterns that you described in the previous problems to write each expression as simply as possible.

a) $x^7 \cdot x^4$

b) $(x^3)^3$

c) $x^6 \div x^3$

d) $8^6 \div 8^3$

e) $(2x^2)^3$

f) $(x^2 y^2)^4$

g) $\dfrac{x^2 y^{11}}{x^5 y^3}$

h) $\dfrac{2x^{12}}{3x^2}$

YS-40. Add this information to your tool kit.

LAWS OF EXPONENTS

The patterns that you have been working with are called the **LAWS OF EXPONENTS**. Here are the basic patterns with examples:

1) $x^a \cdot x^b = x^{a+b}$ examples: $x^3 \cdot x^4 = x^{3+4} = x^7$; $2^7 \cdot 2^4 = 2^{11}$

2) $\dfrac{x^a}{x^b} = x^{a-b}$ examples: $x^{10} \div x^4 = x^{10-4} = x^6$; $\dfrac{2^4}{2^7} = 2^{-3}$

3) $(x^a)^b = x^{ab}$ examples: $(x^4)^3 = x^{4 \cdot 3} = x^{12}$; $(2x^3)^4 = 2^4 \cdot x^{12} = 16x^{12}$

YS-41. How should we simplify x^0?

a) What does your calculator display for 2^0? 17^0? 1000^0?

b) The following should help convince you why it is reasonable that $x^0 = 1$. From the patterns we found when dividing with exponents, what is "t" in $\frac{x^3}{x^3} = x^t$?

c) From arithmetic we know that $\frac{\text{anything}}{\text{itself}} = ?$ Therefore, $x^0 = \underline{}$?

YS-42. How can we write x^{-2} without negative exponents?

a) From the patterns we found when dividing with exponents, what is "t" in $\frac{x^3}{x^5} = x^t$?

b) Remembering that $\frac{x}{x} = 1$, write out $\frac{x^3}{x^5}$ without exponents and then simplify the fraction by finding as many "ones" $\left(\frac{x}{x}\right)$ as you can. What is left?

c) Therefore, since $x^{-2} = \frac{1}{x^2}$, what should $x^{-3} = ?$ What should $3^{-4} = ?$

YS-43. Add this information to your tool kit.

ZERO EXPONENTS AND NEGATIVE EXPONENTS

$x^0 = 1$. Examples: $2^0 = 1$, $(-3)^0 = 1$, $(\frac{1}{4})^0 = 1$.

$x^{-n} = \frac{1}{x^n}$. Examples: $x^{-3} = \frac{1}{x^3}$, $y^{-4} = \frac{1}{y^4}$, $4^{-2} = \frac{1}{4^2} = \frac{1}{16}$.

$\frac{1}{x^{-n}} = x^n$. Examples: $\frac{1}{x^{-5}} = x^5$, $\frac{1}{x^{-2}} = x^2$, $\frac{1}{3^{-2}} = 3^2 = 9$

Rewrite each expression without negative or zero exponents.

a) x^{-5} c) m^0 e) 4^{-1}

b) y^{-3} d) 5^{-2} f) 5^0

YS-44. You know that when your calculator displays $\boxed{2.3 \quad 03}$ it means $2.3 \cdot 10^3$ or 2,300. Recall that 2.3×10^3 is called **scientific notation** and 2,300 is called **standard form**. (Refer to problem KF-96.) Notice that the positive exponent of 3 moved the decimal 3 places to the right when the number was written in standard form. If your calculator displays $\boxed{1.5 \quad -03}$, write the equivalent expression in scientific notation and standard form.

YS-45. Use your scientific calculator and type in $1.5 \cdot 10^{-3}$. How did multiplying by 10^{-3} move the decimal point? How does that compare with multiplying by 10^3?

YS-46. Convert the following scientific notation numbers to standard form.

a) $2.75 \cdot 10^{-4}$
b) $4.5 \cdot 10^{-5}$
c) $1.56 \cdot 10^{-4}$
d) $2.5 \cdot 10^6$

YS-47. Write each of the following expressions in simpler exponential form (<u>not</u> the value).

a) $10^3 \cdot 10^4$
b) $10^5 \div 10^3$
c) $(10^3)^4$
d) $x^2 y \cdot x^3 y^2$
e) $3^2 \cdot 3^5$
f) $3^7 \div 3^3$
g) $(2^3)^{-2}$
h) $(x^2 y)^3$
i) $x^3 \cdot x^5$
j) $x^7 \div x^2$
k) $(x^3)^4$
l) $x^2 \cdot (x^3)^2$

YS-48. Write the equation of:

a) the line with slope $\frac{3}{4}$ and passes through (6, 20).
b) the line passes through (-3, 8) and (-1, 5).

YS-49. Factor the following expressions:

a) $x^2 + 17x + 42$
b) $2x^2 - x - 15$
c) $9x^2 - 6x + 1$
d) $9x^2 - 25$

YS-50. The perimeter of the triangle at right is 86 units.

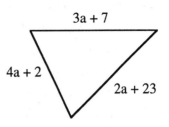

a) Write an equation and find a.

b) Find the lengths of the sides of the triangle.

c) Could this be a right triangle? Justify your answer.

YS-51. Simplify the following expressions.

a) $\dfrac{7659}{7659}$

b) $\dfrac{x}{x}$

c) $\dfrac{x+3}{x+3}$

d) $\dfrac{12x}{3x}$

e) Does $\dfrac{2+3}{2} = 3$?

f) Does $\dfrac{x+3}{x} = 3$?

INTRODUCTION TO RATIONAL EXPRESSIONS

YS-52. Since, for example, $m^3 = m \cdot m \cdot m$, how would you write:

a) x^5 ?

b) $(x+1)^3$?

c) $(x+1)^0$?

YS-53. With your study team, write each of the following expressions in simplest fraction form by looking for factors in fraction form that are equivalent to one.

Example: $\dfrac{15}{6} = \dfrac{5 \cdot 3}{2 \cdot 3} = \dfrac{5}{2} \cdot \dfrac{3}{3} = \dfrac{5}{2} \cdot 1 = \dfrac{5}{2}$

a) $\dfrac{5 \cdot 5 \cdot 5}{5 \cdot 5 \cdot 6}$

b) $\dfrac{10}{10^3}$

c) $\dfrac{8x^3 y^4}{4x^2 y}$

d) $\dfrac{5(x+2)}{2(x+2)}$

e) $\dfrac{(x-1)}{(x-1)^2}$

f) $\dfrac{12(x-2)^2}{3(x-2)}$

g) $\dfrac{6(m+1)^8}{6(m+1)}$

h) In the next several days you will solve problems that will require your factoring, fraction and exponent skills. As you approach these problems, look for the subproblems you need to solve to make these complicated problems easier. Explain in one or two sentences how your team found its solution to part (g).

YS-54. Write each of the following expressions in simplest form.

a) $\dfrac{(x+3)^2}{(x+3)(x-2)}$

b) $\dfrac{8(2x-5)^3}{4(2x-5)^2(x+4)}$

c) $\dfrac{12(x+1)^2(x-2)^3}{6(x+1)^3(x-2)^5}$

d) $\dfrac{x^2+4x-12}{x^2+12x+36}$

(Hint: you must factor first!)

YS-55. Decide whether each of the following statements is correct or not.

a) Does $\dfrac{34}{32} = \dfrac{30+4}{30+2} = 2$?

b) Does $\dfrac{x+7}{x+14} = \dfrac{1}{2}$?

c) Does $\dfrac{x^2}{x}$ always equal 2 ?

YS-56. Factor each of the following expressions. Be sure to look for any common factors.

a) $4x^2 - 12x$

b) $2m^2 + 7m + 3$

c) $3y^2 + 6y + 3$

d) $3x^2 + 4x - 4$

YS-57. For the rectangle at right:

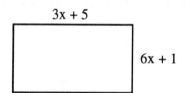

a) write an algebraic expression for this rectangle's area as a sum and as a product.

b) find the area of the rectangle when x = 1 and again when x = 3.

YS-58. Graph the two points (-2, 3) and (2, 1).

a) Determine the slope of the line.

b) Determine the equation of the line.

c) Is the point (30, -17) on the same line? Show how you determined your answer.

d) Find the distance between the two original points.

YS-59. Factor and solve each of the following quadratic equations.

a) $y^2 + 6y - 16 = 0$

b) $2x^2 + 11x - 6 = 0$

c) $5z^2 - 13z = 0$
(Hint: look for a common factor.)

YS-60. Explain your methods in parts (a) and (b) below.

    a) Solve this system for x and y.

$$3x + y = 9$$
$$2x + y = 1$$

    b) Describe another way to solve this problem.

YS-61. Solve for x.

    a) $\sqrt{144} = x$

    b) $x^2 = 144$

    c) How are these problems different?

    d) Compare the graphs of $y = \sqrt{x}$ and then $y = x^2$. You do not need to graph, just describe the comparison in words.

YS-62. A rectangular sign is three times as tall as it is wide. Its area is 588 square centimeters. What are the dimensions of the sign? Write an equation and solve.

SIMPLIFYING RATIONAL EXPRESSIONS

YS-63. Investigate the following problems.

    a) Is $\frac{2+3}{2} = 3$?

    b) Is $\frac{9-3}{3} = 9$?

    c) Is $\frac{2 \cdot 3}{2} = 3$?

    d) Is $\frac{9 \cdot 3}{3} = 9$?

    e) What was the <u>only</u> mathematical operation that allowed us to write the same factor in the numerator and denominator as ONE (i.e. $\frac{1}{1} = 1$)?

    f) Therefore, which of the following is true?

        1) $\frac{x^2 + x + 3}{x + 3} = x^2$

        2) $\frac{(x+2)(x+3)}{(x+3)} = x + 2$

YS-64. Add this information to your tool kit about simplifying rational expressions.

SIMPLIFYING RATIONAL EXPRESSIONS

To simplify rational expressions, both the numerator and denominator must be factored. Then look for factors that make ONE (1).

Example: $\dfrac{x^2+5x+4}{x^2+x-12} = \dfrac{(x+4)(x+1)}{(x+4)(x-3)} = 1 \cdot \dfrac{(x+1)}{(x-3)} = \dfrac{x+1}{x-3}$

Simplify each rational expression. Show all subproblems.

a) $\dfrac{(x+2)(x+2)}{(x-3)(x+2)}$

b) $\dfrac{x^2+5x+6}{x^2+x-6}$

c) $\dfrac{x^2+4x}{2x+8}$

d) $\dfrac{x^2+6x+9}{x^2-9}$

e) $\dfrac{2x^2+x-3}{x^2+4x-5}$

f) $\dfrac{2x^2-x-10}{3x^2+7x+2}$

YS-65. Use what you know about slope and intercepts to graph each line.

a) $y = \dfrac{2}{3}x - 1$

b) $2x + 3y = 6$

c) Describe your process to part (b) above.

YS-66. For the line $2x + 3y = 6$, show how to calculate the coordinates of the x-intercept and the y-intercept. Does the graph in part (b) of the previous problem pass through the points you found?

YS-67. Graph the equation $y = x^2 - 4$ for $-5 \le x \le 5$.

a) Estimate where the graph crosses the x-axis.

b) Solve the equation $x^2 - 4 = 0$ for x by factoring.

c) Write a sentence comparing your solutions to parts (a) and (b). Be sure to note differences, if any, as well as similarities.

YS-68. Solve this system for x and y. Then state precisely where the two lines intersect.

$$x = 3y - 7$$
$$3x - y = 3$$

YS-69. Explain whether each of the following equations is true and how you know.

a) $3^4 + 3^3 = 3^7$

b) $x^4 + x^3 = x^7$

c) $\frac{(x+1)(x+3)}{x+1} = x+3$

d) $\frac{(x+1)+(x+3)}{x+1} = x+3$

YS-70. Use what you know about exponents to write each of the following expressions in a simpler form.

a) $3x^2 \cdot x$

b) $(-2x^2)(-2x)$

c) $\frac{n^{12}}{n^3}$

d) $\frac{-8x^6y^2}{-4xy}$

e) $(x^3)^2$

f) $(2x^3)^3$

YS-71. Copy and complete the table for (x, y), then write the rule.

x	-3	-2	-1	5		1	$\frac{1}{3}$	6
y	-10	-7		14	-1			17

YS-72. Maxine paid $2,379.29 for a computer that was originally priced at $2,500.00. What percent of the original price did Maxine pay? What was the discount percent? Write an equation and solve.

YS-73. Fill in the missing expressions.

a) $\frac{2^3(x^2yw)^2}{(2wx)^4} \cdot \left(\frac{?}{?}\right) = x$

b) $\frac{5^4(xy^2)^3}{(3^2+4^2)y^2} \cdot \left(\frac{?}{?}\right) = x$

Yearbook Sales: Exponents and Quadratics

QUADRATIC EQUATIONS: STANDARD FORM

YS-74. Explore the following quadratic equation.

a) Graph the equation $y = x^2 + 6x + 2$ and estimate the x-intercepts.

b) Substitute the values you found for x in part (a) into the equation $y = x^2 + 6x + 2$ and compute the value.

c) How close to correct were your estimates from part (a)?

d) Is the expression $x^2 + 6x + 2$ factorable? Explain your answer.

e) Are there solutions to the equation $0 = x^2 + 6x + 2$? Explain your answer.

YS-75. Add this information to your tool kit.

STANDARD FORM OF A QUADRATIC EQUATION

Problems such as YS-74 suggest that another method besides the Zero Product Property is needed to solve quadratic equations. There is a formula that accomplishes this task. To use the formula, a quadratic equation must be in what is called **STANDARD FORM**, that is, written as $ax^2 + bx + c$. Then the coefficients **a, b,** and **c** are easily recognized.

EXAMPLE: The quadratic equation $2x^2 - 13x + 21 = 0$ is in **standard form**.

What does a = ? b = ? c = ?.

When a quadratic equation is not in standard form, you first need to convert it to standard form <u>before</u> you can identify a, b, and c.

EXAMPLE: The quadratic equation $3x^2 = 14x - 8$ is <u>not</u> in standard form.

By adding the opposites of 14x and -8, to both sides of our equation, we get:
$$3x^2 - 14x + 8 = 0$$
which <u>is</u> in standard form. The coefficients are a = 3, b = -14, and c = 8.

YS-76. Identify the values of **a, b,** and **c** in each of the following quadratic equations. Some of the equations will need to be written in standard form first.

a) $3x^2 - 5x + 4 = 0$

b) $x^2 + 9x - 1 = 0$

c) $-2x^2 + 9x = 0$

d) $0.017x^2 - 0.4x + 20 = 0$

e) $-6x + 5x^2 = 8$

f) $x(2x + 4) = 7x - 5$

YS-77. a) Find an approximate value of

$$\frac{-2 + \sqrt{(-3)^2 - 4(2)(-3)}}{-3}$$

on your calculator. It will help to list the sub-problems before you use the calculator. Verify your answer before going to part (b).

b) Use a calculator to find the value of each of the following expressions to the nearest hundredth. In parts (c) and (d) the "±" means there are two solutions. One is with addition and one is with subtraction. Find both. Verify your answers with your study team.

1) $8 + \sqrt{8^2 - (4)(5)(3)}$

2) $\frac{-2 + \sqrt{2^2 - 4(3)(-7)}}{2(3)}$

3) $\frac{-2 \pm \sqrt{16}}{2}$

4) $\frac{1 \pm \sqrt{5}}{2}$

YS-78. Solve each equation.

a) $x^2 + 19x + 90 = 0$ b) $2(3x + 1)(x - 7) = 0$ c) $3x^2 + 7x + 2 = 0$

YS-79. Sketch each pair of points and find the distance between them. You may use graph paper or draw generic triangles. Give answers in square root form and the decimal approximation.

a) $(3, -6)$ and $(-2, 5)$

b) $(5, -8)$ and $(-3, 1)$

c) $(0, 5)$ and $(5, 0)$

d) Write the distance you found in part (c) in simplified square root form.

YS-80. Find the area of the shaded region.

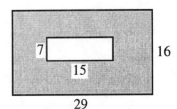

Yearbook Sales: Exponents and Quadratics

YS-81. Use the Laws of Exponents to rewrite the following expressions, if possible.

a) $\dfrac{m^{16}y^{31}}{m^{12}y^{17}}$

b) $(6x^3z)^3$

c) $(5x)^2(3y)^3$

d) $x^2 + y^2$

e) $(3x^2)^2 \div (6x^4)$

f) $(2x-5)^2$

YS-82. Find where the graphs of the lines $y = 2x - 7$ and $y = -x + 5$ intersect. Describe your process and find another way to solve this problem.

YS-83. A right triangle has an area of 40 square centimeters and its shortest side has length 8 centimeters. Draw a diagram and find the length of the hypotenuse.

YS-84. Simplify.

a) $\dfrac{2(x+4)(x-2)}{8(x-1)(x+4)}$

b) $\dfrac{x^2 - 2x - 3}{x^2 - x - 6}$

c) $\dfrac{3x^2 - 2x - 8}{x^2 - 4}$

THE QUADRATIC FORMULA

YS-85. Solve $2x^2 - 9x - 35 = 0$ by factoring and applying the Zero Product Property. How many solutions are there?

YS-86. Problems that do not factor quickly and problems such as YS-74 that do not factor at all create the need for another method of solving quadratic equations. Read the information in the box below. Enter the formula and the example in your tool kit.

USING THE QUADRATIC FORMULA

Once a quadratic equation is in the form $ax^2 + bx + c = 0$, you can use the values for a, b and c to calculate the solutions for the equation; that is, you can find those values of x that make the equation true by using the **QUADRATIC FORMULA**:

$$\text{If } ax^2 + bx + c = 0, \text{ then } x = \frac{-b \pm \sqrt{b^2 - 4ac}}{2a}$$

In words this says:

"The values of x which make the quadratic equation $ax^2 + bx + c = 0$ true are equal to the opposite of **b**, plus or minus the square root of the quantity **b** squared minus the product **4ac**, all divided by the product **2a**."

Remember the $\pm$ is an efficient, shorthand way to write the <u>two</u> solutions to the equation. The formula actually says:

$$x = \frac{-b + \sqrt{b^2 - 4ac}}{2a} \quad \text{or} \quad x = \frac{-b - \sqrt{b^2 - 4ac}}{2a}$$

Unless the quantity $b^2 - 4ac$ is negative or zero, the graph of $ax^2 + bx + c = 0$ crosses the x-axis at <u>two points.</u>

For example, the quadratic equation $x^2 + 5x + 3 = 0$ is in standard form with a = 1, b = 5, and c = 3. When we substitute these values into the Quadratic Formula we get:

$$x = \frac{-5 \pm \sqrt{5^2 - 4(1)(3)}}{2(1)} = \frac{-5 \pm \sqrt{25 - 12}}{2} = \frac{-5 \pm \sqrt{13}}{2},$$

$$\text{which means } x = \frac{-5 + \sqrt{13}}{2} \quad \text{or} \quad x = \frac{-5 - \sqrt{13}}{2}.$$

These square root values of x are the EXACT solutions to the equation $x^2 + 5x + 3 = 0$.

We can use our calculators to find approximate values:

$$x = \frac{-5 + \sqrt{13}}{2} \approx -0.70 \quad \text{OR} \quad x = \frac{-5 - \sqrt{13}}{2} \approx -4.30$$

Here is a summary of the procedure for using the **Quadratic Formula**:

Step 1 Put the equation in standard form (zero on one side of the equation).
Step 2 List the numerical values of the coefficients a, b, and c.
Step 3 Write the Quadratic Formula, <u>even if it is given</u>.
Step 4 Substitute the numerical values for a, b, and c in the Quadratic Formula.
Step 5 Simplify to get the exact solutions.
Step 6 Use a calculator, if necessary, to get approximate solutions

YS-87. Copy and complete the following example in your notebook. Verify the accuracy of each step with your study team.

Use the Quadratic Formula to solve the equation $x^2 - 2x = 4$.

Step 1 $x^2 - 2x - 4 = 0$

Step 2 $a = ? \ b = ? \ c = ?$

Step 3   If $ax^2 + bx + c = 0$, then $x = \dfrac{-b \pm \sqrt{b^2 - 4ac}}{2a}$.

Step 4 $x = \dfrac{-(-2) \pm \sqrt{(-2)^2 - 4(1)(-4)}}{2(1)}$

Step 5 $x = \dfrac{2 \pm \sqrt{4 + 16}}{2} = \dfrac{2 \pm \sqrt{20}}{2}$

Step 6 $x \approx ?$ or $x \approx -1.24$

YS-88. Solve each of the following equations using the six step Quadratic Formula procedure demonstrated in YS-87. Write your answers first in square root form, similar to that in Step 5 of the example, and then in decimal form.

a) $2x^2 - 9x - 35 = 0$ (Compare your answers with YS-85. Did both methods give the same answer?)

b) $3x^2 + 7x = -2$

c) $x^2 - 5x - 2 = 0$

d) $8x^2 + 10x + 3 = 0$

e)   Extension: $(x - 3)(x + 4) = 7x$

YS-89. Use the figure at right to answer the questions below.

a)   Find the equation of line t.

b)   Find the equation of another line parallel to line t.

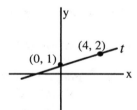

YS-90. The two points (30, -25) and (50, 65) determine a line.

a)   Find the slope of the line.

b)   Find an equation of the line.

YS-91. Find the length of the diagonal of a square with side lengths of 15 inches.

YS-92. Solve the system of equations: $2x + 3y = 5$ and $x - y = 5$. There are two methods that can be used to solve the system: algebraic or graphic. Which method did you choose and why?

YS-93. Use what you know about exponents to write each of the following expressions in a simpler form:

   a) $(x + 3)^2$
   b) $(6x^3y^{-1})^2$
   c) $\dfrac{(x+5)(2x-7)^3}{(x+5)^5(2x-7)^2}$

YS-94. Solve each equation for the variable.

   a) $9x - 5.4 = -2.7x + 42$
   c) $x^2 + 1 = 26$
   b) $8(3x - 5) + 2 = 6$
   d) $\sqrt{x} = 11$

YS-95. Use the Quadratic Formula to solve each of the following equations. Be sure to use all six steps of the procedure. Remember to copy the formula each time you use it. Verify your solutions with your study team.

   a) $-4x^2 + 8x + 3 = 0$
   d) $0.09x^2 - 0.86x + 2 = 0$
   b) $-3x = -x^2 + 14$
   e) $3x^2 + 4x = 0$
   c) $-5 + 11x = 2x^2$
   f) $25x^2 - 49 = 0$
   g) Which of these equations could you have solved by factoring?

YS-96. Look again at the equations in YS-95 and your solutions to them.

   a) List the values of $\sqrt{b^2 - 4ac}$ for each part of problem YS-95.

   b) Find the equations that had rational numbers (integers or fractions) for their solutions. Compare the values of $\sqrt{b^2 - 4ac}$ for these equations to the values of $\sqrt{b^2 - 4ac}$ for the equations whose solutions were not integers or fractions. What do you notice?

   c) Can $3x^2 + 8x + 5 = 0$ be solved by factoring? Factor if possible and solve.

   d) Calculate the value of $\sqrt{b^2 - 4ac}$ for the equation $3x^2 + 8x + 5 = 0$. What kind of number do you get? Write one or two sentences that explains the relation of this result to your conclusion in part (b).

   e) Explain what the value of $\sqrt{b^2 - 4ac}$ tells you about the solutions of the quadratic equation $ax^2 + bx + c = 0$. What does it tell you about the factorability of the polynomial $ax^2 + bx + c$?

Yearbook Sales: Exponents and Quadratics

YS-97. A right triangle has one leg two centimeters longer than the other. The hypotenuse is 17 centimeters long.

    a) Draw a diagram of this right triangle and label the length of each side.

    b) Write an equation to find the lengths of the legs.

    c) Find the length of each leg of the triangle.

YS-98. Use the slope and y-intercept to graph each of the following equations.

    a) $y = 2x - 4$

    b) $y = -0.5x + 2$

    c) Find the point of intersection for the lines in parts (a) and (b).

YS-99. Find an equation of the line:

    a) with slope $\frac{2}{3}$ passing through the point (-6, -1).

    b) passing through the points (60, 400) and (50, 500).

YS-100. Test to see if each of the following quadratic equations can be solved by factoring by applying your conclusion from problem YS-96. Remember to put each equation in standard form first!

    a) $x^2 + 4x - 5 = 0$

    b) $5x = -x^2 + 6$

    c) $x^2 + 7x + 5 = 0$

    d) $5x^2 = 6 + 14x$

YS-101. Use the Laws of Exponents to rewrite each of the following expressions.

    a) $(3x^{11}z^5)^2$

    b) $(2b)^5(3k^2)^2$

    c) $x^{13} \cdot x^{-16} \cdot x^9$

    d) $(3x - 4)^2$

YS-102. Simplify the following rational expressions.

    a) $\dfrac{(x-3)(x+5)}{(x-2)(x-3)}$

    b) $\left(\dfrac{3x^2}{6x^5}\right)^3$

    c) $\dfrac{3x^2 + x - 2}{2x^2 + 7x + 5}$

    d) $(6x)^2 \div (24x^3)$

UNIT SUMMARY AND REVIEW

YS-103. **UNIT 10 SUMMARY**

Select 3 or 4 big ideas that you feel best represent Unit 10. Write a description of each main idea. For each topic, choose a problem that best demonstrates your understanding. Explain your solution as in the previous unit summaries. Show all your work.

YS-104. **YEARBOOK SALES**

Last year, the yearbook at Central High School cost $50 and only 500 books were sold. It cost so much because so few were sold and so few were sold because it cost so much! The student body officers want more people to be able to buy yearbooks this year. A student survey discovered that, on the average, for every $5 reduction in price, 100 more students will buy yearbooks. The yearbook company has also raised the minimum order for any school to $27,500. What price should the school charge this year so that more students will buy the yearbook and the school can achieve the minimum sales amount?

a) Graph this situation. Let x represent the cost of the book and let y represent the number of books sold.

b) Use the points on your line to find an equation relating the number of books sold to cost.

c) The total sales revenue is (cost of one book)(number of books sold). Write this expression in terms of x.

d) Since the minimum school order is $27,500, take your answer to part (c) and make it equal to $27,500 and solve the quadratic equation.

e) What do these answers represent? What is the cost and expected number sales associated with each answer?

f) What price should the school charge and why?

g) Extension: What price should the student body charge to have the greatest amount of revenue under the same conditions?

Yearbook Sales: Exponents and Quadratics

YS-105. Solve each of the following equations either by factoring or by using the Quadratic Formula, whichever method you prefer.

a) $x^2 + 8x + 5 = 0$

b) $2x^2 + 5x + 3 = 0$

c) $x^2 + 7x + 5 = 0$

d) $x^2 = 10 - 3x$

YS-106. Karen Camero drives straight south on Interstate 405 from the Santa Monica Freeway interchange going 10 miles per hour faster than Melinda Mustang, who leaves the same interchange at the same time going straight east on Interstate 10. After one hour they are 108 miles apart, measured by a straight line drawn between them.. If the highways were actually straight, how fast would each driver be going? (Hint: use a solution procedure similar to what you did the day before in problem YS-97.)

YS-107. A certain rectangle has an area of 50 square meters and its length is five more than twice its width. Find the lengths of the sides of the rectangle.

a) Draw a diagram and label the length of each side. Think about whether you want to use "x" for the length or for the width.

b) Write an equation for the area in terms of the width and length.

c) Find the lengths of each side of the rectangle.

YS-108. Write each of the expressions below in a simpler exponential form.

a) $x^2y^3 \cdot x^3y^4$

b) $-3x^2 \cdot 4x^3$

c) $(x^3)^4$

d) $(2x^2)^3$

e) $\dfrac{6x^2y^3}{2xy}$

f) $(x^3y)^2(2x)^3$

YS-109. Simplify.

a) $\dfrac{x^2 + 7x + 10}{2x^2 + 11x + 5}$

b) $\dfrac{x^2 + 3x - 10}{x^2 - 7x + 10}$

c) $\dfrac{2x + 1}{4x^2 + x - 1}$

d) $\dfrac{3x^2 + 21x + 30}{3x^2 + 3x - 6}$

YS-110. TOOL KIT CHECK-UP

Your tool kit contains reference tools for algebra. Return to your tool kit entries. You may need to revise or add entries.

Be sure that your tool kit contains entries for all of the items listed below. Add any topics that are missing to your tool kit NOW, as well as any other items that will help you in your study of algebra.

- Definition of Base, Exponent, and Value
- Laws of Exponents
- Zero Exponents and Negative Exponents
- Extended Factoring of $ax^2 + bx + c$ when $a \neq 1$

- Standard Form of Quadratic Equations
- Using the Quadratic Formula
- Simplifying Rational Expressions

GM-19. PYTHAGOREAN TRIPLES

$$3, 4, 5 \qquad\qquad 7, 24, 25$$

Above are two examples of a Pythagorean Triple. Pythagorean Triples are made up of whole numbers. These two examples are special because in both cases the hypotenuse is 1 greater than the longer leg.

Your Task:

- How many Pythagorean Triples can you find where the hypotenuse is one more than the longer leg? How many do you think there are? Explain your answer.

- Find one where one of the legs is 37 and the hypotenuse is one more than the longer leg. Explain how you got your answer.

Yearbook Sales: Exponents and Quadratics

GM-20. COOKIES FOR DESSERT

a) Three algebra students were doing their homework together. As a treat, one mom offered to bake some cookies. While waiting for them to cool, all three students fell asleep. After a while, Latisha woke up, ate her equal share of cookies and went back to sleep. A little while later, Susan woke up, ate what she  thought was her equal share and fell asleep again. Then Hieu woke up, ate what she thought was her equal share, and went back to sleep. Later, all three kids woke up and discovered 8 cookies left. How many cookies were baked originally?

b) The next day, four students got together to study. Another mom baked cookies. Again the four students fell asleep. As before, they woke up one at a time and each ate her equal share. When they all awakened, they discovered 81 cookies remained.

How many cookies were baked originally?

Your Task:

- Solve both problems.
- Compare the original cookie numbers to the final numbers of cookies left. What is their relationship?
- **What would happen if there were 5 students? As before, each student ate her equal share of what was left. When they all awakened, how many cookies remained? How many cookies were baked originally?**
- Explain how this problem would work for any number of students.

Unit 11
The Cola Machine

Functions and Equality

Unit 11 Objectives
The Cola Machine: **FUNCTIONS AND EQUALITY**

This unit is intended to help you focus and formalize your study of equations. You will study two more types of equations, learn another way to solve linear systems, and tie work you have done with equations to the laws and properties that govern the real numbers.

You will study some formal structures of algebra, namely, relations and functions and the properties of real numbers that are the basis for the work you did this year with variables.

In this unit you will have the opportunity to:

- explore the nature of relations and functions.
- add the Elimination (addition) Method to your list of strategies to solve systems of linear equations.
- work with absolute value in the context of distance from a reference point.
- solve equations with absolute value and square roots.
- extend your work with rational expressions to include multiplying and dividing them.
- explore the properties of real numbers and identify them in the context of algebraic expressions.

Look for patterns in the investigations of the new ideas you encounter in this unit. Also look for ideas that help you see the "big picture" for all the work you have done with equations this year.

Read the following problem carefully. **Do not try to solve it now**. During this unit, you will gain more proficiency and skills to solve this problem.

CM-0.	The cola machine at your school offers several types of soda. Your favorite drink, *Blast!*, has two buttons dedicated to it, while the other drinks (*Slurp, Lemon Twister,* and *Diet Slurp*) each have one button.
	Your task is to determine whether the machine models the mathematical idea known as a "function." You must determine whether the machine can be depended upon to deliver the same kind of soda each time a specific button is pushed.
	Think about this model for a few days, especially when you are buying a soda from a vending machine. We will return to it later.

Unit 11
The Cola Machine: FUNCTIONS AND EQUALITY

RELATIONS

CM-1. Examine the table of input (x) and output (y) values below. Is there a relationship between the input and output values? If so, state the relationship.

x	-3	-2	-1	0	1	2	3
y	8	3	0	-1	0	3	8

CM-2. Use the table from problem CM-1 to graph $y = x^2 - 1$. Using the table and graph, answer the questions below:

a) Each equation that relates inputs to outputs is called a **RELATION**. This is easy to remember because the equation helps us know how all the y-values (outputs) on our graph are related to their corresponding x-values (inputs).

A relation works like a machine, as shown in the diagram at right. Numbers are put into a relation (in this case, $y = x^2 - 1$), the relation performs operations on each input, and determines an output. For example, for the relation $y = x^2 - 1$, when $x = 2$ is put into the relation, the output is $y = 3$. Find the output when the input is $x = 4$.

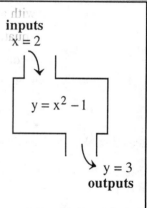

b) The set of all possible inputs of a relation is called the **DOMAIN** of the relation. If <u>any</u> number can go into the equation for x, we called the domain "all real numbers," meaning that all real numbers can be put into our relation. Explain why the domain for the relation $y = x^2 - 1$ is the set of all real numbers.

c) Likewise, the set of all possible outputs of a relation is called the **RANGE** of the relation. Examine your table and graph and decide which outputs are possible for the relation $y = x^2 - 1$.

d) If the output of the relation is 24, what was the input? Is there more than one possible input?

The Cola Machine: Functions and Equality

CM-3. Find the outputs for the following relations and the given inputs. If there is no possible output for the given input, explain why not.

a) x = -3
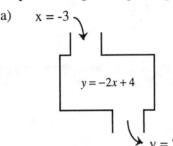

c) x = 11
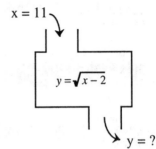

b) x = 5
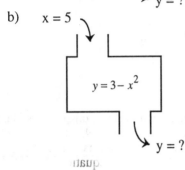

d) x = 2
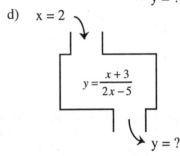

CM-4. In problem CM-14, we will focus on solving system of equations. Review what you know now by completing parts (a) through (c) to solve the system of equations at right.

$2x + y = 11$
$x - y = 4$

a) Change each equation into y-form.

b) Solve the system by graphing. (Where do the lines cross?)

c) Solve the system using substitution.

CM-5. The incidence of rabies in skunks was recently reported to be three of every 10 skunks in a particular county. Sampling studies (just like fish sampling) revealed that there are about 22,400 skunks in the county. About how many carry rabies?

CM-6. Start at the point (-4, 5). Move (slide) ten units down and six units to the right.

a) Write the coordinates of the resulting point and find the distance between the two points.

b) What is the slope of the line containing these two points?

c) Compare your answers to parts (a) and (b). Are the length and slope of the segment connecting the two points the same?

CM-7. Simplify by factoring and looking for fractions that are equivalent to one.

a) $\dfrac{(x+4)^2}{(x+4)(x-2)}$

b) $\dfrac{6(3x-1)^3}{3(3x-1)^2(x+4)}$

c) $\dfrac{8(x+2)^2(x-3)^3}{4(x+2)^3(x-3)^5}$

d) $\dfrac{x^2+3x}{x^2+6x+9}$

CM-8. Write each expression without negative or zero exponents. Use your tool kit to verify your answers.

a) 4^{-1}

b) 5^{-2}

c) 7^0

d) x^{-2}

CM-9. Arturo has earned 134 points so far this quarter in algebra. To get an A, he needs to earn 90% of the total possible points. One more test, worth 45 points, is scheduled. It will bring the total number of points to 195. How many points does Arturo need to get on this test in order to get an A for the quarter? Write an equation and solve the problem.

CM-10. Factor each polynomial completely.

a) $2x^2 - 5x + 3$

b) $36x^2 - 25$

CM-11. Find the distance between each pair of points by drawing a generic right triangle. Write the distances you calculate in parts (a) and (b) in square root form, simplified square root form, and decimal form rounded to the nearest 0.01. Write the distances you calculate in parts (c) and (d) only in decimal form to the nearest 0.01.

a) (0, 0) and (4, 4)

b) (-2, 4) and (4, 7)

c) (12, 18) and (-16, -19)

d) (0, 0) and (25, 25)

The Cola Machine: Functions and Equality

SOLVING LINEAR SYSTEMS BY ELIMINATION

CM-12. With your study team, examine the balanced scales in figures 1 and 2. Determine what could be placed on the right side of the scale in figure 3 to balance with the left side. Draw a picture on your paper to represent this balance and justify your solution in complete sentences.

figure 1   figure 2   figure 3

CM-13. Rianna thinks that if

              $a = b$
and if $c = d$
then $a + c = b + d$

Is she correct? Explain why or why not.

CM-14. We will use the same concept as in problems CM-12 and CM-13 to solve systems of equations. Read the following information and summarize it in your tool kit. Then complete parts (a) through (c) on your paper.

THE ELIMINATION METHOD
Part One

In problem CM-4, you solved one system of equations by both graphing and substitution. There is a third algebraic method that can be used to solve systems of equations. This method uses subproblems.

We know how to solve single-variable equations. When we have a pair of two-variable equations, we can often **ELIMINATE** one of the variables to obtain one single variable equation. We can do this by adding, as shown below.

Solve the system:
$$2x + y = 11$$
$$x - y = 4$$

To eliminate the y terms,
ADD the two equations together,

$$\begin{array}{r} 2x + y = 11 \\ \underline{x - y = 4} \\ 3x = 15 \end{array}$$

then solve for x.

$$3x = 15$$
$$x = 5$$

Once we know the x-value we can substitute it into <u>either</u> of the original equations to find the corresponding value of y.

Using the first equation:
$$2x + y = 11$$
$$2(5) + y = 11$$
$$10 + y = 11$$
$$y = 1$$

We can check our solution by substituting both the x-value and y-value into the other original equation, $x - y = 4$, and $5 - 1 = 4$ checks!

a) What happens if you first substitute the x-value into the second original equation? Try it and check your answer.

b) Now write the solution to the system as an ordered pair and compare this answer to the one you got in problem CM-4.

c) Which of the three methods for solving the given system of equations (graphing, substitution, or elimination) was easiest for you? Write one or two sentences to explain your response.

CM-15. Solve the following pairs of equations by eliminating one of the variables. Check your solutions as shown in the tool kit example above.

a) $4x + 2y = 14$
    $x - 2y = 1$

b) $5x + 3y = 25$
    $7x - 3y = -1$

c) $x + 3y = 13$
    $-x + 2y = 2$

d) $5x + y = 20$
    $2x + y = 8$

CM-16. Find the outputs for the following relations using the given inputs. If there is no possible output for the given input, explain why not.

a)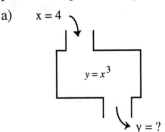

b)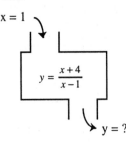

CM-17. Examine the table of input (x) and output (y) values below. Is there a relationship between the input and output values? If so, state the relationship.

x	-3	-2	-1	0	1	2	3
y	-11	-7	-3	1	5	9	13

CM-18. The points (-23, 345) and (127, 311) are on a line. Find the coordinates of a third point.

CM-19. Factor each polynomial completely. (Hint: Look for a common factor first.)

a) $3x^2 - 6x + 3$

b) $2xy - 4y^2$

CM-20. Simplify each expression using the Laws of Exponents.

a) $(x^2)(x^2y^3)$

b) $(2x^2)(-3x^4)$

c) $\dfrac{x^3y^4}{x^2y^3}$

d) $(2x)^3$

CM-21. Joe and Jahi are cutting a deck of cards to see who will mow the lawn this week. Joe cuts the deck and shows Jahi his card, a ten. He then replaces it and reshuffles the deck. What is the probability that Jahi will cut to a higher card? (A Jack, Queen, King, and Ace beat a 10.)

CM-22. Simplify by factoring and looking for fractions that are equivalent to one.

a) $\dfrac{(x-9)(x-5)^4}{(x-5)(x-9)^2}$

b) $\dfrac{6(x-2)^3(3x-2)^4}{2(x-2)(3x-2)^5}$

CM-23. Complete the following problem on the same set of axes. Label each graph with its equation.

a) Graph $y = \frac{1}{2}x - 3$.

c) How are the equations the same? How are they different?

b) Find the equation of the line parallel to part (a) that passes through the point (4, 4).

CM-24. Use the Elimination Method to find x and y.

a) $3x + 2y = 11$
$4x - 2y = 3$

b) $3x + 2y = 11$
$4x + 2y = 3$

c) Discuss the difference between parts (a) and (b) with your teammates. What must be done to systems like the one in part (b)?

CM-25. Consider what happens when you multiply an entire equation by some number.

a) If $2x - 5y = 13$, then does $4x - 10y = 13$ or does $4x - 10y = 26$? Why or why not?

b) Jovan thinks that if $a = b$, then $a \cdot c = b \cdot c$. Is he correct? Explain why or why not.

CM-26. Samantha used Elimination in the problem below but got stuck with an equation she did not know how to solve. Examine her work below:

$$3x + 4y = 1$$
$$\underline{4x + y = 10}$$
$$7x + 5y = 11 \;??$$

a) With your study team, describe what went wrong.

b) Compare Samantha's problem and part (a) of problem CM-24. In order to eliminate a variable, what must be true about the equations? Explain using complete sentences.

c) Use Jovan's observation (problem CM-25 (b)) to alter the second equation so that the y-values will meet the requirement you stated in part (b) above. Then use Elimination to solve for the point of intersection.

The Cola Machine: Functions and Equality

CM-27. Here is another example of solving a system of equations by first eliminating one of the variables. Continue your tool kit entry for The Elimination Method with the following example. Then use the Elimination Method to solve parts (a) through (c).

SOLVING SYSTEMS BY ELIMINATION
Part Two

Suppose we want to solve this system of equations:

$$3x + 2y = 11$$
$$4x + 3y = 14$$

Again, one subproblem is to <u>eliminate</u> either x or y when we add the equations together. In this case we need to do something to BOTH equations before we add them. To eliminate y we can multiply the first equation by 3 and multiply the second equation by -2 to get:

$$9x + 6y = 33$$
$$-8x - 6y = -28$$

We can eliminate the y terms by adding the two new equations:

$$9x + 6y = 33$$
$$\underline{-8x - 6y = -28}$$
$$x = 5$$

Now we know $x = 5$ and can substitute to find that $y = -2$. Therefore, the solution to the system of equations is (5, -2).

a) Discuss the example in your study team. Could you solve the system by multiplying the first equation by 4 and the second equation by -3? Try this and show your subproblems.

Solve each of the following systems of equations. There may be more than one way to do them, so expect a variety of <u>methods</u> in your study team, but the <u>same</u> <u>solutions</u>, (x, y). Remember to check your solution.

b) $2x + 3y = -1$
   $5x - 2y = -12$

c) $2x + 3y = 17$
   $x + 3y = 16$

CM-28. SOLVING ANY SYSTEM OF EQUATIONS

In this course we have solved systems of equations with a variety of methods. We started with graphing, moved to the Substitution Method, and have now learned the Elimination Method. Each method works best with certain forms of equations.

For each system below, determine which method would be best (easiest) to use. Then solve the system to find the point of intersection.

a) $x = 4y - 7$
   $3x - 2y = 1$

b) $x + 2y = 16$
   $x - y = 2$

c) $y = \frac{3}{4}x - 1$
   $y = -\frac{1}{3}x - 1$

d) $x + 3y = 4$
   $3x - y = 2$

CM-29. Examine the graph at right for the relation $y = \sqrt{x-2}$. Then answer the following questions.

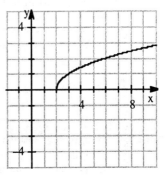

a) Find the outputs for the inputs $x = 2, 3,$ and 6.

b) What happens when the input is $x = 0$?

c) What do all the outputs of this relation have in common?

d) Find the domain and range of $y = \sqrt{x-2}$.

e) Use the graph to find x when $y = 1$ and when $y = 2$.

CM-30. Find the inputs for the following relations with the given outputs. If there is no possible input for the given output, explain why not.

a)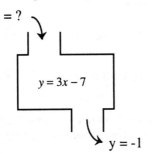

b)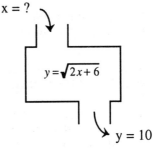

CM-31. Simplify by factoring and looking for fractions that are equivalent to one.

a) $\dfrac{x^2 + 2x}{x + 2}$

b) $\dfrac{x^2 + 2x + 1}{x^2 + x}$

CM-32. On the same set of axes, use the slope and y-intercept to sketch a graph of the following lines.

a) $y = 2x + 3$ b) $y = \dfrac{1}{2}x + 3$ c) $y = -\dfrac{1}{2}x + 3$

CM-33. Felice works as a sales clerk at Stacy's Department Store. She is paid $5.00 per hour plus a $7.50 meal allowance when she works evenings.

a) How much does she make for a typical eight hour day that ends at 9:00 p.m.? Write an expression to represent how much money she would be paid on this day if she worked n hours.

b) Felice's hours last week included just one evening shift. If she made $190 last week, how many hours did Felice work? Write an equation and solve it.

The Cola Machine: Functions and Equality

CM-34. Use the Laws of Exponents to rewrite the following expressions.

a) $(2x^2)(3x^2y^3)$

b) $(-3x^4)^3$

c) $\dfrac{20x^3y^4}{5xy^4}$

d) $x^{-2}x^4$

e) $(x-2)^2$

f) $(2x+1)^2$

MULTIPLYING AND DIVIDING RATIONAL EXPRESSIONS

CM-35. Without your calculator, complete each part below. Pay particular attention to explaining how you do each problem.

a) Multiply $\dfrac{2}{3} \cdot \dfrac{9}{14}$ and reduce your result. Using complete sentences, describe your method for multiplying fractions.

b) Divide $\dfrac{3}{5} \div \dfrac{12}{25}$ and reduce your result. Using complete sentences, describe your method for dividing fractions.

CM-36. Use your method for multiplying and dividing fractions to simplify these expressions.

a) $\dfrac{x+2}{x-1} \cdot \dfrac{x-1}{x-6}$

b) $\dfrac{(4x-3)(x+2)}{(x-5)(x-3)} \cdot \dfrac{(x-1)(x-3)}{(x-1)(x+2)}$

c) $\dfrac{3x-1}{x+4} \div \dfrac{x-5}{x+4}$

d) $\dfrac{(x-6)^2}{(2x+1)(x-6)} \cdot \dfrac{x(2x+1)(x+7)}{(x-1)(x+7)}$

e) $\dfrac{(x+3)(2x-5)}{(3x-4)(x-7)} \div \dfrac{(2x-5)}{(3x-4)}$

f) $\dfrac{x-3}{x+4} \cdot \dfrac{3x-10}{x+11} \cdot \dfrac{x+4}{3x-10}$

CM-37. Find the output for the following relation with the given input. If there is no possible output for the given input, explain why not.

$x = 1$, $y = \dfrac{x}{x-1}$, $y = ?$

CM-38. What input number(s) would cause the same problem for these machines as the one illustrated in the previous problem?

a) 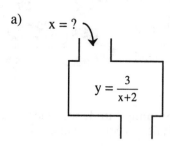 $x = ?$, $y = \dfrac{3}{x+2}$

b) 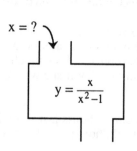 $x = ?$, $y = \dfrac{x}{x^2-1}$

CM-39. Read the following information and summarize it in your tool kit. Then complete the problems below it on your paper.

MULTIPLYING AND DIVIDING RATIONAL EXPRESSIONS

Just as we can multiply and divide fractions, we can multiply and divide rational expressions. Study the following examples and answer the summary questions below.

Problem A: Multiply $\dfrac{x^2+6x}{(x+6)^2} \cdot \dfrac{x^2+7x+6}{x^2-1}$ and simplify your result.

After factoring, our expression becomes: $\dfrac{x(x+6)}{(x+6)(x+6)} \cdot \dfrac{(x+6)(x+1)}{(x+1)(x-1)}$

After multiplying, reorder the factors: $\dfrac{(x+6)}{(x+6)} \cdot \dfrac{(x+6)}{(x+6)} \cdot \dfrac{x}{(x-1)} \cdot \dfrac{(x+1)}{(x+1)}$

Since $\dfrac{(x+6)}{(x+6)} = 1$ and $\dfrac{(x+1)}{(x+1)} = 1$, simplify: $1 \cdot 1 \cdot \dfrac{x}{x-1} \cdot 1 \Rightarrow \dfrac{x}{x-1}$.

Problem B: Divide $\dfrac{x^2-4x-5}{x^2-4x+4} \div \dfrac{x^2-2x-15}{x^2+4x-12}$ and simplify your result.

First, change to a multiplication expression: $\dfrac{x^2-4x-5}{x^2-4x+4} \cdot \dfrac{x^2+4x-12}{x^2-2x-15}$

After factoring, we get: $\dfrac{(x-5)(x+1)}{(x-2)(x-2)} \cdot \dfrac{(x+6)(x-2)}{(x-5)(x+3)}$

After multiplying, reorder the factors: $\dfrac{(x-5)}{(x-5)} \cdot \dfrac{(x-2)}{(x-2)} \cdot \dfrac{(x+1)}{(x-2)} \cdot \dfrac{(x+6)}{(x+3)}$

Since $\dfrac{(x-5)}{(x-5)} = 1$ and $\dfrac{(x-2)}{(x-2)} = 1$, simplify: $\dfrac{(x+1)(x+6)}{(x-2)(x+3)}$

Thus, $\dfrac{x^2-4x-5}{x^2-4x+4} \div \dfrac{x^2-2x-15}{x^2+4x-12} = \dfrac{(x+1)(x+6)}{(x-2)(x+3)}$ or $\dfrac{x^2+7x+6}{x^2+x-6}$

a) Verify that $\dfrac{x^2+6x}{(x+6)^2} \cdot \dfrac{x^2+7x+6}{x^2-1} = \dfrac{x}{x-1}$ by substituting two different values for x.

b) There are three values of x that cannot be used when verifying our equality. What are these restrictions? Why can x not assume these values?

CM-40. For each system below, decide to use graphing, Substitution or Elimination. Then solve the system to find the point of intersection.

a) $y = -3x + 5$
    $y = x - 11$

b) $7x - 2y = 2$
    $y = x + 4$

c) $2x + y = -2$
    $2x + 3y = 14$

d) $9x - 5y = 4$
    $3x + 2y = 5$

CM-41. Find the outputs for the following relations with the given inputs. If there is no possible output for the given input, explain why not.

a) $x = -3$ 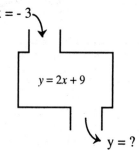 $y = 2x + 9$, $y = ?$

b) $x = 13$ 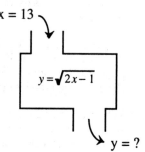 $y = \sqrt{2x - 1}$, $y = ?$

CM-42. Solve the following equations for x.

a) $2x^2 + x - 6 = 0$

b) $5 - 2(3x - 4) = 17$

c) $3x^2 - 11 = 5x$

d) $x(x - 7)(3x - 1) = 0$

CM-43. Use your knowledge of exponents to rewrite the following expressions:

a) $\dfrac{5}{x^{-1}}$

b) $(3x^5)^2(2x^{-2})^3$

c) $\dfrac{32x^3y^7}{18x^5y^2}$

d) $(x^4)(x^{-1})(x^{-3})$

CM-44. Find the product: $(3x - 1)(2x^2 + 6x + 1)$

CM-45. An isosceles right triangle has two legs of equal length. If the length of each leg is increased by two, the length of the hypotenuse becomes $5\sqrt{2}$. Find the dimensions of the original triangle.

INTRODUCTION TO FUNCTIONS

CM-46. THE COLA MACHINE

The cola machine at your school offers several types of soda. Your favorite drink, *Blast!*, has two buttons dedicated to it, while the other drinks (*Slurp, Lemon Twister,* and *Diet Slurp*) each have one button.

a) Explain how the soda machine is a relation.

b) Describe the **domain** and **range** of this soda machine.

c) While buying a soda, Mr. Hagen pushed the button for *Lemon Twister* and got a can of *Lemon Twister*. Later he went back to the same machine but this time pushing the *Lemon Twister* button got him a can of *Blast!* Is the machine functioning consistently? Why or why not?

d) When Karen pushed the top button for *Blast!* she received a can of *Blast!* Her friend, Miguel, decided to be different and pushed the second button for *Blast!* He, too, received a can of *Blast!* Is the machine functioning consistently? Why or why not?

e) When Loufti pushed a button for *Slurp*, he received a can of *Lemon Twister*! Later, Tayeisha also pushed the *Slurp* button and received a can of *Lemon Twister*. Still later, Tayeisha noticed that everyone else who pushed the *Slurp* button received a *Lemon Twister*. Is the machine functioning consistently? Explain why or why not.

f) When a relation is functioning consistently, we call that relation a **function**. What is the main difference between a relation that is a function and a relation that is not a function?

CM-47. Using your own words, write a definition of a function.

CM-48. Examine each of the relations below. Compare the inputs and outputs of each relation and decide if the relation is a function. Explain why or why not. Use your definition of a function (from problem CM-47) to help you justify your conclusion.

a)
x	7	-2	0	4	9	-3	6
y	6	-3	4	2	10	-3	0

b)
x	1	-5	3	8	1	-4	5
y	3	-7	2	3	3	10	5

c)
x	3	-1	2	2	1	0	9
y	4	-5	9	7	4	-8	2

The Cola Machine: Functions and Equality

CM-49. Examine the graph of lines a and b at right.

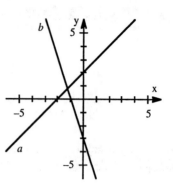

a) Find the equation for each line.

b) Are lines a and b functions? Explain how you know.

c) Are all lines functions? If not, draw an example of a line that is not a function.

CM-50. Solve each of the following systems of equations for x and y.

a) $x + 2y = 5$
$x + y = 5$

b) $2x + 3y = 5$
$x + 3y = 4$

c) $2x + y = 7$
$x + 5y = 12$

d) $3x + 2y = 11$
$2x - y = 1$

CM-51. Simplify the expressions below.

a) $\dfrac{(x+5)(2x-3)}{x+4} \cdot \dfrac{(x+1)(x+4)}{(x+5)(x-3)}$

b) $\dfrac{(3x+4)(x+5)}{(3x+15)(x+4)} \div \dfrac{(3x-4)}{(x+4)}$

CM-52. Use the given information to find the equation of the line that:

a) has slope $-\dfrac{2}{3}$ and passes through (0, -6).

b) passes through (-1, 7) and (3, 7).

CM-53. Find the x-intercepts of the parabola $y = 3x^2 - x - 11$.

CM-54. Write a shorter expression for each of the following polynomials.

a) $(3x^2 + 0.8) - (7x^2 - 5x)$

b) $(12x^2 - 3x) + (2x^2 - 4x + 3)$

c) $(7x^2 + 5) - (3x^2 + 2x - 4)$

d) $(x^2 + 3x + 4) + (3x^2 - 4x - 7)$

e) $(x^3 + 2x^2 + 4) - (5x^2 - 4)$

CM-55. Solve each quadratic equation by the method of your choice.

a) $x^2 + 3x - 5 = 0$

b) $4x^2 + 19x - 63 = 0$

c) What do your answers to (a) and (b) tell you about whether the problems can be factored?

FUNCTION NOTATION

CM-56. Using complete sentences, explain the meaning of **relation** and **function**. Be sure to offer examples to demonstrate your understanding.

CM-57. Examine the graph of the relation at right. Then find the output(s) of the relation when:

a) the input is 3.

b) the input is 1.

c) the input is -4.

d) Is this relation a function? Why or why not?

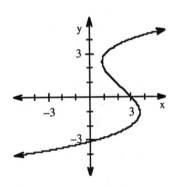

CM-58. The graphs of several relations are shown below. Decide if each is a function. If the relation is not a function, explain why not.

a)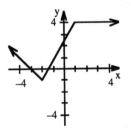

c)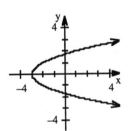

e)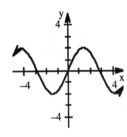

b)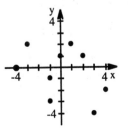

d)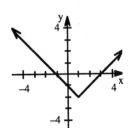

f)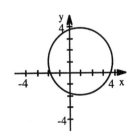

The Cola Machine: Functions and Equality

CM-59. Review the definition of domain and range in your tool kit. The domain and range of a relation can be determined by carefully examining its graph. For example, the relation graphed in problem CM-58 part (a) above has a domain of all reals and a range of all numbers $y \geq -1$.

Find the domain and range for the relations in parts (b) through (f) in problem CM-58.

CM-60. Carefully read the following definitions regarding relations and functions and add them to your tool kit.

RELATIONS AND FUNCTIONS
Definitions and Vocabulary

A **RELATION** is an equation which establishes the relationship between two variables and helps determine one variable when given the other. Some examples of relations are:

$$y = x^2, \quad y = \frac{x}{x+3}, \quad y = -2x + 5$$

Since y depends on the value assigned to x, y is often referred to as the **DEPENDENT VARIABLE**, while x is called the **INDEPENDENT VARIABLE**.

The set of possible inputs of a relation is called the **DOMAIN**, while the set of all possible outputs of a relation is called the **RANGE**.

A **relation** is called a **FUNCTION** if there exists <u>no more than one</u> output for each input. If a relation has two or more outputs for a single input value, it is not a function. For example, problem CM-58 part (c) is not a function because there are two outputs for each input value greater than -3.

Functions are often given names, most commonly "f," "g" or "h". The notation **f(x)** represents the output of a function, named f, when x is the input. It is pronounced "f of x." The notation g(2), pronounced "g of 2," represents the output of function g when x = 2.

Similarly, the function y = 3x + 4 and f(x) = 3x + 4 represent the <u>same function</u>. Notice that this notation is interchangeable, that is, y = f(x). In some textbooks, 3x + 4 is called the **RULE** of the function.

$x = 2$
$f(x) = 3x + 4$
$f(2) = 10$

CM-61. If $f(x) = x^2$, then $f(4) = 4^2 = 16$. Find $f(1)$, $f(-3)$, and $f(t)$.

CM-62. If $g(x) = \sqrt{x - 7}$, find $g(8)$, $g(32)$ and $g(80)$.

CM-63. Before class, Jethro and Janice were chatting on the stairs. Jethro was two steps from the bottom, while Janice was five steps from the bottom.

    a) How many steps was Janice from Jethro?

    b) How did you determine your answer to part (a)?

    c) How many steps from Jethro was Janice?

    d) Your answer to parts (a) and (c) represent a <u>distance</u>. Can distance be negative? Why or why not?

CM-64. Solve each system for x and y:

    a) $y = 3x - 4$
        $y = \frac{1}{2}x + 7$

    b) $2x - 5y = 5$
        $2x + 5y = -25$

    c) $2x + y = 6$
        $y = 12 - 3x$

    d) $y = 0$
        $y = x^2 - 2x + 1$

CM-65. If $f(x) = -x^2$, find $f(1)$, $f(-1)$ and $f(4)$.

CM-66. Solve each of the following quadratic equations by factoring or by using the Quadratic Formula.

    a) $2x^2 + 6x = 8$

    b) $x^2 - 7x + 10 = 0$

    c) $3x(x - 5) = 17 - x$

    d) $(x - 1)(x - 5) + 4 = 0$

CM-67. Simplify the expressions below.

    a) $\dfrac{x^2+5x+6}{x+2} \div \dfrac{x+3}{4x-1}$

    b) $\dfrac{2x^2+7x-30}{3x^2-20x-7} \cdot \dfrac{x-7}{2x-5}$

CM-68. In Unit Four, you used a Guess and Check table to solve the following problem and write an equation in one variable. Now that you know how to solve equations with two variables, write two equations with two different variables, and solve the problem.

A stick 152 centimeters long is cut up into six pieces: four short pieces, all the same length, and two longer pieces (both the same length). A long piece is 10 centimeters longer than a short piece. What are the lengths of the pieces? (Hint: let S represent the length of a short piece and let L represent the length of a long piece.)

The Cola Machine: Functions and Equality

ABSOLUTE VALUE

CM-69. **WHAT'S THE DIFFERENCE?**

Examine the following situations in which we need to find the difference between two amounts.

a) Rocio has $298 saved in the bank, while Thomas has $314. What is the difference between their bank balances? How did you get your answer?

b) Herneisha is 24 years old, while her brother, Jason, is 17. What is the difference of their ages? How did you get your answer?

c) Urban High School has 1850 students while Saint Ignatius has 1490 students. What is the difference of their student populations?

d) Explain why these differences are all positive.

CM-70. **ARE YOU POSITIVE?**

Problems CM-63 and CM-69 offer examples of situations in which we want to calculate a positive difference. This difference helps us compare two quantities, regardless of which is larger. To find these differences, we subtract the smaller amount from the larger. However, what if we want to compare two amounts we do not know?

For example, what if Herneisha is x years old and Jason is y years old?

a) If Herneisha is older, what is the difference of their ages?

b) If Jason is older, what is the difference of their ages?

CM-71. Read the following information and summarize it in your tool kit.

ABSOLUTE VALUE DEFINITION AND NOTATION

An **ABSOLUTE VALUE**, represented by two vertical bars, "| |", determines the positive value of a number. Numerically, it represents a distance between the number and zero. Since a distance is always positive, the absolute value always returns a positive value or zero.

For example, the number -3 is a distance of 3 units from 0, as shown on the number line at right. The absolute value of -3 is 3. This is written

$$|-3| = 3$$

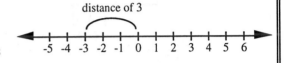

Likewise, the number 5 is a distance of 5 units from 0. The absolute value of 5 is 5, written:

$$|5| = 5$$

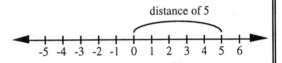

In the case of Herneisha and Jason in problem CM-70, we wanted to find a positive difference between their ages. The absolute value offers us another way to write this difference:

$$|x - y|$$

In this notation, the vertical bars indicate that we want the result to be positive. It does not matter if x or y is larger. Once we subtract, the absolute value (which is always positive) will be given.

CM-72. Find the following absolute values.

a) $|-7|$ b) $|4|$ c) $|0|$ d) $|3-8|$

CM-73. Explain why there are two solutions to the equation $|x| = 10$.

CM-74. The absolute value can be used in relations as well. Find the corresponding inputs or outputs for the following relations. If there is no solution, explain why not. Be careful: in some cases, there is more than one solution.

a) x = -3
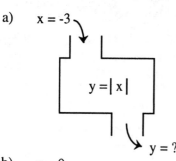
y = ?

c) x = ?
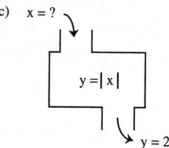
y = 2

b) x = 9
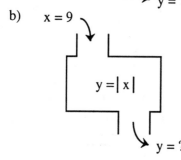
y = ?

d) x = ?
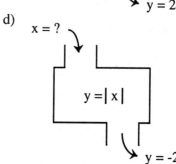
y = -2

CM-75. Complete this table to graph the function $f(x) = |x+2| - 4$. Then answer the questions below.

x	-4	-3	-2	-1	0	1	2
y							

a) Find f(6), f(-10), f(-5) and $f(\frac{1}{2})$.

b) Determine the domain and range of f(x).

c) Find the x-value(s) at which the y-value, or f(x), is 3.

CM-76. Make a table and graph $y = |x|$ for $-5 \le x \le 5$. Use your graph to answer the following questions.

a) Describe the shape of the graph.

b) Find the **domain** and **range** of the relation.

c) Is this relation a function? Explain.

d) Find all x-values for which y = 2.

CM-77. Make a table and graph $f(x) = |x - 2|$ on the same set of axes as problem CM-76. Use your graph to answer the following questions.

    a) Compare this graph with the graph from CM-76. Write down any observations.

    b) On your graph, label the x- and y-intercepts.

    c) Find all x-values for which $y = 1$.

    d) Find $f(2)$, $f(-6)$ and $f(10)$.

CM-78. Solve each of the following equations.

    a) $\frac{x}{4} = x - 1$ c) $\frac{x}{4} = \frac{x-2}{3}$

    b) $\frac{2x}{3} = x - 4$ d) $\frac{x}{2} + \frac{x}{5} = 7$

CM-79. Solve for x in the triangle:

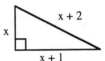

CM-80. Solve each system for x and y.

    a) $5x - y = 3$ b) $2x + y = 6$
        $3x + y = 5$ $y = 12 - 3x$

CM-81. Simplify using only positive exponents

    a) $(3x^2y)(5x)^2$ c) $\frac{x^3}{x^{-2}}$

    b) $(x^2y^3)(x^{-2}y^{-2})$ d) $(2x^{-1})^3$

SOLVING ABSOLUTE VALUE EQUATIONS

CM-82. Find the following absolute values.

    a) $|0.75|$ b) $|-99|$ c) $|4 - 2 \cdot 3|$ d) $|\pi|$

CM-83. Solve for x. Be sure to find all possible solutions. Check your solutions by substituting them back into the original equation. If there is no solution, explain why not.

    a) $|x| = 2$ d) $|x| + 6 = 7$

    b) $|x| = 5$ e) $|x - 4| = 1$

    c) $|x| - 3 = 8$ f) $|x + 8| = 2$

CM-84. Since absolute value is defined as the distance from zero, Sherry wondered if the solutions to the equation $|x|=5$ are related to distance. She plotted her solutions (5 and -5) on a number line and noticed that both points were 5 units away from zero.

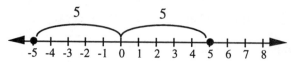

This result interested her! She decided to try the same thing with her solutions to $|x-2|=3$. Her solutions, -1 and 5, are plotted below.

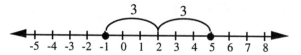

a) Examine her number line carefully. Using complete sentences, explain how the solutions $x = -1$ and $x = 5$ are related to the equation $|x-2|=3$.

b) For part (a), the central number 2 is called a **REFERENCE POINT**, since the solutions -1 and 5 are an equal distance away from it. On a number line, locate the reference point for the equation $|x-10|=2$. Then use the idea of distance to find the solutions to $|x-10|=2$.

c) Translate the equation $|x-4|=3$ into a sentence describing distance. Then find the solutions to the equation.

d) Using complete sentences, describe how to find the solutions of any absolute value equation of the form $|x-a|=b$.

CM-85. Use Sherry's method to find the solutions to $|x-1|=5$. Compare with your answer to problem CM-83 part (e). Did it work?

CM-86. When solving $|x+3|=1$, Sherry noticed that there was no longer a difference in the absolute value. In order to find her reference point, she needed to first rewrite her expression as $|x-(-3)|=1$. Using this expression:

a) What is the reference point for the equation $|x+3|=1$?

b) Use the reference point and a number line to solve for x.

CM-87. Draw a number line on your paper. Then use Sherry's method to find the solutions to the following equations. You may need to re-write the problem first. Check your solutions by substituting them into your original equation.

a) $|x|=3$

b) $|x-4|=2$

c) $|x-5|+2=3$

d) $|x+1|=6$

UNIT 11

CM-88. Make a table and graph $f(x) = |x+3|$ for $-4 \leq x \leq 4$. Compare this graph with your graph from CM-76. Write down any observations.

CM-89. Find the corresponding inputs for the following relations. If there is no solution, explain why not. Be careful: in both cases, there is more than one solution. Find both.

a) x = ?, $y = |x|$, y = 5

b) x = ?, $y = |x-2|$, y = 5

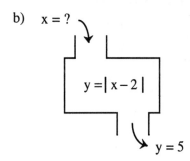

CM-90. Use the Elimination Method to solve for x and y:

a) x + y = 3
    2x - y = -9

b) 2x + 3y = 13
    x - 2y = -11

CM-91. For the equation $|x| = 4$,

a) what is the reference point?

b) find the solutions for x.

CM-92. Solve each of the following equations.

a) $x^2 - 1 = 15$

c) $2x^2 = -x + 7$

b) $x^2 - 2x - 8 = 0$

CM-93. For the equation: $4x + 2y = 8$:

a) Write the coordinates of the x-intercept and the y-intercept.

b) Draw a graph of the equation.

c) Solve the equation for y.

d) Is the point (23, -42) on the graph? Explain how you know.

CM-94. For the parabola $y = 2x^2 - 7x + 3$,

    a) give the coordinates of the y-intercept.

    b) give the coordinates of the two x-intercepts. Explain how you found your solution.

CM-95. Explain why $|-2| = |2|$.

CM-96. Draw a number line on your paper. Then use Sherry's method to find the solutions to the following equations. Check your solutions by substituting them into each original equation.

    a) $|x-9| = 2$ b) $|x+5| = 4$

CM-97. Let's re-examine the equation $|x-9| = 2$ from problem CM-96 part (a).

    a) Explain why the quantity inside the absolute value (x - 9) must equal 2 or - 2.

    b) We can use the reasoning from part (a) above to write two new equalities, as shown below.

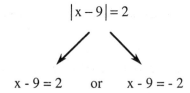

Solve each of these new equalities to find solutions for x. Check these solutions by substituting them into the original equation $|x-9| = 2$. Do your solutions match those found in problem CM-96 part (a)?

    c) Use similar reasoning to solve the equation $|x+5| = 4$. Check your solutions with those you found in problem CM-96 part (b).

CM-98. Explain why $|x+5| = -4$ has no solution.

CM-99. Read the following information, then summarize it in your tool kit.

SOLVING ABSOLUTE VALUE EQUATIONS ALGEBRAICALLY

To solve an equation with an absolute value algebraically, first determine the possible values of the quantity inside the absolute value.

For example, if $|2x+3|=7$, then the quantity $(2x + 3)$ must equal 7 or -7.

With these two values, set up new equations and solve as shown below.

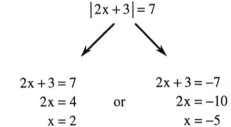

$$|2x+3|=7$$

$2x + 3 = 7$		$2x + 3 = -7$
$2x = 4$	or	$2x = -10$
$x = 2$		$x = -5$

Always check your solutions by substituting them into the original equation:

Test x = 2: $|2(2)+3|=7$ True   Test x = -5: $|2(-5)+3|=7$ True

---

CM-100. Solve for x. Use any method. Check your solution(s) by testing them in the original equation.

a) $|x-3|=5$

b) $|x+1|=2$

c) $5|x|=35$

d) $|x+3|-6=-4$

CM-101. Plot the points A(5, 2), B(0, -1), C(3, -6) on graph paper.

a) Find the slope of line AB.

b) Draw a line through point C that is parallel to the line through points A and B. What is the slope of this parallel line?

c) Draw the line through points B and C. Line BC is **perpendicular** to the lines from parts (a) and (b) above because the lines meet at right angles. What is the slope of line BC?

d) Compare the slopes of the perpendicular lines AB and BC. Record your observation in your tool kit.

The Cola Machine: Functions and Equality

CM-102. Solve for x. Use any method. Check your solution(s) by testing them in the original equation.

a) $8 = |x - 4|$

b) $|x| + 7 = 8$

c) $|x - 1.5| = 2$

d) $|x + 7| + 1 = -1$

CM-103. Solve each of the following systems for x and y. Use any method you choose.

a) $y = 3x - 5$
$y = 5x - 9$

b) $x + 2y = 1$
$3x - 2y = -5$

c) $y = 2x + 5$
$3x + 2y = 31$

d) If $y = 3x - 5$ and $y = 5x - 9$ were graphed on the same set of axes, at what point would the lines intersect?

CM-104. Solve each of the following equations.

a) $\dfrac{3x}{5} = \dfrac{x - 2}{4}$

b) $\dfrac{2x}{5} - \dfrac{1}{3} = \dfrac{137}{3}$

c) $\dfrac{4x - 1}{x} = 3x$

d) $\dfrac{4x - 1}{x + 1} = x - 1$

CM-105. Find each of the values below if $g(x) = 2x^2 - 3$.

a) $g(0)$

b) $g(-1)$

c) $g(14)$

CM-106. THRIFTY PEOPLE

Janet has $290 and saves $5 a week. David has $200 and saves $8 a week. In how many weeks will they both have the same amount of money?

a) Let x represent the number of weeks Janet saves her money. Write an equation which describes how much money Janet will save, using y to represent her total savings. Label the equation "Janet's savings."

b) Use the same variables to write and label a similar equation for David.

c) Use your two equations to determine the number of weeks needed for Janet and David to save the same amount of money.

d) If you had graphed the lines, where would they intersect?

CM-107. Compute each of the following products.

a) $(\sqrt{36})^2$

b) $(\sqrt{5})^2$

c) $(\sqrt{x+2})^2$

d) $(\sqrt{x^2+4})^2$

e) $(x+3)^2$

f) Based on your answers to (a) through (d), what is the effect of "squaring a square root?"

CM-108. Using the patterns from the previous problem, discuss in your team how to rewrite this equation without square roots: $\sqrt{x+1} = 4$. Solve the equation and check your answer.

CM-109. Bradford was given the equation $\sqrt{x^2+6} = x+2$ and did not know what to do. His teacher suggested that eliminating the square root symbol was a subproblem.

a) Describe how you can eliminate the square root.

b) Solve for x.

CM-110. Sean was given this problem: Solve the equation $\sqrt{5x+3} + 7 = 10$.

a) Compare Sean's equation with Bradford's in the previous problem. To solve Sean's equation, what is the first subproblem?

b) What would be the next subproblem? Show how to do this.

c) Find x.

CM-111. Use the graph at right for $y = -|x-2|+3$ to explain why the equation below has no solution for x.

$$4 = -|x-2|+3$$

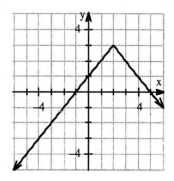

CM-112. Solve for x. Use any method. Check your solution(s) by testing them in the original equation.

a) $|2x+1| = 5$

b) $|x-1|-2 = -3$

The Cola Machine: Functions and Equality

CM-113. Solve each of the following systems of equations.

  a) $2x - y = 16$
      $x + y = 14$

  b) $2x - y = 16$
      $3x + 4y = 24$

  c) $x^2 - y = 9$
      $3x + y = 19$

CM-114. Simplify each expression using the Laws of Exponents, if possible.

  a) $(x^5)(y^2 x^3)$

  b) $(-6x^2)^2(-x^5)$

  c) $\dfrac{x^6 y^8}{x^{11} y^{-4}}$

  d) $x^2 - x^3$

CM-115. Make a table and graph $f(x) = |2x| - 3$ for $-4 \le x \le 4$. Compare this graph with your graph from problem CM-76. Write down any observations.

CM-116. For each part below:

  a) Solve for x: $\sqrt{x^2 + 5} = 5$

  b) Solve for (x, y): $\frac{1}{2}x + \frac{1}{3}y = 6$
      $-\frac{1}{2}x + y = 2$

  c) Solve for m: $m(m - 7) = 3$

  d) Solve for (x, y): $x + y = 16$
      and $2y = x - 4$

CM-117. Determine the corresponding input(s) or output(s) for the given function. If there is no possible input or output, explain why not.

  a)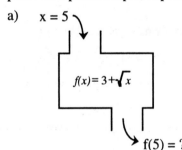

  b)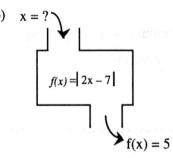

CM-118. Find the slope of the line $4x + 2y = 9$. Then find the slope of a line perpendicular to it. (Hint: you may want to review problem CM-101.)

PROPERTIES OF EQUALITY

CM-119. ALGEBRAIC PROPERTIES

Throughout this course we have focused on the meaning of equality and have studied how to solve equations. During this process, several important algebraic properties have naturally emerged. Although most have not been formally introduced and labeled, they should all be familiar from your experiences once we state them.

Two of these properties, the Distributive and the Substitution Properties, were stated explicitly in Unit Two. These properties are summarized below. Notice that along with a generalization of the property, a description and several examples are given.

Your task today is to similarly describe another property in a presentation to the class. Once your team has been assigned a property, design a presentation that demonstrates the multiple uses of the property. Be sure to include a numerical example and a generalization of the property and be prepared to answer questions about it.

During the team presentations, take careful notes about the properties presented by other study teams. Your notes should include a numerical example for each property. Your notes from these presentations will become a tool kit entry entitled, "Algebraic Properties of Real Numbers."

THE DISTRIBUTIVE PROPERTY

Generalization: $a(b + c) = ab + ac$

The Distributive Property basically states that if you multiply a quantity with two or more terms, such as $(b + c)$, by a number "a," then <u>each</u> term in the quantity is multiplied by "a."

For example, simplify:
$$2(5x^2 - 4x + 11) = 2 \cdot 5x^2 - 2 \cdot 4x + 2 \cdot 11$$
$$= 10x^2 - 8x + 22$$

Even multiplying binomials can be viewed as a form of the Distributive Property:

Multiply and simplify:
$$(4x + 5)(2x - 7) = 4x(2x - 7) + 5(2x - 7)$$
$$= 8x^2 - 28x + 10x - 35$$
$$= 8x^2 - 18x - 35$$

THE SUBSTITUTION PROPERTY

Generalization: If $a = b$, then either a or b can be replaced by the other.

The Substitution Property allows us to replace one expression with another that is equal to it. Substitution is used most often to rewrite an equation in a form that can be solved or simplified. For example, values can be substituted for variables as shown this example:

Evaluate $\frac{1}{2}x^2 - 7$, if $x = 8$.

Since $x = 8$, then $\frac{1}{2}x^2 - 7 = \frac{1}{2}(8)^2 - 7 = \frac{1}{2}(64) - 7 = 32 - 7 = 25$

Substitution also provides us with a method for solving systems of equations. For example, variable expressions can be substituted as follows:

Find the point of intersection for the system:
$$x = 3y - 2$$
$$5x + y = 22$$

Since $x = 3y - 2$,
x can be <u>replaced</u> with $3y - 2$ in the second equation.
Therefore, $5x + y = 22$ becomes:

$$5(3y - 2) + y = 22$$
$$15y - 10 + y = 22$$
$$16y - 10 = 22$$
$$16y = 32$$
$$y = 2$$

Substituting $y = 2$ back into the first equation:

$$x = 3y - 2$$
$$x = 3(2) - 2$$
$$x = 4$$

Therefore, the point of intersection is $(4, 2)$.

ALGEBRAIC PROPERTIES OF REAL NUMBERS

Commutative Property

The Commutative Property states that if two terms are added or multiplied, the order is reversible.

$$a + b = b + a$$
$$ab = ba$$

Multiplicative and Additive Identity

The Multiplicative Identity Property states that any term multiplied by one (1) remains unchanged, while the Additive Identity Property states that any term added to zero (0) remains unchanged.

$$a \cdot 1 = a$$
$$a + 0 = a$$

Substitution Property

The Substitution Property allows us to replace one expression with another that is equal to it.

If $a = b$, then a can be replaced by b.

Multiplicative Inverse

The Multiplicative Inverse Property states that when multiplying a term by its reciprocal, the result is always one.

$$\frac{a}{b} \cdot \frac{b}{a} = 1$$

Symmetric Property

The Symmetric Property states that if two terms are equal, it does not matter which is stated first.

If $a = b$, then $b = a$

Additive Inverse

The Additive Inverse Property states that when opposites are added, the result is always zero.

$$a + (-a) = 0$$

Reflexive Property

The Reflexive Property states that a term is always equal to itself.

$$a = a$$

Transitive Property of Equality

The Transitive Property of Equality states that if two terms (a and c) are both equal to a third term (b), then they must also be equal to each other.

If $a = b$ and $b = c$, then $a = c$

Distributive Property

The Distributive Property states that if you multiply a quantity (b + c) by a number "a," then <u>each</u> term in the quantity is multiplied by "a."

$$a(b + c) = ab + ac$$

Associative Property

The Associative Property states that if a sum or product contains terms that are grouped, then the sum or product can be grouped differently with no affect on the sum or product.

$$a + (b + c) = (a + b) + c$$
$$a \cdot (b \cdot c) = (a \cdot b) \cdot c$$

Additive Property of Equality

The Additive Property of Equality states that equality is maintained if you add the same amount to both sides of an equation.

If $a = b$, then $a + c = b + c$

Multiplicative Property of Equality

The Multiplicative Property of Equality states that equality is maintained if you multiply both sides of an equation by the same amount.

If $a = b$, then $a \cdot c = b \cdot c$

The Cola Machine: Functions and Equality

CM-120. Solve each of the following equations for x.

a) $x^2 - 4 = 16$

b) $\sqrt{x^2 - 4} = 16$

c) $\sqrt{x - 4} + 2 = 16$

d) $x^2 - 4x + 2 = 0$

CM-121. Solve each of the following systems of equations.

a) $3x - 2y = 4$
$4x + 2y = 10$

b) $p + q = 4$
$-p + q = 7$

c) $y = x + 2$
$x + y = -4$

d) $2x - 5y = 16$
$3x + y = 11$

CM-122. Find each of the values below if $h(x) = 2x + 3$.

a) $h(0)$

b) $h(0.5)$

c) $h(-11)$

CM-123. Determine the corresponding input(s) or output(s) for the given function. If there is no possible input or output, explain why not.

a)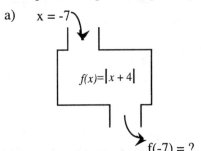

b)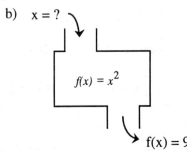

CM-124. Find the area of a right triangle with a hypotenuse of length 13 centimeters and one leg of length 5 centimeters. What subproblem did you need to solve?

CM-125. A rectangle has one side of length 7 centimeters and a **diagonal** 10 centimeters long. Draw a diagram and find its area.

CM-126. Simplify the expressions below.

a) $\dfrac{3x^2 + 8x + 5}{x^2 - 5x - 6} \cdot \dfrac{2x - 5}{3x + 5}$

b) $\dfrac{x^2 + x - 12}{x^2 - x - 6} \div \dfrac{x - 5}{x^2 - 3x - 10}$

UNIT SUMMARY AND REVIEW

CM-127. UNIT 11 SUMMARY

With your study team, list 3 or 4 big ideas that you feel best represent Unit 11. Write a description of each main idea. For each topic, choose a problem that best demonstrates your understanding. Explain your solution as in previous unit summaries. Show all your work.

CM-128. Review the Algebraic Properties in your tool kit. Then identify which property is being used below.

a) $\dfrac{(x+4)^2}{(x+4)(x-2)} = \dfrac{x+4}{x-2}$

b) $16 + 2x = 2x + 16$

c) If $x + y = 16$ and $16 = m^2$, then $x + y = m^2$.

d) $4 + (-4) = 0$

e) $3 + (6 + 2x) = (3 + 6) + 2x$

f) $18x^2y^3 + 0 = 18x^2y^3$

g) $-4x(3x - 16x^2) = -12x^2 + 64x^3$

h) If $\frac{3}{4}x = 21$, then $x = 21 \cdot \frac{4}{3}$.

CM-129. Solve for x using any method. Check your solution(s) by testing them in the original equation.

a) $|2 - x| = 4$

b) $|x - 2| = 4$

c) Explain why the equations in parts (a) and (b) above had to have the same solution.

CM-130. Find the equation of the line perpendicular to $y = \frac{4}{3}x + 7$ and which passes through the point (-4, 5).

CM-131. Simplify the expressions below.

a) $\dfrac{2x^2 - x - 15}{x + 4} \cdot \dfrac{x^2 + 4x}{2x + 5}$

b) $\dfrac{2x - 3}{x^2 + 2x - 35} \div \dfrac{x - 6}{x^2 - 11x + 30}$

CM-132. Solve for x. Use any method. Check your solution(s) by testing them in the original equation.

a) $|x-2|=3$

b) $|x-2|+3=6$

c) Explain why the equations in parts (a) and (b) above had to have the same solution.

CM-133. Solve each of the following equations.

a) $\frac{x}{3} = x + 4$

b) $\frac{x+6}{3} = x$

c) $\sqrt{3x-1} = 18$

d) $\frac{2x+3}{6} + \frac{1}{2} = \frac{x}{2}$

CM-134. Find the area of each triangle described below:

a) The triangle enclosed by the line $y = 2x + 6$, the x-axis and the y-axis.

b) The triangle enclosed by the line $y = 7x - 56$, the x-axis and the y-axis.

CM-135. Solve each system for x and y:

a) $3x + 3y = 15$
$x - y = 6$

b) $x = 6 - 2y$
$3 - 2x = -5y$

c) $y = 2x + 4$
$y = 4x - 3$

d) Extension: $2x^2 - y = 5$
$x + 2y = -5$

CM-136. TOOL KIT CHECK-UP

Your tool kit contains reference tools for algebra. Return to your tool kit entries. You may need to revise or add entries.

Be sure that your tool kit contains entries for all of the items listed below. Add any topics that are missing to your tool kit NOW, as well as any other items that will help you in your study of algebra.

- Algebraic Properties
- The Elimination Method, Parts One and Two
- Solving Equations with Absolute Value Algebraically
- Relations and Functions
- Definition of Absolute Value
- Using a Reference Point to Solve Equations with Absolute Value

GM-21. HOW IS THAT POSSIBLE?

Ms. Speedi needs your help! She wrote a test question (without making a key) and does not know the answer. To her surprise, her class arrived at two different solutions and Ms. Speedi does not know which answer is correct. The question was:

"Using scissors, cut out the shapes at right, form a large triangle, and find the total area of all the pieces. Be sure to sketch your large triangle and justify your solution."

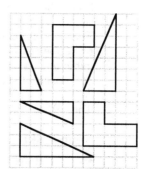

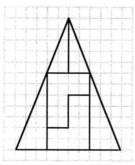

Figure 1

Half of the class put the pieces together as shown in figure 1 at left.

By using the triangle formula, the total area of the pieces in Figure 1 is 60 un^2.

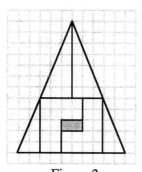

Figure 2

However, the other half of her class put the pieces together as shown in Figure 2. This triangle had the same overall area, but there was a gap of 2 square units inside!

Subtracting the missing area, the total area of the pieces in Figure 2 is 58 un^2.

Ms. Speedi is not sure which answer is correct and needs to know so she can grade these tests. She's provided you with a Resource Page so you can cut out the pieces for yourself and test these two solutions.

<< problem continues on next page >>

Your Task:

- Cut out the pieces on the Resource Page and arrange them as shown in both figures.

- Calculate the areas of both configurations and check the solutions given above.

- Write a letter to Ms. Speedi explaining to her what happened. Give her advice on how to grade these tests.

GM-22. **STAIRCASES**

Study the following staircases.

The first staircase is made of one cube (its volume is one cubic unit) and its surface is composed of six squares.

  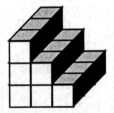

The second staircase has a volume of 6 cubic units and a surface area of 22 square units.

Your Task:

- Find the volume and surface area of the third staircase shown above.

- Describe what the 50th staircase will look like. Use geometric patterns, tables, and sketches to find its volume and the surface area.

- In your explanation, describe as many **patterns** as you can and explain how you arrived at your answers and why you think they are correct.

Unit 12
The Grazing Goat
Problem Solving and Inequality

Unit 12 Objectives
The Grazing Goat: **PROBLEM SOLVING AND INEQUALITY**

Throughout the year you have learned to solve various kinds of equations. This unit will use those skills as the basis for learning how to solve inequalities. In addition, this unit will give you more practice with your problem solving strategies for problems that involve distance, rate, and time as well as rate of work, mixture, and area problems.

You will use the **subproblem** approach to "uncomplicate" many of the problems in this unit. The Grazing Goat problems (GG-0 and GG-127) are examples of a complicated situation that can be most easily approached by looking for subproblems (smaller, more manageable problems).

In this unit you will have the opportunity to:

- review and extend the various problem solving strategies that have been introduced during the year.

- extend your study of solving equations to solving inequalities.

- extend your study of linear systems to linear inequalities.

- solve more challenging problems that involve area and subproblems.

- add and subtract rational expressions.

You will review solving equations and systems of equations while extending them to inequalities. Your work with rational expressions will help you practice factoring. The entire unit will challenge you to apply the problem solving strategies you have learned this year.

Read the following problem carefully. **Do not try to solve it now**. During this unit, you will gain more proficiency and skills to solve this problem.

GG-0.	THE GRAZING GOAT
	A barn 15 meters by 25 meters stands in the middle of a large grassy field. Zoe the goat, who is always hungry, is tied by a rope to one corner of the barn. Over what area of the field can Zoe graze if the rope is x meters long?

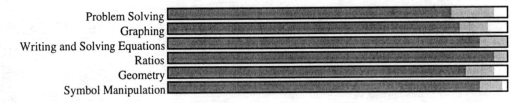

Unit 12
The Grazing Goat: PROBLEM SOLVING AND INEQUALITY

PROBLEM SOLVING WITH DISTANCE, RATE AND TIME

GG-1. Carol and Jan leave from the same place and travel on the same road. Carol walks at a rate of two miles per hour. Carol left five hours earlier than Jan, but Jan bikes at a rate of six miles per hour. When will Jan catch up? To solve this problem, answer the questions below:

a) We know that they left from the same place, are traveling in the same direction, and that Jan catches up with Carol, so draw a diagram like the one at right on your paper.

b) Carol has a head start. How many miles has Carol traveled before Jan leaves? Place that distance in the proper place on the diagram.

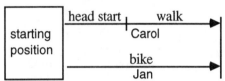

c) Let x represent the time Carol travels <u>after</u> her head start. On your diagram, write an expression using x to represent the distance Carol travels <u>after</u> her head start. Then write an expression that represents Carol's <u>total</u> distance. On your paper, write "Carol's total distance = _____ " so that you know what the expression represents.

d) On your diagram, write an expression using x to represent the distance Jan travels. Be sure to label your expression with "Jan's distance = _____ ."

e) Use your diagram to write an equation that says Carol and Jan travel the <u>same</u> distance, and then solve the equation.

f) How long does it take Jan to catch up with Carol?

GG-2. Matilda and Nancy are 60 miles apart, bicycling <u>toward</u> each other on the same road. Matilda rides 12 miles per hour and Nancy rides eight miles per hour. In how many hours will they meet?

GG-3. Two trucks leave a rest stop at the same time. One heads due east; the other heads due north and travels twice as fast as the first truck. The trucks lose radio contact when they are 47 miles apart. (They are obviously using an illegal power amplifier since FCC regulations limit CB's to five watts, a normal range of about five miles.) How far has each truck traveled when they lose contact?

GG-4. Two cars start together and travel in the same direction. One car goes twice as fast as the other. After five hours, they are 275 kilometers apart. How fast is each car traveling?

GG-5. Two cars leave a parking lot at the same time. One goes south and one goes west. One car is traveling at 55 miles per hour and the other at 45 miles per hour. How long will it take before the cars are 150 miles apart?

GG-6. A cheetah spots a gazelle 132 meters away. The cheetah starts towards the gazelle at a speed of 18 meters per second. At the same instant, the gazelle starts moving away at 11 meters per second. How long will it take the cheetah to get dinner?

GG-7. Two trucks leave the same rest stop at the same time traveling at the same speed. One heads south, and the other travels west. When the two trucks lose CB contact, they are 53 miles apart. How far has each traveled?

GG-8. Without a calculator, add $\frac{5}{12} + \frac{1}{12}$. Simplify your solution. Describe your method.

GG-9. Simplify the expressions below.

a) $\dfrac{x+3}{x^2+6x+9}$

b) $\dfrac{2x-10}{x^2-x-12} \cdot \dfrac{x-4}{x-5}$

GG-10. Solve each of the following systems of equations for x and y.

a) $4x + y = -4$
$2x - 3y = 19$

b) $2x - 5y = 7$
$x + 3y = 5$

GG-11. Solve for x. Use any method. Check your solution(s) by testing them in the original equation.

a) $|x+4| = 2$

b) $|x-5| = -5$

c) $|x-0.5| = 0.5$

d) $\frac{1}{2}|x-2| = 6$

GG-12. Solve for r in parts (a) and (b).

a) $A = \pi r^2$

b) $C = 2\pi r$

c) What part of the circle is associated with the length 2r?

GG-13. Write an absolute value equation that represents the solutions shown on the number line below.

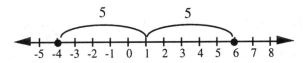

GG-14. Review the Algebraic Properties in your tool kit. Then identify which property is being used below.

a) $x + (-x) = 0$

b) $5x(2y - 3x) = 10xy - 15x^2$

c) If $y = 4x$ and $y = 2x - 1$, then $4x = 2x - 1$.

d) $-7 + 0 = -7$

ADDING AND SUBTRACTING RATIONAL EXPRESSIONS

GG-15. Without your calculator, complete each part below. Pay particular attention to explaining how you do each problem.

a) Add $\frac{3}{8} + \frac{1}{8}$ and reduce your result, if possible. Using complete sentences, describe your method for adding fractions with a common denominator.

b) Subtract $\frac{1}{3} - \frac{3}{4}$ and reduce your result if possible. Using complete sentences, describe your method for subtracting fractions with unlike denominators.

GG-16. Add the following fractions. Simplify your solution, if possible.

a) $\frac{2}{x+3} + \frac{x}{x+3}$

b) $\frac{x}{(x+2)(x+3)} + \frac{2}{(x+2)(x+3)}$

The Grazing Goat: Problem Solving and Inequality

GG-17. Subtract $(3x + 4) - (x - 5)$.

    a) What did you have to do to the second polynomial?

    b) Subtract and simplify, if possible: $\dfrac{3x+4}{x+3} - \dfrac{x-5}{x+3}$.

    c) Subtract and simplify, if possible: $\dfrac{8x+3}{2x+3} - \dfrac{2x-6}{2x+3}$.

GG-18. Add or subtract, then simplify if possible.

    a) $\dfrac{x^2}{x-5} - \dfrac{25}{x-5}$ c) $\dfrac{a^2}{a+5} + \dfrac{10a+25}{a+5}$

    b) $\dfrac{x^2}{x-y} - \dfrac{2xy - y^2}{x-y}$ d) $\dfrac{x}{x+1} - \dfrac{1}{x+1}$

GG-19. Add $\dfrac{3}{4} + \dfrac{1}{6}$. Reduce your answer.

    a) What did you do to add these fractions? Describe each subproblem thoroughly.

    b) What would be the least common multiple of 4x and 6x?

    c) Add $\dfrac{3}{4x} + \dfrac{11}{6x}$.

Recall that for a circle of radius r, its circumference, C, is $2\pi r$ and its area, A, is πr^2. Find the π key on your scientific calculator and use it for computations.

GG-20. A picture of an ice-cream cone is shown at right. Find the area of the picture. Assume the shape of the ice-cream creates a semi-circle. Show all subproblems.

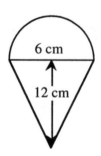

6 cm

12 cm

GG-21. The diameter of a flattened Frisbee is 9.2 inches. Find its area and circumference. Be sure to draw pictures and identify any subproblems.

GG-22. Add or subtract and simplify, if possible.

a) $\frac{x}{2} + \frac{x-4}{2}$

b) $\frac{x}{x^2 - 2x - 3} - \frac{3}{x^2 - 2x - 3}$

GG-23. Solve for x. Check your solution.

a) $3x^2 = 12$

b) $\sqrt{x-5} = 7$

c) $|x-5| = 7$

d) $\sqrt{3-x} + 5 = 8$

GG-24. Find the roots (x-intercepts) of the parabola $y = \frac{1}{2}x^2 - 6$. Verify your solution by graphing.

GG-25. If $f(x) = \frac{2}{3}x + 11$, find f(-6), f(5) and f(300).

GG-26. Using complete sentences, write down what you know about adding and subtracting rational expressions. Use examples to demonstrate your knowledge.

GG-27. What is the effect of multiplying a number by one? By $\frac{3x}{3x}$? By $\frac{2-x}{2-x}$? Explain your reasoning.

GG-28. Use your results in part (a) below to help solve part (b).

a) What is the least common multiple of (2x + 6) and (x + 3) ?

b) Add and simplify: $\frac{4x}{2x+6} + \frac{3}{x+3}$

GG-29. Use your results in part (a) below to help solve part (b).

a) What would be the least common multiple of $x^2 - 1$ and $x + 1$? (Hint: first, factor $x^2 - 1$.)

b) Subtract and simplify: $\frac{x}{x^2-1} - \frac{2}{x+1}$.

The Grazing Goat: Problem Solving and Inequality

GG-30. The following box contains examples of adding and subtracting algebraic fractions with unlike denominators. Add this information to your tool kit.

ADDING AND SUBTRACTING RATIONAL EXPRESSIONS

The Least Common Multiple of $(x+3)(x+2)$ and $(x+2)$ is $(x+3)(x+2)$. **Why?**

$$\frac{4}{(x+2)(x+3)} + \frac{2x}{x+2}$$

The denominator of the first fraction already is the Least Common Multiple. To get a common denominator in the second fraction, multiply the fraction by $\frac{x+3}{x+3}$, a form of one (1).

$$= \frac{4}{(x+2)(x+3)} + \frac{2x}{x+2} \cdot \frac{(x+3)}{(x+3)}$$

Multiply the numerator and denominator of the second term:

$$= \frac{4}{(x+2)(x+3)} + \frac{2x(x+3)}{(x+2)(x+3)}$$

Distribute the numerator.

$$= \frac{4}{(x+2)(x+3)} + \frac{2x^2 + 6x}{(x+2)(x+3)}$$

Add, factor, and simplify.

$$= \frac{2x^2 + 6x + 4}{(x+2)(x+3)} = \frac{2(x+1)(x+2)}{(x+2)(x+3)} = \frac{2(x+1)}{(x+3)}$$

GG-31. Some of the following algebraic fractions have common denominators and some do not. With your team, add or subtract the expressions and simplify them if possible.

a) $\dfrac{3}{(x-4)(x+1)} + \dfrac{6}{x+1}$

b) $\dfrac{x}{x^2 - x - 2} - \dfrac{2}{x^2 - x - 2}$

c) $\dfrac{y^2}{y+4} - \dfrac{16}{y+4}$

d) $\dfrac{5}{2(x-5)} - \dfrac{3x}{x-5}$

e) $\dfrac{3x}{x^2 + 2x + 1} + \dfrac{3}{x^2 + 2x + 1}$

f) $\dfrac{x+2}{x^2 - 9} - \dfrac{1}{x+3}$

GG-32. Cleopatra rode an elephant to the outskirts of Rome at two kilometers per hour and then took a chariot back to camp at 10 kilometers per hour. If the total traveling time was 18 hours, how far was it from camp to the outskirts of Rome?

GG-33. Consider the following equations:

a) Solve for x. Check your solution. 2x - 5 = 3

b) Clifford thinks x = 4 is a solution to 5x - 11 = 9. Is he correct? Show why or why not.

GG-34. Fernando solved 4(3x - 12) = 16 for x below. Justify each step of his solution with the appropriate Algebraic Property.

The problem: 4(3x - 12) = 16

Step 1: 12x - 48 = 16

Step 2: 12x - 48 + 48 = 16 + 48

Step 3: 12x = 64

Step 4: $\frac{1}{12} \cdot 12x = \frac{1}{12} \cdot 64$

Step 5: $x = \frac{16}{3}$

GG-35. Ms. Gerek recently bought a scooter for $6000. She estimates that over time the value of the scooter will depreciate (decrease) by $20 per month. Mr. Armentrout, on the other hand, bought a car for $8700. Studies show his car will depreciate at a rate of $80 per month. When will their vehicles be worth the same amount?

GG-36. Factor each of the following expressions.

a) $x^2 - 25$

b) $x^2 + 25$

c) $4x^2 - 49$

d) $7x^2 - 49$

e) $100x^2 - 49y^2$

f) $16y^2 - 1$

g) Which of the above problems are examples of the "Difference of Squares" ?

h) Compare part (b) with part (a). What was different?

i) Compare parts (c) and (d). They appear very similar, yet their answers are very different. Why?

GG-37. Find each of the following sums.

a) $\frac{1}{2} + \frac{1}{3}$

b) $\frac{1}{2} + \frac{x}{3}$

c) $\frac{x}{2} + \frac{x}{3}$

d) $\frac{1}{2} + \frac{1}{x}$

e) $\frac{1}{3} + \frac{1}{2x}$

GG-38. Solve for x. Check your solution.

a) $x - 6 = 4$

b) $\sqrt{x - 6} = 4$

c) $(x - 6)^2 = 4$

d) $|x - 6| = 4$

INTRODUCTION TO INEQUALITIES

GG-39. We introduced the symbols $\leq$ and $\geq$ in Unit 3. Review the definitions and then determine if the statements below are true or false.

INEQUALITY NOTATION	
Symbol	**Translation**
<	less than
>	greater than
$\leq$	less than or equal to
$\geq$	greater than or equal to

a) $3 < 5$ c) $2 \leq -5$ e) $4 > 4$ g) $-1 \leq -1$

b) $9 > -3$ d) $-2 \geq -6$ f) $|-4| < |6|$ h) $7 \geq |-10|$

GG-40. Clifford is at it again! He thinks $x = 3$ is a solution to $2x + 1 > 5$.

a) Is he correct? Show why or why not.

b) Are there any other solutions? If so, name at least three. If not, explain why not.

GG-41. This time, Clifford thinks $x = 7$ is a solution to $3(x - 2) \leq 4$. Is he correct? Show why or why not.

GG-42. With your study team, find at least <u>five</u> x-values that make this inequality true:
$$2x - 5 \geq 3$$

a) How many solutions are there?

b) What is the smallest solution for x?

c) What is the significance of the smallest solution from part (b)?

d) On a number line, plot a point for each of your solutions for x.

GG-43. Why did $2x - 5 = 3$ only have one solution, while $2x - 5 \geq 3$ had an infinite number of solutions? Discuss this with your study team and record all ideas on your paper.

GG-44. Find an x-value that makes $x - 4 > 2$ true.

a) Is $x = 6$ a solution? Why or why not?

b) Is $x = 7$ a solution? Is $x = 6.1$ a solution? Is $x = 6.01$ a solution? What about $x = 6.001$?

c) In your own words, describe which x-values are solutions to $x - 4 > 2$.

d) Using mathematical notation, write an expression that represents all solutions to the inequality $x - 4 > 2$.

GG-45. Simplify each expression:

a) $\dfrac{x^2}{x-2} \cdot \dfrac{x^2 - 3x + 2}{x}$

b) $\dfrac{x^2 - 3x - 10}{x^2 - 8x + 15} \div \dfrac{x^2 + 8x + 12}{x^2 + 3x - 18}$

GG-46. Find the equation of the line:

a) with slope $-\dfrac{3}{5}$ passing through the point $(-6, 2)$.

b) perpendicular to the line in part (a) and passing through $(-6, 2)$.

GG-47. Add the following fractions using the same process you used in problem GG-31. Simplify your solution, if possible.

a) $\dfrac{2-x}{x+4} + \dfrac{3x+6}{x+4}$

b) $\dfrac{x^2-4}{x-5} - \dfrac{x^2-4}{x-5}$

c) $\dfrac{3}{(x+2)(x+3)} + \dfrac{x}{(x+2)(x+3)}$

d) $\dfrac{3}{x-1} - \dfrac{2}{x-2}$

e) $\dfrac{8}{x} - \dfrac{4}{x+2}$

GG-48. Solve each of the following quadratic equations. Although the Quadratic Formula will always work, factoring is sometimes faster.

a) $x^2 - 7x + 6 = 0$

b) $y^2 + 6 = 7y$

GG-49. Solve the following systems of equations using any method.

a) $2x + y = 4$
$3x - y = 16$

b) $x + y = 4$
$2y + 3x = -39$

GG-50. Find the area and perimeter of this trapezoid. Show all subproblems.

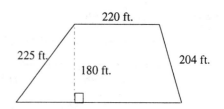

GG-51. Jethro and Mac were in a 24-hour bicycle race. Jethro biked at an average speed of 17 miles per hour and finished 60 miles ahead of Mac. What was Mac's average speed? If you need help getting started, draw a diagram.

GG-52. Farmer Fran is building a rectangular pig pen alongside her barn. She has 100 feet of fencing, and she wants the largest possible area in which the pigs can muck around. What should the dimensions of the pig pen be?

SOLVING LINEAR INEQUALITIES

GG-53. Using complete sentences, explain why $x = 9$ is a solution to $x \leq 9$ but <u>not</u> a solution to $x < 9$.

GG-54. Write the inequality that represents the x-values highlighted on the number line below. On a number line graph, the **open endpoints** represent < or >, while **closed endpoints** represent ≤ and ≥.

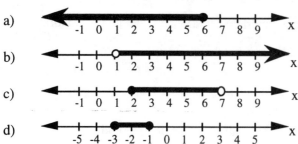

GG-55. Read the information in the box below, add it to your tool kit, then complete parts (a) through (c).

> In the solutions of the inequality $2x - 5 \geq 3$ (problem GG-42), $x = 4$ is called the **DIVIDING POINT** because it divides the number line into two parts: the set of numbers less than four and the set of numbers greater than four.
>
> The dividing point is the solution to the **equality** $2x - 5 = 3$. Once we find the dividing point, we can test numbers on each side of our dividing point to determine the solution to our **inequality**. For $2x - 5 \geq 3$, the solution is $x \geq 4$, shown above right on the number line.
>
>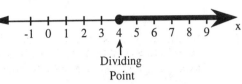
>
> a) Find the dividing point for the inequality $2 + 3x \leq -4$ by solving the equation $2 + 3x = -4$. Plot this point on a number line.
>
> b) Your solution to part (a) divides the set of numbers into two parts. Which part is the solution for this inequality? With your study team, test points on both side of your dividing point to help determine the solution to $2 + 3x \leq -4$.
>
> c) In writing your solution, decide if your dividing point is part of your solution. If it is not part of your solution, < and > are the appropriate inequality symbols to use. Otherwise, $\leq$ and $\geq$ will indicate that you are <u>including</u> your dividing point. Write the solution as an algebraic inequality.

GG-56. Use the method from problem GG-55 to solve the following inequalities for x.

   a) $4x - 1 \geq 7$

   b) $2(x - 5) \leq 8$

   c) $3 - 2x < x + 6$

   d) $\frac{1}{2}x > 5$

GG-57. For the following problem, set up an inequality that shows the various numbers of cans Mr. Drucker could add to the machine, then solve it. Using an inequality allows for the possibility that he might restock the machine at less than its full capacity.

As Mr. Drucker prepared to restock the soda machine, he noticed there were already 20 cans of soda in the machine. If the soda machine can hold a maximum of 140 cans of soda, how many six-packs can he load in the machine?

The Grazing Goat: Problem Solving and Inequality

GG-58. Carefully read the following information. Then add "Solving Inequalities" to your tool kit.

SOLVING INEQUALITIES

To solve an inequality, we first treat the problem as an equality. The solution to the equality is called the **DIVIDING POINT**. For example, x = 12 is the Dividing Point of the inequality $3 + 2(x - 5) \leq 17$, as shown below:

Problem: $3 + 2(x - 5) \leq 17$

First change the problem to an equality and solve for x:

$$3 + 2(x - 5) = 17$$
$$3 + 2x - 10 = 17$$
$$2x - 7 = 17$$
$$2x = 24$$
$$x = 12$$

Since our original inequality <u>included</u> x = 12, we place our Dividing Point on our number line as a solid point. We then test one value on either side in the <u>original</u> inequality to determine which set of numbers makes the inequality true.

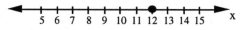

Test: x = 8	Test: x = 15
$3 + 2(\mathbf{8} - 5) \leq 17$	$3 + 2(\mathbf{15} - 5) \leq 17$
$3 + 2 \cdot 3 \leq 17$	$3 + 2 \cdot 10 \leq 17$
$3 + 6 \leq 17$	$3 + 20 \leq 17$
TRUE!	**FALSE!**

Therefore, the solution is $x \leq 12$.

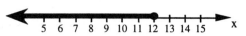

When the inequality is "<" or ">," then the Dividing Point is <u>not</u> included in the answer. On a number line, this would be indicated with an open circle at 12.

GG-59. Solve the following inequalities for x. Graph your solutions on a number line.

a) $5 < 8x - 3$

b) $9 + 4x \geq 6x + 15$

c) $\frac{x}{3} \leq \frac{6}{7}$

d) $5(x - 2) > -10$

GG-60. Complete each problem below. Look for similarities and differences in these two problems.

a) Add: $\frac{1}{x} + \frac{2}{3x}$

b) Solve for x: $\frac{1}{x} + \frac{2}{3x} = 10$

GG-61. Tony managed to drive from Pittsburgh to Philadelphia at a speed of 70 miles per hour. About how long (in seconds) did it take Tony to drive one mile?

GG-62. Add or subtract, then simplify if possible.

a) $\dfrac{3}{8} - \dfrac{1}{8}$

b) $\dfrac{2x}{5} - \dfrac{3y}{4}$

c) $\dfrac{x^2}{x-2} - \dfrac{2x+7}{x-2}$

d) $\dfrac{4}{2a} + \dfrac{-2}{6a}$

GG-63. Solve for x. Check your solution.

a) $|x| = 19$

b) $|x - 52| = 12$

c) $|x + 10| = 0$

d) $|2x - 1| = 7$

GG-64. When a family with two adults and three children bought tickets for a movie, they paid a total of $27.75. The next family in line, with two children and three adults, paid $32.25 for the same movie. Find the adult and child ticket prices by writing a system of equations with two variables.

GG-65. A grass playing field at Josie Smith Middle School is shaped as shown at right. The rounded pieces are each circular arcs of radius 35 yards. If two pounds of fertilizer are needed for every 100 square feet of grass, about how many pounds of fertilizer will be needed?

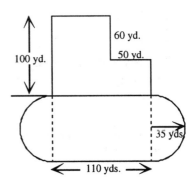

The Grazing Goat: Problem Solving and Inequality

SOLVING QUADRATIC INEQUALITIES

GG-66. In problem GG-55, we focused on linear inequalities. How can we use those ideas with non-linear inequalities? In this problem, we will focus on the quadratic inequality at right: $x^2 - 3x - 18 \leq 0$

a) To find our dividing point(s), solve the equality $x^2 - 3x - 18 = 0$.

b) In the case of this parabola, we seem to have **two** dividing points! Plot these dividing points on a number line.

c) How many regions (sections or "pieces") do two different points create on the number line? Identify each region.

d) For what x-values is $x^2 - 3x - 18 \leq 0$? Test a number in each region. Be sure to consider whether the dividing points themselves are part of the solution.

e) Suppose we change the direction of the inequality symbol. Solve: $x^2 - 3x - 18 > 0$.

GG-67. Solve each inequality for x by finding the dividing points and testing a point in each region.

a) $(x + 2)(x - 4) < 0$
c) $x^2 - 4 \leq 0$
b) $x^2 + 5x + 4 > 0$

GG-68. For the following problem, set up an inequality and solve it.

Ms. Speedi's convertible can travel 32 miles per gallon of gasoline. If her gas tank already has 4 gallons of gas, how many gallons could she add in order to travel at least 280 miles?

GG-69. Solve the inequality for x: $x^2 - 4x + 8 \geq 2$

GG-70. Examine the graph of the line at right. Write the equation of the line.

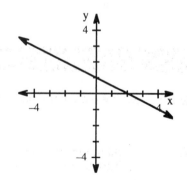

GG-71. Solve for x.

a) $\frac{2}{3}x \geq 16$
c) $4(6x - 1) < 40$
b) $1 - 3(x - 4) = -11$
d) $-12 < 3 + 8x$

GG-72. A piece of metal at 20°C is warmed at a steady rate of two degrees every minute. At the same time, another piece of metal at 240°C is cooled at a steady rate of three degrees every minute.

    a) Write equations to describe how the temperature of each piece of metal is changing.

    b) After how many minutes is the temperature of each piece of metal the same?

GG-73. Add or subtract and simplify. You may need to start by finding the common denominator.

    a) $-\dfrac{1}{3} - \dfrac{1}{6}$

    b) $\dfrac{x}{3} + \dfrac{-4}{5}$

    c) $\dfrac{x+3}{4} + \dfrac{x-1}{4}$

    d) $\dfrac{x^2 - 2x + 1}{x^2 - 1} + \dfrac{x^2 + 2x - 3}{x^2 - 1}$

    e) Extension: $\dfrac{7}{x} - \dfrac{x-1}{y}$

GG-74. Solve for x. Check your solution.

    a) $|x-11| = 2$

    b) $5|x-2| = 15$

    c) $|x-2| - 6 = -3$

    d) $|3x+1| + 4 = 11$

GG-75. Review the Algebraic Properties in your tool kit. Then identify which property is being used below. Name two properties in parts (b) and (d).

    a) $(8x + 5y) - 3y = 8x + (5y - 3y)$

    b) $\dfrac{4}{3} \cdot (\dfrac{3}{4} x) = 24 \cdot \dfrac{4}{3}$

    c) If $x = 7$ and $y = 3x - 9$, then $y = 3(7) - 9$

    d) If $3x - 8 = 27$, then $3x - 8 + 8 = 27 + 8$

SOLVING INEQUALITIES WITH ABSOLUTE VALUE

GG-76. Solve the inequality for x: $x^2 - x - 6 > 0$

GG-77. What happens when inequalities include absolute values? With your team, investigate the solutions for $|x| \leq 4$. How many dividing points does this inequality have?

The Grazing Goat: Problem Solving and Inequality

GG-78. Find the dividing points for the inequality $|x-3|>2$. Then, on a number line, test each region to determine the solution.

GG-79. Using Sherry's method from Unit 11, problem CM-84, $|x-3|=2$ translates as "The distance between x and three is equal to two units."

a) Similarly translate the inequality $|x-3|>2$.

b) The solution to $|x-3|>2$ is shown on the number line below. Using complete sentences, describe how the solution to this inequality relates to your translation from part (a).

c) Translate $|x-7|\geq 3$ and use a number line to quickly find its solution.

d) Write an inequality that represents the set of numbers that are less than 4 units away from 5. Then draw a number line that shows the solution to this inequality.

GG-80. Solve each inequality for x:

a) $|x|<3$

b) $x^2<9$

c) Compare parts (a) and (b) and their solutions. What do you notice? Is there any connection?

GG-81. Solve the following inequalities for x. Graph your solutions on a number line.

a) $3(x-6) \leq 12 - 2x$

b) $x^2 - 2x - 15 > 0$

c) $2x^2 - 5x - 3 < 0$

GG-82. Below is a mixture of rational expressions. Perform the indicated operation and simplify your solution, if possible.

a) $\dfrac{6}{x^2+8x+12} \cdot \dfrac{x+6}{3x-3}$

b) $\dfrac{4x}{5} - (x+1)$

c) $\dfrac{(x-2)(x+5)}{4(x+3)(x-2)} \div \dfrac{(x+1)(x+5)}{10(x+1)}$

d) $\dfrac{x^2-5x+10}{x^2+x-2} - \dfrac{x^2-8x+4}{x^2+x-2}$

e) $\dfrac{x+4}{2x^3} + \dfrac{5}{10x}$

GG-83. Solve each inequality for x:

a) $|x|-4>1$

b) $|x-4|>1$

c) Compare parts (a) and (b) and their solutions. What do you notice? Is there any connection?

GG-84. Review the Algebraic Properties in your tool kit. Then identify which property is being used below.

a) If y = 4 and x = 2y - 1, then x = 2(4) - 1.

b) $\frac{2}{3} \cdot \frac{3}{2} = 1$

c) 3x · 1 = 3x

d) $x^3 = x^3$

GG-85. One leg of a right triangle is five inches shorter than the hypotenuse. The other leg is 15 inches long. How long is the hypotenuse?

GG-86. Solve for x: $\frac{4}{3x} + \frac{6}{x} = 9$

GRAPHING LINEAR INEQUALITIES

GG-87. For the line $y = \frac{3}{2}x + 2$:

a) Does (-4, -4) lie on the line?

b) Does (6, 10) lie on the line? How can you tell?

c) Use graph paper and graph the line.

GG-88. With your study team, decide if the points listed below make the inequality $y > \frac{3}{2}x + 2$ true or false. If true, plot the point on your graph from the previous problem.

a) (2, 6)   b) (0, 0)   c) (4, 10)   d) (-1, 3)

GG-89. A point is called a **solution** to an inequality if it makes the inequality true.

a) Find three more solutions (in addition to those you found in problem GG-88) that make the inequality $y > \frac{3}{2}x + 2$ true.

b) How many solutions exist for $y > \frac{3}{2}x + 2$?

c) What do all the solutions to $y > \frac{3}{2}x + 2$ have in common?

d) How can we represent all the solutions to the equation?

The Grazing Goat: Problem Solving and Inequality

GG-90. Read the information in the box below, add it to your tool kit, then complete parts (a) through (d).

> To graph the linear inequality $y > \frac{3}{2}x + 2$, we must first graph $y = \frac{3}{2}x + 2$, referred to as the **DIVIDING LINE**. This dividing line is similar to a dividing point since it divides the graphing region into two parts.

The dividing line represents the points which satisfy the **equality** $y = \frac{3}{2}x + 2$. Once we graph the dividing line, we can test points on each side to determine the solution to our **inequality**. For $y > \frac{3}{2}x + 2$, the solution turned out to be all the points that lie above the line, as shown shaded at right.

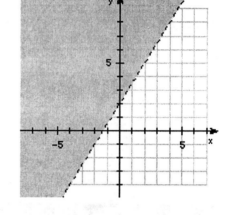

a) Why was the dividing line graphed at right dashed and not solid?

b) Find the dividing line for the inequality $y \leq -2x - 3$ by graphing the line $y = -2x - 3$.

c) Your line from part (b) divides the graph into two parts. However, which part is the solution for $y \leq -2x - 3$? With your study team, test points on both sides of the dividing line to help determine the solution to $y \leq -2x - 3$.

d) Decide if your dividing line is part of your solution. If it is not part of your solution, your dividing line should be represented with a dashed line. Otherwise, your dividing line should be solid.

GG-91. On a new set of axes, graph and shade the region represented by the inequality at right: $y > \frac{3}{4}x - 5$

GG-92. Solve for x. Check your solution.

a) $8 - 3x = 15$

b) $\frac{x-1}{5} = \frac{3}{x+1}$

c) $|x - 1| - 2 = 0$

d) $4 - 2(3x + 1) = 9x - 13$

GG-93. Solve the system of equations at right by graphing.

$y = 2x - 3$

$y = -\frac{1}{3}x + 4$

GG-94. Find the equation of the line parallel to $y = \frac{3}{4}x - 5$, and passing through the point (12, -2).

GG-95. Solve each inequality for x:

a) $6 - 3x > 2(3 + x)$

b) $x(x - 5) \geq 0$

c) $|x - 6| + 2 \leq 3$

d) $|x - 1| < 0$

GG-96. Simplify:

a) $\dfrac{3x^2 + 6x + 3}{x + 2} \cdot \dfrac{x + 3}{x^2 + 4x + 3}$

b) $\dfrac{(2x+1)(3x-7)}{(4x+11)(x-12)} \div \dfrac{(9x+7)(3x-7)}{(4x+11)(9x+7)}$

GG-97. Find the area of the shaded region in each of the following figures.

a)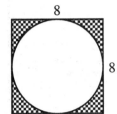

b)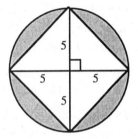

GG-98. Cassie walked for awhile at three miles per hour and then continued her journey on a bus at 15 miles per hour. Her time on the bus was twice as long as her time walking. How long did she ride on the bus if the total distance she covered was 66 miles? If you need help getting started, draw a diagram.

GRAPHING SYSTEMS OF LINEAR EQUATIONS

GG-99. Match each graph below with the correct inequality.

a)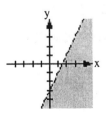

b)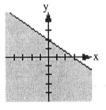

c)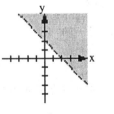

d)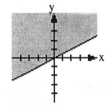

1) $y > -x + 2$

2) $y < 2x - 3$

3) $y \geq \dfrac{1}{2}x$

4) $y \leq -\dfrac{2}{3}x + 2$

GG-100. What does a solution for a system of equations represent?

The Grazing Goat: Problem Solving and Inequality

GG-101. We have much experience with graphing systems of equations such as the one shown at right from problem GG-93. However, what would a solution to a system of inequalities look like?

Notice that we have slightly altered our equations from problem GG-93 to create the system of inequalities below.

$$y \geq 2x - 3$$
$$y \leq -\frac{1}{3}x + 4$$

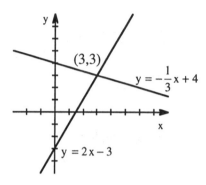

a) Sketch a copy of the graph (above right) on your paper. Then modify the graph (using the method in problem GG-90) to show the solution for each inequality.

b) Since we have two dividing lines, the graph is divided into four regions. With your study team, test at least two points in each region of the graph. Decide which region(s) satisfy both inequalities.

c) Compare your results from part (b) with the shadings in part (a). How does the shading of each inequality lead to the solution for the system?

GG-102. On graph paper, graph and shade the solution for each of the systems of inequalities below. Describe each resulting region.

a) $y \leq \frac{2}{5}x$
$y > 5 - x$

b) $y \leq \frac{1}{4}x + 3$
$y \geq \frac{1}{4}x - 2$

GG-103. Write a system of inequalities represented by the graph at right.

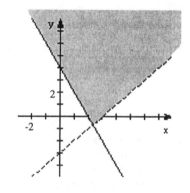

GG-104. Using complete sentences, describe your method for graphing systems of inequalities for a student who has missed class for the last couple of days. Be sure to include examples and important details.

GG-105. On graph paper, graph and shade the solution for the system of inequalities at right. Describe the resulting region.

$y > 3x + 1$
$y < 3x - 4$

420   UNIT 12

GG-106. Add the following fractions using the same process you used in problem GG-31. Simplify your solution, if possible.

a) $\dfrac{6x}{x-3} - \dfrac{1}{x+2}$

b) $\dfrac{12x-11}{3x-1} - \dfrac{3x-8}{3x-1}$

c) $\dfrac{2x-18}{(2x-1)(x+8)} + \dfrac{2}{(2x-1)}$

d) $\dfrac{3}{x} - \dfrac{x}{x+6}$

e) $\dfrac{x^2-3x-1}{4} + \dfrac{7x+5}{4}$

GG-107. Using complete sentences, explain why $|x|<-4$ cannot have a solution.

GG-108. Solve the inequality for x:

a) $11 + 4x - 3 < 6x$

b) $x^2 - 6x + 8 \leq 0$

c) $|x-1| > 5$

d) $|x-9| \geq -6$

GG-109. Graph and shade the solution for the system of inequalities at right.

$y \geq \dfrac{3}{4}x - 2$

$y < -\dfrac{1}{2}x + 3$

GG-110. Find the equation of the line perpendicular to $y = -2x$ and which passes through the point (6, 0).

GG-111. What diameter pizza would you have to buy to get at least 100 square inches of pizza? Be sure to draw pictures and identify any subproblems.

GG-112. A square with sides of length x centimeters has a circle of <u>diameter</u> x centimeters cut out of it. Find the fraction of the square that is left for the given values of x below. Hint: if you need help, review your solution to problem GG-97 (a).

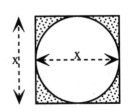

a) x = 2

b) x = 20

c) x = 10

d) For any x (in terms of x)

APPLYING PROBLEM SOLVING STRATEGIES

GG-113. You have solved a great variety of problems using several different strategies presented to you, as well as using strategies you and your study team developed. Some of these strategies are listed below in no particular order. Add these problem solving strategies, as well as any others you developed, to your tool kit as necessary.

PROBLEM SOLVING STRATEGIES

- Making a Guess and then Checking It (Guess And Check)
- Using Manipulatives such as Algebra Tiles
- Making Systematic Lists
- Graphing a Situation
- Drawing a Diagram
- Breaking a large problem into smaller Subproblems
- Writing and Solving an Equation

In problems GG-114 through GG-120, you will encounter a variety of problems that will require a problem solving strategy to solve. Remember to add new problem solving strategies to your tool kit as they are developed.

GG-114. Steven can look up 20 words in the dictionary in an hour. His teammate, Mary Lou, can look up 30 words per hour. Working together, how long will it take them to look up 100 words?

GG-115. Susan can paint her living room in 2 hours. Her friend, Jaime, estimates it would take him 3 hours to paint the same room. If they work together, how long will it take them to paint Susan's living room?

GG-116. How much coffee costing $6 a pound should be mixed with 3 pounds of coffee costing $4 per pound to create a mixture costing $4.75 per pound?

GG-117. A truck going 70 miles per hour passes a parked highway patrol car. When the truck is half a mile past the patrol car, the officer starts after it going 100 miles per hour. How long does it take the patrol car to overtake the truck?

GG-118. A rectangle is three times as long as it is wide. If the length and width are each decreased by four units, the area is decreased by 176 square units. What are the dimensions of the original rectangle?

GG-119. A square cake plate has area 150 square inches. Regina is baking a round cake. What is the maximum area the bottom of a round cake can have and still fit on the plate?

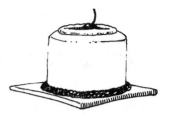

GG-120. Ms. Speedi's favorite recipe for fruit punch requires 12% apple juice. How much pure apple juice should she add to 2 gallons of punch that already contains 8% apple juice to meet her standards?

GG-121. Find the area of the remaining pizza if the figure at right was originally a 16" pizza.

GG-122. Find the equation of the line perpendicular to $y = \frac{1}{3}x - 5$ and which passes through the point (2, 1).

GG-123. On a new set of axes, graph and shade the region represented by the system at right.

$$y < -\frac{1}{3}x + 3$$
$$y \leq x + 1$$

GG-124. For the following problem, set up an inequality and solve it.

Baking dishes are three inches longer than they are wide. Susan is going to bake lasagna for six people and estimates that each person will want at least 18 square inches of lasagna. Help her decide which baking dishes she can use.

GG-125. Review the Algebraic Properties in your tool kit. Then identify which property is being used below.

a) $3 + (5 + x) = (3 + 5) + x$

c) $19 + 4x = 4x + 19$

b) $8(9 - 4x) = 72 - 32x$

UNIT SUMMARY AND REVIEW

GG-126. Solve for x and y.

a) $x = 3y + 1$
    $5y - x = 3$

b) $x + 3y = 4$
    $4x - 6y = 1$

GG-127. THE GRAZING GOAT

Zoe the goat is tied by a rope to one corner of a 15 meter by 25 meter barn in the middle of a large, grassy field. Over what area of the field can Zoe graze if the rope is:

a) 10 meters long?
b) 20 meters long?
c) 30 meters long?
d) 40 meters long?

e) Zoe is happiest when she has at least 400 m^2 to graze. What possible lengths of rope could be used?

GG-128. UNIT 12 SUMMARY

With your study team, list 3 or 4 big ideas that you feel best represent Unit 12. Write a description of each main idea. For each topic, choose a problem that best demonstrates your understanding. Explain your solution as in previous unit summaries. Show all your work.

GG-129. Using complete sentences, describe the different problem solving techniques you used during this unit. It may help to look back through the unit, review your homework, or reflect on classwork.

GG-130. On graph paper, graph and shade the solution for the system of inequalities at right.

$y \leq 2x + 5$
$y > 3$

GG-131. Solve the following inequalities for x. Graph your solutions on a number line.

a) $3x - 5 \leq 7 + 2x$
b) $5(2 - x) + 6 > 16$
c) $|x| - 3 < 7$
d) $|x + 2| > 3$

GG-132. Add the following fractions using the same process you used in problem GG-31. Simplify your solution, if possible.

a) $\dfrac{5}{x-8} + \dfrac{3+x}{x-8}$

b) $\dfrac{x+2}{3x+6} - \dfrac{5}{3}$

c) $\dfrac{2(x-1)}{x^2+3x-10} + \dfrac{2(2x-5)}{x^2+3x-10}$

d) $\dfrac{x}{x+6} + \dfrac{6}{x-2}$

e) $\dfrac{4}{2x-1} + 5$

GG-133. In the figure at right, the curves are quarter circles cut out of a rectangle. The radius of the smaller circle in the lower left corner is one unit.

a) Find the dimensions of the original rectangle.

b) Find the area of the figure.

c) Find the perimeter of the figure.

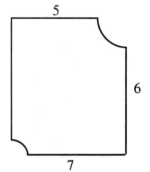

GG-134. If two 10-inch pizzas cost the same as one 15-inch pizza, which choice gets you more pizza for your money? What if the crust is your favorite part of a pizza? Would you still make the same choice? Explain your answers.

GG-135. **TOOL KIT CHECK-UP**

Your tool kit contains reference tools for algebra. Return to your tool kit entries. You may need to revise or add entries.

Be sure that your tool kit contains entries for all of the items listed below. Add any topics that are missing to your tool kit NOW, as well as any other items that will help you in your study of algebra.

- Problem Solving Strategies
- Solving and Graphing Inequalities
- Adding and Subtracting Rational Expressions
- Graphing Systems of Linear Inequalities

GM-23. BASIL'S BACKYARD (or THE CITY PERMIT)

Jerome has a problem. His pet schnauzer, Basil, has recently behaved very badly and your mail delivery will stop unless you can prevent Basil from reaching the mailbox. To make matters worse, a local city ordinance requires that you obtain a permit to leave a dog leashed during the day. In order to receive the permit, Jerome must fulfill the following requirements:

- The dog must have water.
- The dog must have access to shade.
- The dog must have at least 1,600 square feet to roam.

Jerome can attach the leash anywhere along the outside of his house as long as the requirements listed above are met. Basil's water dish is located below the water faucet, located two feet from the right front corner of Jerome's house, as shown in the diagram at right. The porch off the left side of Jerome's house provides shade. Jerome's mailbox is on the front of his house, 5 feet from the right-hand corner.

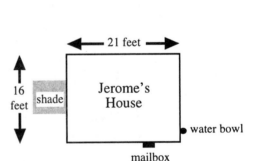

Your Task:

- Decide where Jerome can attach Basil's leash, and how long the leash should be to meet the city's requirements, as well as please the mail deliverer. Since Jerome likes to spoil Basil, try to maximize the space Basil has to roam.
- Obtain a city permit from your teacher and fill it out for Jerome.

GM-24. **THE COOKIE CUTTER**

A cookie baker has an automatic mixer that turns out a sheet of dough 12 inches wide and $\frac{3}{8}$ inch thick. His cookie cutter cuts 3-inch diameter circular cookies as shown at right. The supervisor complained that too large a fraction of the dough had to be re-rolled each time and ordered the baker to use a 4-inch diameter cookie cutter.

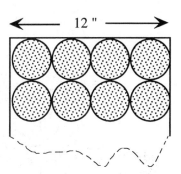

Your Task:

- Analyze the percentage of dough that needs to be re-rolled for different sizes of cookies.

- Write a note to the supervisor explaining your results. Justify your conclusion.

The Rocket Show

More About Quadratic Equations

Unit 13 Objectives
The Rocket Show: MORE ABOUT QUADRATIC EQUATIONS

In past units, we were given linear and non-linear equations to graph and analyze. With the graph or the equation, we could answer questions about intercepts, slope and points of intersection. However, in this unit, we will start with data instead of equations. Finding the line or parabola that fits a trend will create a **model** with which to analyze the data.

We will also learn more about **parabolas**. We will study the connection between a graph and its equation, bringing together all of our skills from previous units.

In this unit you will have the opportunity to:

- work backwards from the x-intercepts of a parabola to write its equation and sketch its graph.
- use Algebra Tiles to learn the technique known as Completing the Square.
- examine the shape, location, and direction of parabolas.
- consider data points and trends for straight lines and parabolas.
- derive the Quadratic Formula using several of the skills you have learned this semester.

The course concludes with an opportunity to connect many of the concepts and skills you have learned all year, including data organization skills, graphing of data, detecting patterns and trends, all leading to writing equations that are reasonable representations of the data.

Read the problem below. **Do not try to solve it now.** Over the next few days you will learn what is needed to solve it.

> RS-0.   THE ROCKET SHOW
>
> It is the end of the year and the Math Club is going to have one last event. The students plan to have a rocket show. Jairo suggests that the rocket fly over the school building and land near the parking lot.
>
> June, the Math Club president, needs your help to program the launch mechanism so that the rocket not only misses the building, but lands in a safe spot: the pond. June knows that a rocket that flies extremely high would miss the building, but she wants it as low as possible so it will be visible to the spectators. What path should the rocket follow?

Unit 13
The Rocket Show: MORE ABOUT QUADRATIC EQUATIONS

TREND LINES

RS-1. The Math Club has been fundraising for months in order to buy the rocket. Bomani, the treasurer, expects an excellent rocket to cost roughly $250 and wants to plan ahead. Use the information shown below from the club's bank statement for the first 10 weeks of school to estimate when they will have enough money to buy the rocket.

a) Plot the data as ten ordered pairs (x, y) with x representing the week. Scale the axes carefully.

week number	balance
0	$ 7.50
1	$ 53.00
2	$ 60.22
3	$ 85.64
4	$ 92.88
5	$ 99.41
6	$ 116.67
7	$ 122.72
8	$ 134.60
9	$ 150.53

b) Use a ruler to draw the line that best fits the data. Your line may not actually pass through any of the data points. In particular, <u>do not</u> connect the dots with short segments.

c) Write an equation for your line of best fit.

d) Use your equation from part (c) to predict when the Math Club will have enough money to buy the rocket.

RS-2. Is it always possible to fit a line to a set of data points? Sketch a graph for which you think it would be impossible to fit a line with any accuracy. Describe the situation that might produce data that would create your graph.

RS-3. On graph paper, graph the parabola $y = x^2 - 4x + 3$. Label the x-intercepts.

RS-4. Is (4, -6) on the graph of $y = x^2 - 3x - 10$? Explain how you know.

RS-5. Simplify using positive exponents only.

a) $(x^{-3})(y^0)$

b) $(2xy^3)(-4x^2)$

c) 2^{-4}

d) $(x^2)(2xy)^3$

RS-6. Multiply or factor the following.

a) $x^2 - 4x + 4$

b) $3x^2 + 19x - 14$

c) $(x - 9)^2$

d) $x^2 - 5x + \frac{25}{4}$

e) $(x + 31)^2$

f) $(x - 7)(2x + 1)$

RS-7. Solve each of the following equations. You may have more than one correct value of x.

a) $|x - 6| = 12$

b) $\frac{2}{5x} + \frac{x+1}{2x} = 1$

c) $\sqrt{4x + 3} = 1$

d) $4x^2 + 3x = 1$

RS-8. One of the most important skills needed to work with graphs is the ability to write an equation for a given line. Write an equation in the form $y = mx + b$ for each of the graphs shown. You may need to estimate to the nearest 0.5.

a)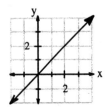

d)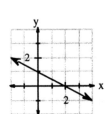

f)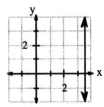

b)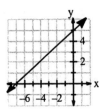

e)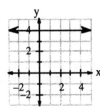

g)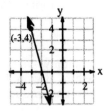

RS-9. Solve the following inequalities for x. Graph your solutions on a number line.

a) $3x - 5(4 - x) \geq 4$

b) $|x - 6| < 3$

RS-10. MATCH-A-GRAPH

One of the graphs at right is for the equation $y = -2x(x - 6)$. Determine which graph corresponds to $y = -2x(x - 6)$. Explain how you discovered your solution.

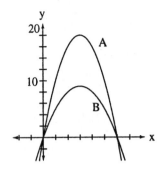

RS-11. RETURN TO THE BEEBOPPER SHOE STORE

As she started to clear out her desk at the end of the school year, Ms. Speedi found the data one of her classes compiled for the Beebopper Shoe Store problem from Unit One, shown below. The shoe sizes were adjusted for the differences between women's and men's shoe sizing.

Student	height (inches)	shoe size
Dina	70	$11\frac{1}{2}$
Max	66	9
Maxine	68	9
Janelle	62	7
Karen	62	5
Thu	68	9
Elizabeth	60	4
Arturo	61	6
Elaine	67	$7\frac{1}{2}$
Judy	65	7

Student	height (inches)	shoe size
Tom	76	14
Bob	69	12
José	65	$8\frac{1}{2}$
Alicia	58	5
Manuel	70	$10\frac{1}{2}$
Henry	75	14
Brian	67	10
Leslie	60	$6\frac{1}{2}$
Pat	62	6
Lim	71	11

a) Plot the data for Ms. Speedi's class. Use height as the x variable.

b) Draw a line of best fit for the data.

c) Write an equation for the line you drew in part (b).

d) Ms. Speedi, who is 6'3" tall, was absent the day the data was compiled. Predict her shoe size.

e) Are you <u>sure</u> that Ms. Speedi's <u>actual</u> shoe size is what the graph indicated in part (d)?

The Rocket Show: More About Quadratic Equations

RS-12. Nine students did a science experiment by rolling a marble down a slanted board and timing how long it took.

Adrián found that the marble took 10 seconds to roll 50 cm, but Mícheál's took only 8.9 seconds to roll 40 cm. Monica's results were 14.1 seconds to roll 100 cm, while Karla's were 9.5 seconds to roll 45 cm. Tammy got 12.6 seconds to roll 80 cm. Allan reported 0 seconds for 0 cm and George found the ball took 4.5 seconds to go 10 cm. Marni and Rebecca reported 11 seconds to roll 60 cm and 13.4 seconds to roll 90 cm, respectively.

a) Organize the information.

b) Plot the data as nine ordered pairs (x, y) with x representing time and y representing distance. Scale the axes carefully.

c) Draw a parabola of best fit for the data.

d) Use your graph to predict how long it will take the ball to roll 120 cm, 5 cm, and 15 cm.

e) Use your graph to predict how far the ball will travel after 3 seconds, 6 seconds, and 15 seconds.

f) Write an equation for the parabola you drew in part (c).

RS-13. Ms. Speedi's class measured some circular items to determine a relationship between the diameter and circumference. Their data is recorded in the table below.

diameter (cm)	circumference (cm)
3	10
5	16
10.8	32.8
13	40
10	32.3
6.8	21
4.5	18

a) Plot the data, fit a line, and write an equation for the line.

b) Compare your equation for the line of best fit to the equation which truly relates circumference and diameter. How close is your equation? Other than measuring more accurately, what else could Ms. Speedi's class do to get a more precise description for the relationship from their data?

c) Explain the significance of the slope in this problem.

RS-14. At 10:40 a.m., the time the math club meeting was scheduled to start, only four members of the math club had arrived! If an average of 7 students arrived every three minutes after that time, predict when all 53 members will be in attendance.

RS-15. Solve each of the following equations. Some equations have **two** solutions.

a) $2x + 1 = 7$

b) $2x^2 + 1 = 7$

c) $5(x - 1) + 2[x + 3(x + 1)] = 25$

d) $\sqrt{x - 3} - 7 = 8$

e) $x^2 = 100$

RS-16. Solve each of the following systems for x and y.

a) $0.7x - 0.3y = 1$
   $2x + y = 9$

b) $y = x^2 - 7$
   $y = 8 + 2x$

RS-17. For the relation graphed at right:

a) Find the domain and range.

b) Is this relation a function? Explain why or why not.

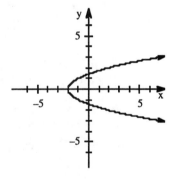

RS-18. Solve the following inequalities for x. Graph your solutions on a number line.

a) $5(9 - 3x) \leq -6(x + 9)$

b) $|x + 2| + 2 > 1$

RS-19. Identify which algebraic property allows you to change the expression on the left side of the equal sign to an equivalent expression on the right side.

a) $8x + 1 = 1 + 8x$

b) $32(2 - 9x) = 64 - 288x$

c) $3 + (6 + 2x) = (3 + 6) + 2x$

The Rocket Show: More About Quadratic Equations

FINDING THE EQUATION OF A PARABOLA FROM ITS X-INTERCEPTS

RS-20. On graph paper, graph $y = x^2 - x - 12$.

   a) Find the x-intercepts.

   b) Solve the equation: $x^2 - x - 12 = 0$.

   c) Using complete sentences, describe how you can find the x-intercepts of a parabola <u>without</u> graphing.

RS-21. Find an equation of a parabola with x-intercepts:

   a) $(-2, 0)$ and $(1, 0)$.

   b) $(6, 0)$ and the origin.

   c) $(b, 0)$ and $(c, 0)$.

RS-22. Find a possible equation of a parabola with x-intercepts $(0, 0)$ and $(5, 0)$. Is there more than one answer?

RS-23. Find an equation for a parabola with x-intercepts $(-10, 0)$ and $(6, 0)$.

   a) Latisha knows that the vertex of a parabola is $(6, -10)$. If one x-intercept is at $(4, 0)$, then where is the other?

   b) Is there more than one solution? Explain how you know.

RS-24. The two solutions to $x^2 - 6x - 8 = 0$ are $3 + \sqrt{17}$ and $3 - \sqrt{17}$. Substitute each of them into the equation to show that they work.

RS-25. Solve the following problem. In complete sentences, describe your method, and be prepared to share with the class tomorrow.

For a food sale, the math club sold egg rolls and won tons. Ms. Speedi bought a combination plate of 2 egg rolls and 3 won tons for $1, while Latisha spent $2.30 for 4 egg rolls and 9 won tons. How much did the math club charge for an egg roll?

RS-26. Use algebraic methods to convert each of the following equations to slope-intercept form. State the slope.

   a) $2y = 4x + 8$

   b) $5x - y = 8$

   c) $y - 8x = 7$

   d) $x = \frac{2y - 1}{3}$

   e) $6x + 3y - 5 = 0$

   f) $\frac{2}{y} = \frac{6}{x + 3}$

   g) Use the y-intercept and the slope to graph the equations in parts (a), (e) and (f) <u>without</u> making a table of values.

RS-27. Find the equation of the line with y-intercept (0, 6) through the point (2, -1).

RS-28. The slope of line AB is $\frac{5}{9}$ and the coordinates of point B are (18, 6).

  a) What is the y-intercept of line AB?

  b) Find the equation of the line perpendicular to line AB through the point B.

RS-29. Factor each of the following expressions as far as possible.

  a) $x^2 + 4xy + 4y^2$

  c) $x^3 + 4x^2 + 4x$

  b) $6x^2 + 18x + 12$

  d) Extension: $x^2 + 2y^2 + 3xy$

RS-30. Solve each of the following equations for x.

  a) $11 = |x+1|$

  c) $x(x-1)(x-2) = 0$

  b) $\frac{10}{x} + \frac{20}{x} = 5$

  d) $\sqrt{x+4} = 10$

RS-31. Ventura claims that if he knows the x-intercepts of a parabola, he also knows the x-coordinate of the vertex. Is he correct? If so, explain how.

RS-32. What are the x-intercepts of $y = 3(x - 1)(x + 3)$? Plot them on axes using $-5 \le x \le 3$.

  a) Find the x-coordinate of the vertex for $y = 3(x - 1)(x + 3)$.

  b) Use the x-coordinate you discovered in part (b) to find the y-coordinate of the vertex. State the vertex as a point (x, y).

  c) Sketch the parabola using the x-intercepts and the vertex.

RS-33. Make a large pair of coordinate axes on a sheet of graph paper by placing the origin at the center of the paper. Label your graph clearly. Graph $y = x^2$.

  a) Each member of your study team should choose one of the equations from the list below. Neatly graph the parabolas on the same set of axes as above by making tables and assigning values to x for $-4 \le x \le 4$. Share the results of your graph with your team members.

  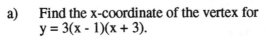

  1) $y = -x^2$ 2) $y = -\frac{1}{3}x^2$ 3) $y = \frac{1}{3}x^2$ 4) $y = 3x^2$

  b) After verifying their accuracy, copy all the parabolas onto your graph. Label each parabola with its equation.

  c) Write a few sentences to describe some patterns you see in the relationship between the quadratic equations and their graphs.

RS-34. Add or subtract the following fractions. Simplify your solution, if possible.

a) $\dfrac{2x}{3x-1} + \dfrac{x-1}{3x-1}$

b) $\dfrac{6}{x+3} + \dfrac{2x}{x+3}$

c) $\dfrac{x+3}{5} - \dfrac{x+4}{2}$

d) $\dfrac{2}{2x-1} + \dfrac{x+2}{3}$

RS-35. Bomani loves squares! In fact, he's trying to build one with Algebra Tiles. He has four large squares and 8 rectangles. How many small squares does he need to complete his square?

RS-36. Find the x-intercept, the y-intercept, and the slope of the line for each of the following equations.

a) $2x - 3y = 6$

b) $y = 4$

c) $y - 3x = 5$

d) $x + 44 = 0$

RS-37. Find the roots (x-intercepts) of the parabola $y = x^2 + 6x + 2$.

RS-38. How **long** is the line segment from the point (4, 3) to the point (-2, 7)? Write your answer in simplified square root form and as a decimal. Then determine the slope of the line that connects these two points.

RS-39. Solve each of the following systems of equations.

a) $3x + 2y = 12$
$2x + 2y = 13$

b) $x + y = 1$
$2(x + y) = 5 + x$

RS-40. Use your conclusions from RS-33 to decide with your team how each graph differs from x^2.

a) $y = \dfrac{1}{2}x^2$

b) $y = -2x^2$

RS-41. Find an equation of a parabola with x-intercepts at (0, 0) and (6, 0).

a) Find the x-coordinate of the vertex.

b) Find a coefficient of your parabola so that the y-coordinate of the vertex is 3.

c) Sketch this parabola using the x-intercepts and the vertex.

RS-42. Add this information to your tool kit:

THE EQUATION OF A PARABOLA FROM ITS X-INTERCEPTS

A parabola with x-intercepts (b, 0) and (c, 0) can be written in the form:

$$y = a(x - b)(x - c)$$

The coefficient, "a", determines the shape (wide or narrow) and direction (opens upward or downward) of the parabola. If we know the x-intercepts and one other point on the parabola, we can determine the equation.

For example: suppose you have x-intercepts (2, 0) and (4, 0), and point (1, 18) on the parabola.

Working backwards from the Zero Product Property, we know the equation is of the form $y = a(x - 2)(x - 4)$.

Substitute any other point on the parabola, in this case (1, 18), into the equation for x and y. Then solve for "a".

$y = a(x - 2)(x - 4)$
$18 = a(1 - 2)(1 - 4)$
$18 = a(-1)(-3)$
$18 = 3a$
$\frac{18}{3} = a$
$6 = a$

Therefore, the equation of the parabola with x-intercepts (2, 0) and (4, 0) and passing through (1, 18) is:

$$y = 6(x - 2)(x - 4)$$

We can next find the **vertex** (the point at the top or bottom) of the parabola by using the average of (that is, halfway between) the x-intercepts. In the above example, x = 3 is midway between x = 2 and x = 4. Substitute x = 3 in the equation to find the y-coordinate of the vertex.

$$y = 6(3 - 2)(3 - 4) = -6$$

The vertex of the parabola is (3, -6).

RS-43. Find the equation for the parabola with x-intercepts (3, 0) and (15, 0) that goes through (6, -162).

RS-44. Find the vertex of the parabola $y = \frac{1}{2}(x - 2)(x + 6)$.

RS-45. What are the x-intercepts of $y = -2x(x - 5)$? Plot them on axes using $-2 \leq x \leq 7$.

a) Find the x-coordinate of the vertex for $y = -2x(x - 5)$.

b) Use the x-coordinate you discovered in part (a) to find the y-coordinate of the vertex. State the vertex as a point (x, y).

c) Sketch the parabola using the x-intercepts and the vertex.

RS-46. Find an equation of a parabola with x-intercepts (0, 0) and (10, 0). Find the value of "a" so that the vertex is at (5, 100). Refer to problem RS-42 if you need help.

RS-47. If a parabola has its vertex at (1, 6) and one x-intercept at (5, 0), find its equation.

RS-48. Suppose you are given one x^2 tile and 16 small square tiles.

a) Is it possible to use some rectangular tiles together with the square tiles you are given to get a composite rectangle which is a square? Use a diagram to show how this can be done.

b) Write the area of the composite square you formed in part (a) as a product and as a sum.

RS-49. For the graph shown at right:

a) Estimate the x-intercept.

b) Find the slope of the line.

c) Find the equation of the line at right.

d) Use your equation to determine if your estimate of the x-intercept was accurate.

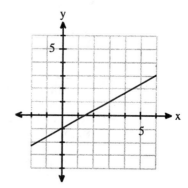

RS-50. Find the equation of the parabola graphed at right.

RS-51. Solve each of the following equations for x.

a) $4x^2 = 25$

b) $(x - 1)^2 + 3 = 7$

c) $x^2 - 2 = 14$

d) $2[2 - 3(5x + 4)] + 1 = 11$

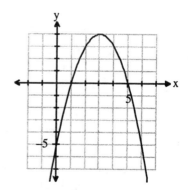

RS-52. The math club purchased a 52" big screen TV for the math department from a catalog and it will be arriving soon. Alfredo built the cabinet, as shown at right, to store the TV. Now he is worried that it might not fit! His friend, Kate, has heard that the width of the screen is always $\sqrt{2}$ times as wide as it is tall. Will the new TV fit? (Note that the 52" refers to the diagonal of the television screen.)

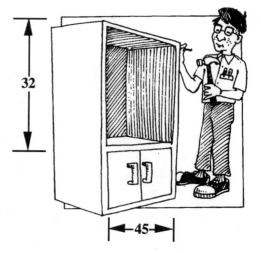

RS-53. Find the area of the shaded region. Show all subproblems.

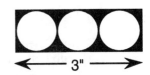

COMPLETING THE SQUARE

RS-54. Solve for x if $\sqrt{x} - 3 = 7$. How did you eliminate the square root?

RS-55. Solve for x if $x^2 = 25$. Explain why there are two solutions.

RS-56. Find all possible solutions to each of the following equations. Show your work clearly for each solution.

   a) $x^2 = 9$
   b) $x^2 = 7$
   c) $(x - 5)^2 = 36$
   d) $(x + 3)^2 = 49$
   e) $x^2 + 8 = 51$
   f) $(x - 4)^2 + 9 = 12$

RS-57. Suppose you are given one x^2 tile and twelve rectangles.

   a) How many small squares will you need to make a composite rectangle which is a square?
   b) Sketch the composite square you formed in part (a).
   c) Write the area of the composite square as a product and as a sum.
   d) Now do parts (a), (b) and (c) if you are given one x^2 tile and 10 rectangles.

RS-58. COMPLETING A SQUARE

The expression $x^2 + 6x$ can be represented by tiles as shown at right. To complete the diagram and make a **complete composite square**, we need to add nine unit (small) squares.

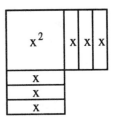

What number would you need to add to each of the following expressions to make each one represent the area of a complete composite square? Also write the dimensions of each square.

Example: $x^2 + 6x + \underline{9} = (x + 3)^2$.

   a) $x^2 + 10x + \underline{}$
   b) $x^2 + 4x + \underline{}$
   c) $x^2 - 2x + \underline{}$
   d) $x^2 - 8x + \underline{}$

RS-59.   **GOLD DIGGER'S GULCH**

During a field trip to the Amusement Park, Alfredo noticed that a portion of the roller coaster was shaped like a parabola. Near the base of the parabola, the track goes underground into a gulch. A car on the track, 300 feet above the ground, sped toward the gulch and emerged above ground 100 feet later. How far below ground did the car travel?

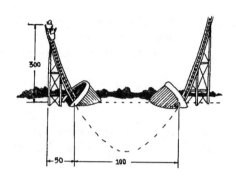

RS-60.   Factor each of the following polynomials completely. (Hint: remember to look for the largest common factor first.)

a) $3x^2 - 243$

b) $x^2 + 7x - 98$

c) $2x^2 + 9x + 10$

d) $2x^2 + 9xy + 10y^2$

RS-61.   Solve each of the following systems for x and y.

a) $2x + y = 7$
    $3x - 2y = 7$

b) $y = 0.5x + 4$
    $2x - 4y = 1$

c) If you graphed the lines in part (b), where would they intersect?

RS-62.   To encourage club spirit, June decided to order T-shirts and sweatshirts with the Math Club logo. She called a T-shirt company and found that she could order 40 T-shirts and 13 sweatshirts for $425.35. After surveying the club membership, she changed her order to 30 T-shirts and 23 sweatshirts, for a total of $477.35. What is the price of a single T-shirt?

RS-63.   Carefully graph $f(x) = |x - 3| + 1$.

a) Evaluate f(3), f(0) and f(-5).

b) Use your graph to solve the equation for x: $|x - 3| + 1 = 5$.

RS-64. On a thermometer, 20° Celsius is equal to 68° Fahrenheit. Also, 10° Celsius is equal to 50° Fahrenheit. The conversion from Celsius to Fahrenheit degrees is linear.

    a) Use the data to plot the points. (Let the horizontal axis represent Celsius.)

    b) Draw a line through the points.

    c) Use the data to write an equation for the line.

    d) Use the equation to determine the temperature in degrees Fahrenheit for 30° Celsius. Check this with your graph. You have just written a conversion formula, which is simply a linear equation.

RS-65. For the equation $2y - 4x = 6$:

    a) write the coordinates of the x- and y-intercepts.

    b) graph the intercepts and draw the line.

    c) find the slope of the line.

    d) find the distance between the x- and y-intercepts.

RS-66. Solve these equations for x.

    a) $x^2 = 36$ c) $(x + 4)^2 = 4$

    b) $(x + 2)^2 = 36$ d) $x^2 + 8x + 12 = 0$

    e) Compare your answers for parts (c) and (d). They should be the same. Which part was easier to solve? Why?

RS-67. Taking notes is always an important study tool. Take careful notes and record sketches as your read this problem with your study team.

COMPLETING THE SQUARE

In problem RS-58, we added tiles to form a square. This changed the value of the original polynomial. However, by using a neutral field, we can take any number of tiles and create a square without changing the value of the original expression. This technique is called **COMPLETING THE SQUARE**. For example, start with the polynomial: $x^2 + 8x + 12$:

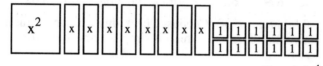

First, put these tiles together in the usual arrangement and you can see a "square that needs completing."

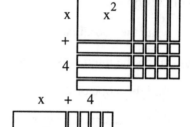

a) How many small squares are needed to complete this square?

b) Draw a neutral field beside the tiles. Does this neutral field affect the value of our tiles? The equation now reads:

$$x^2 + 8x + 12 + 0$$

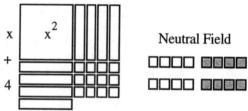

c) To complete the square, we are going to need to move tiles from the neutral field to the square. When we take the necessary four positive tiles that complete the square, what is the value of the formerly neutral field?

$$(\underbrace{x^2 + 8x + 12 + 4}_{\text{complete square}}) + \underbrace{(-4)}_{\text{neutral field}}$$

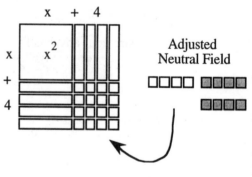

d) Combining like terms,

$$x^2 + 8x + 16 + -4$$

e) Factoring the trinomial square,

$$(x + 4)^2 - 4$$

So, $x^2 + 8x + 12 = (x + 4)^2 - 4$

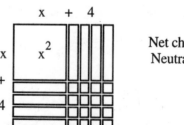

RS-68. You noted in problem RS-66 that parts (c) and (d) were the same equation in different forms. Completing the square is used to transform quadratic equations like $x^2 + 8x + 12 = 0$ into perfect squares like $(x + 4)^2 - 4 = 0$ and then into $(x + 4)^2 = 4$. Once the square is completed, we can solve by taking the square root of both sides. Complete the square and solve for x in each part below.

a) $x^2 + 10x + 21 = 0$

b) $x^2 - 2x = 0$

c) $x^2 + 4x + 3 = 0$

d) $x^2 - 8x = -7$

RS-69. Find the equation of a parabola with x-intercepts at (5, 0) and the origin that passes through (2, -2).

RS-70. Explain how to determine what number to add and subtract to both sides of an equation of the form $x^2 + bx + c = 0$ when you complete the square. Use drawings and examples.

RS-71. Complete the square to solve for x.

a) $x^2 + 2x - 3 = 0$

b) $x^2 + 8x = 5$

RS-72. Welded wire fencing is sold in rolls of 100 linear feet. A farmer intends to build a rectangular rabbit pen with an area of 481 square feet. The pen is two feet less than three times its width, and uses exactly one roll of wire.

a) Draw a diagram of the pen and label its dimensions.

b) Develop two equations to find the dimensions of the pen, one using area and one using perimeter.

c) Find the dimensions of the pen.

RS-73. If the hundred-foot length of wire in the previous problem weighs 62 pounds, about how much will 240 feet of the same wire weigh?

RS-74. Solve the following inequalities for x. Graph your solutions on a number line.

a) $-2(x + 6) < 10$

b) $4 \geq -5 + |x|$

RS-75. On graph paper, graph and shade the solution for the systems of inequalities at right. Describe the resulting region.

$y \geq 3x - 5$

$y < -\frac{1}{3}x + 2$

The Rocket Show: More About Quadratic Equations

RS-76. Kristen has a box full of ping-pong balls numbered from 1 to 53. If she reaches in and pulls out one ball, find the probability that the ball has a number on it which is:

a) even.

b) less than or equal to 10.

c) a perfect square (1, 4, 9, …).

d) two digits.

RS-77. THE ROCKET SHOW

It is the end of the year and the Math Club is going to have one last event. To celebrate, the club members plan to have a rocket show. Jairo suggested that the rocket fly over the school building and land near the parking lot. For safety reasons, the principal wants assurances that the students can control where it lands.

Since the rocket will be a fireball when it lands, the Math Club needs to aim the rocket so that it lands in the pond near the parking lot. Here are some facts about the school grounds. Use this data to sketch the path of the rocket on the resource page provided by your teacher.

The launcher will be placed on the football field, 60 meters from the school building. The school building is 40 meters wide and 30 meters tall. The parking lot on the other side of the school building is 15 meters wide. At the edge of the parking lot is a pond which is 10 meters wide.

June, the Math Club president, needs your help to program the launch mechanism so that it not only misses the building, but lands in a safe spot: the pond. June knows that a rocket that flies extremely high would miss the building, but she wants it as low as possible so that it will be visible to the spectators.

She has enlisted your help in programming the launcher. Find an equation that will tell the launcher where to send the rocket.

a) On your resource page, draw the lowest path possible for the rocket that flies at least 5 meters over the building and lands in the pond.

b) Name the intercepts of your parabola.

c) Find a possible equation for the parabola using the x-intercepts from part (b).

d) If a = -1, what will be the maximum height of the rocket?

e) Does your equation of the rocket ensure that you will miss the building? Substitute an x-coordinate into your launching equation to find out if the rocket clears the building.

f) June wants the lowest possible rocket trajectory so that it is visible to the crowd and wants at least a 5 meter clearance over the building. Double check your rocket's height on the top corner of the building nearest the parking lot. Find the x-coordinate that will help you determine whether you have proper clearance. What "a" value should your equation have to miss the building and yet be as low as possible? (Hint: figure out another point on the parabola you want to hit and use it in the equation to solve for "a".)

g) Elmo, a reporter for the school newspaper, wants to know how high the rocket flies. Use your graph to estimate the height, then verify your estimate using your equation.

RS-78. Elizabeth fixed her tricycle and wants to race Leslie again. She now races 5 meters every 2 seconds, while Leslie rides 2 meters every second. If they start at the same place and time, and the race is 25 meters long, by how much of a margin does Elizabeth win?

RS-79. Factor the polynomial $4x^2 + 12x + 9$ by drawing a rectangle.

a) Label its dimensions and area(s).

b) The polynomial $4x^2 + 12x + 9$ is called a "perfect square" trinomial. Explain why this name is appropriate.

RS-80. During a camping trip, Cami forgot the poles to her tent! June suggested that the club search the vicinity for a tree branch that would work instead. When Cami measured the tent, she found that the base was 5 feet wide and the sides were 4.75 feet long. Determine how long the branch should be to properly hold up the tent.

RS-81. The slope of line AB is 5, with points A (-3, -1) and B (2, n). Find the value of n.

RS-82. Write each of the following expressions as simply as possible.

a) $\dfrac{(x - 1)(x + 2)(x - 3)(x + 4)}{(x + 2)(x + 3)(x + 4)(x + 5)}$

b) $\dfrac{(x^2 - 1)(x^2 - 2x - 3)}{(x - 1)^2(x - 2)(x - 3)}$

RS-83. Complete the square. Solve parts (c) and (d).

a) $x^2 + 12x + 11$

b) $x^2 - 14x + 33$

c) $x^2 + 14x - 10 = 0$

d) $x^2 - 12x + 20 = 0$

RS-84. A rectangular garden has a perimeter of 100 meters and a diagonal of 40 meters. Find its dimensions.

DERIVING THE QUADRATIC FORMULA

> We will now see how the Quadratic Formula is developed. We will approach the derivation of the Quadratic Formula as a series of subproblems. Look for <u>patterns</u> as you solve each problem and discuss the patterns you observe with your study team.

RS-85. **SUBPROBLEM 1**

Solve each of the following equations for y. Each equation has **two** solutions.

a) $y^2 = 25$

b) $y^2 = 19$

c) $y^2 = 45$

d) $y^2 = 16b^2$

e) $y^2 = 13b^2$

f) $y^2 = \frac{36}{49}$ (write solutions as fractions)

> Stop to discuss with your study team any patterns you see developing in parts (a) through (f). Use that pattern to solve parts (g), (h), and (i).

g) $y^2 = 4a^2$

h) $y^2 = \frac{b^2}{4a^2}$

i) $y^2 = \frac{b^2 - 4ac}{4a^2}$

RS-86. **SUBPROBLEM 2: COMPLETING A SQUARE**

Remember that the expression $x^2 + 6x$ can be represented by tiles.

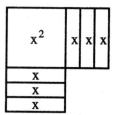

To complete the diagram and make a complete composite square, we would need to add unit (small) squares.

What number would you need to add to each of the following expressions to make each one represent the area of a composite square?

a) $x^2 + 2x +$ _____

b) $x^2 - 4x +$ _____

c) $x^2 + 7x +$ _____

d) $x^2 - 5x +$ _____

e) $x^2 + bx +$ _____

f) $x^2 + \frac{b}{a}x +$ _____

RS-87. **SUBPROBLEM 3**

What are the dimensions of each of the completed squares in RS-86? (Hint: think of a completed square, then use the factoring method you learned for rectangles.)

We know $x^2 + 6x + 9 = (x + 3)^2$, so the dimensions of the composite square are $x + 3$ by $x + 3$.

RS-88. **SUBPROBLEM 4**

Write each of the following sums as a single expression. (Hint: common denominators are necessary in order to add fractions.)

a) $\frac{3}{4a} + \frac{5}{a}$

b) $\frac{b^2}{4a} + \frac{2}{a}$

c) $\frac{7}{4a^2} - \frac{c}{a}$

d) $\frac{b^2}{4a^2} - \frac{c}{a}$

RS-89. DERIVATION OF THE QUADRATIC FORMULA

We are ready to derive the Quadratic Formula. To do this we will use the last part of each of the subproblems in problems RS-85 through RS-88 to solve the equation $ax^2 + bx + c = 0$ for x. Our goal is to transform the equation into:

$$x = \frac{-b \pm \sqrt{b^2 - 4ac}}{2a}.$$

Fold a piece of lined paper in half vertically, make a crease, then unfold the paper. Copy the algebraic steps shown below onto the left-hand side of your paper. Write your answer to each question to the right of the corresponding algebraic step.

We want to solve the equation $ax^2 + bx + c = 0$.

$$x^2 + \frac{b}{a}x + \frac{c}{a} = 0$$

What did we do to get this?

$$x^2 + \frac{b}{a}x = -\frac{c}{a}$$

What did we do to get this?

What do we do to get the next step?

$$x^2 + \frac{b}{a}x + \frac{b^2}{4a^2} = \frac{b^2}{4a^2} - \frac{c}{a}$$

Why do you think we chose $\frac{b^2}{4a^2}$?

Now we make a major replacement for the whole left side of the equation:

$$\left(x + \frac{b}{2a}\right)^2 = \frac{b^2}{4a^2} - \frac{c}{a}$$

Why is that possible? What subproblem did we use?

This time we replace the right-hand side:

$$\left(x + \frac{b}{2a}\right)^2 = \frac{b^2 - 4ac}{4a^2}$$

What subproblem shows we can do this?

What operation do we need to do to get the next result?

$$x + \frac{b}{2a} = \pm\frac{\sqrt{b^2 - 4ac}}{2a}$$

On what subproblem are we relying?

$$x = -\frac{b}{2a} + \frac{\sqrt{b^2 - 4ac}}{2a}$$

What did we do to get this result?

Finally, we get to the long awaited solution:

$$x = \frac{-b \pm \sqrt{b^2 - 4ac}}{2a}$$

Which subproblem do we use to get this?

The Rocket Show: More About Quadratic Equations

RS-90. Find the equation of the line through the points (-6, 2) and (8, 9).

RS-91. Solve the system of equations:

a) $x = 7y - 3$
$2x - 3y = 8$

b) $3x - 5y = 4$
$x + 2y = 5$

RS-92. Complete the square and solve for x.

a) $x^2 - 6x = -5$

b) $x^2 - 6x + 7 = 0$

RS-93. Solve the equation $2x^2 - 7x + 3 = 0$ by:

a) factoring

b) using the Quadratic Formula

c) How could you use your result in part (b) to write the factors of the equation?

RS-94. Use the Quadratic Formula to find the roots of each equation. Use the roots to find the factors of each of the following polynomials.

a) $6x^2 - 7x - 10 = 0$

c) $9x^2 + 11x + 2 = 0$

b) $36x^2 - 37x - 48 = 0$

RS-95. Simplify using positive exponents only.

a) $(2x^3)(-3x)^2$

c) $\dfrac{x^3 y^3}{xy}$

b) $(-2x^2 y)^3$

d) $\dfrac{x^3 y^3}{x^{-1} y^{-1}}$

UNIT SUMMARY AND REVIEW

RS-96. UNIT THIRTEEN SUMMARY

With your study team, list 3 or 4 big ideas that you feel best represent Unit 13. Write a description of each main idea. For each topic, choose a problem that best demonstrates your understanding. Explain your solution as in previous unit summaries. Show all your work.

RS-97. **MEMBERSHIP DRIVE**

Each year, the math club has steadily grown in membership and members predict it will soon be the biggest club on campus! The science club, currently the largest, is growing as well. Some of the membership records for the last 17 years are shown below.

Math Club	
Year	Membership
1985	15
1987	27
1989	32
1992	36
1993	45
1995	45
1999	59

Science Club	
Year	Membership
1983	34
1986	35
1988	43
1990	50
1994	49
1997	61
1999	63

a) Plot the points on the same set of axes. Let x represent the year, and x = 0 for the year 1983. Use a different marking (or color) to distinguish the points for each club on your graph.

b) Find the equation of a line of best fit for each set of data.

c) Based on your graphs, estimate when the membership of the math club will surpass that of the science club.

d) Using the equations from part (b), solve the system of equations algebraically to determine when the math club will become the largest club on campus.

RS-98. Complete the square and use the resulting expression to solve for x. Then check your answer by factoring or by using the Quadratic Formula.

a) $x^2 - 4x + 3 = 0$ b) $x^2 - 10x + 14 = 0$

RS-99. Find the equation of the parabola shown in the graph at right.

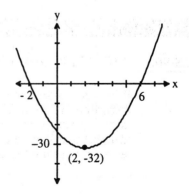

RS-100. Below is a mixture of rational expressions. Perform the indicated operation and simplify your solution, if possible.

a) $\dfrac{x}{2} + \dfrac{x}{6}$

b) $\dfrac{3}{2x-1} \div \dfrac{x+5}{4x-2}$

c) $\dfrac{2x-2}{x^2+4x-5} \cdot \dfrac{x^2+2x-15}{x^2+x-12}$

d) $5 - \dfrac{3}{2x}$

e) $\dfrac{5}{x^2-4} + \dfrac{2}{x-2}$

RS-101. **THE CAMPAIGN POSTER**

Judy was running for math club president. She used an enlargement of a 3" by 5" photograph for her campaign poster. She surrounded the enlarged photo with a two-foot border. Including the border, the area of her poster was three times the area of the enlarged photograph. How large was Judy's poster?

RS-102. The point (3, -7) is on line k with slope $\tfrac{2}{3}$.

a) Find the equation of line k.

b) Find the equation of the line perpendicular to line k that passes through (3, -7).

RS-103. Solve each of the following equations.

a) $x^2 - 2x - 4 = 0$

b) $3 - 2[3 - (x - 4)] = 10$

c) $3x^2 + 6x + 4 = 0$

d) $2x^2 - 7x - 4 = 0$

e) Extension: $(x^2 - 5)^2 - 3 = 13$

RS-104. Simplify each of the following expressions.

a) $(x^2)^3 \cdot (3x^4)^2$

b) $\dfrac{4x^3 y}{12xy^3}$

c) $\dfrac{2x^2 - x}{x} \cdot \dfrac{1}{2}$

RS-105. Use what you know about the slopes of lines to show that points A, B, and C are on the same line:

A (4, 2)   B (17, 28)   C (7, 8)

(Hint: What is true about the slopes of line segments which are on the same line?)

RS-106. Solve the equation $2x^2 + 3x - 9 = 0$.

a) If you graph the equation $y = 2x^2 + 3x - 9$, where will the graph cross the x-axis?

b) What equation would you have to solve if you wanted to know where the graph of $y = 2x^2 + 3x - 9$ would cross the line $y = 11$?

c) Solve the equation in part (b).

RS-107. A track around a soccer field is shown at right. All measurements represent distance in feet.

a) Find the distance a jogger runs for one lap around the track.

b) Find the area inside the track.

RS-108. COURSE REFLECTION

You have arrived at the end of this algebra course. Reflect back on how much you have learned throughout this year. Write a paragraph addressing your year in this course. The questions below are merely suggestions to get you started.

What topics were difficult for you to learn at first? What learning are you most proud of? Is algebra what you thought it would be? Have the study teams helped your understanding? What talent are you most proud of? How have you helped others learn? Have your study habits changed? Have your feelings about math changed? What topics are still difficult for you? What are your goals for your next math course?

RS-109. A LETTER OF ADVICE

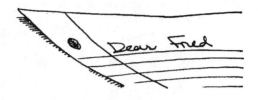

Write a letter of advice to a student entering CPM algebra for the first time. Include in this letter how they can be successful in this course. If you could start over, what would you do differently? What types of things helped you succeed in this class? What tools were important? What general advice would you give this student?

RS-110. TOOL KIT CLEAN-UP

Examine the elements in your tool kit from Units 7 - 13. You should create a consolidated tool kit representing the entire course of algebra.

Include items you value the most in your tool kit. Spend time on this. It is a good way to prepare for your final exam. You may also want to use it as a resource tool kit for next year!

a) Examine the list of tool kit entries from this unit. Check to be sure you have all of these entries. Add any you are missing.

b) Identify which concepts in your complete tool kit (Units 7 - 13) you understand well.

c) Identify the concepts you still need to work on to master.

d) Choose entries to create a Unit 0 - 13 tool kit that is shorter, clear, and useful. You may want to consolidate or shorten some entries.

- Equation of a Parabola from its X-Intercepts
- Finding the Vertex of a Parabola
- Completing the Square

GM-25. CANDY SALES

For a fund-raiser, each math club member must sell 30 candy bars each day for a week. Although they all sell the same type of candy, the members may choose the sales price to help them compete for top sales member.

Alfredo decided to sell three candy bars for $1, earning $10 per day, while June priced hers at two for $1, earning her $15 per day.

One day, both Alfredo and June were on a field trip, so they asked Bomani to sell their candy bars for them. Bomani agreed and promised he would not change their prices. He decided that instead of offering three for $1 and two for $1, he would put them together and sell the 60 candy bars for five for $2.

When Alfredo and June returned, Bomani handed them the money he earned for the day, $24. Alfredo and June were angry and demanded the dollar they were sure Bomani stole! Bomani is now confused... what happened?

Your Task:

- Write a letter to Alfredo and June to explain what happened to the missing dollar.

ALGEBRA 1 SKILL BUILDERS
(Extra Practice)
Introduction to Students and Their Teachers

Learning is an individual endeavor. Some ideas come easily; others take time--sometimes lots of time--to grasp. In addition, individual students learn the same idea in different ways and at different rates. The authors of this textbook designed the classroom lessons and homework to give students time--often weeks and months--to practice an idea and to use it in various settings. This section of the textbook offers students a brief review of 27 topics followed by additional practice with answers. Not all students will need extra practice. Some will need to do a few topics, while others will need to do many of the sections to help develop their understanding of the ideas. This section of the text may also be useful to prepare for tests, especially final examinations.

How these problems are used will be up to your teacher, your parents, and yourself. In classes where a topic needs additional work by most students, your teacher may assign work from one of the skill builders that follow. In most cases, though, the authors expect that these resources will be used by individual students who need to do more than the textbook offers to learn an idea. This will mean that you are going to need to do some extra work outside of class. In the case where additional practice is necessary for you individually or for a few students in your class, you should not expect your teacher to spend time in class going over the solutions to the skill builder problems. After reading the examples and trying the problems, if you still are not successful, talk to your teacher about getting a tutor or extra help outside of class time.

Warning! Looking is not the same as doing. You will never become good at any sport just by watching it. In the same way, reading through the worked out examples and understanding the steps are not the same as being able to do the problems yourself. An athlete only gets good with practice. The same is true of developing your algebra skills. How many of the extra practice problems do you need to try? That is really up to you. Remember that your goal is to be able to do problems of the type you are practicing on your own, confidently and accurately.

Another source for help with the topics in this course is the *Parent's Guide with Review to Math 1 (Algebra 1)*. Information about ordering this resource can be found inside the front page of the student text. It is also available free on the Internet at *www.cpm.org*.

Skill Builder Topics

1. Arithmetic operations with numbers
2. Combining like terms
3. Order of operations
4. Distributive Property
5. Substitution and evaluation
6. Tables, equations, and graphs
7. Solving linear equations
8. Writing equations
9. Solving proportions
10. Ratio applications
11. Intersection of lines: substitution method
12. Multiplying polynomials
13. Writing and graphing linear equations
14. Factoring polynomials
15. Zero Product Property and quadratics
16. Pythagorean Theorem
17. Solving equations containing algebraic fractions
18. Laws of exponents
19. Simplifying radicals
20. The quadratic formula
21. Simplifying rational expressions
22. Multiplication and division of rational expressions
23. Absolute value equations
24. Intersection of lines: elimination method
25. Solving inequalities
26. Addition and subtraction of rational expressions
27. Solving mixed equations and inequalities

ARITHMETIC OPERATIONS WITH NUMBERS   #1

Adding integers: If the signs are the <u>same</u>, add the numbers and keep the same sign. If the signs are <u>different</u>, ignore the signs (that is, use the absolute value of each number) and find the difference of the two numbers. The sign of the answer is determined by the number farthest from zero, that is, the number with the greater absolute value.

same signs   different signs

a) $2+3=5$ or $3+2=5$ c) $-2+3=1$ or $3+(-2)=1$

b) $-2+(-3)=-5$ or $-3+(-2)=-5$ d) $-3+2=-1$ or $2+(-3)=-1$

Subtracting integers: To find the difference of two values, change the subtraction sign to addition, change the sign of the number being subtracted, then follow the rules for addition.

a) $2-3 \Rightarrow 2+(-3)=-1$ c) $-2-3 \Rightarrow -2+(-3)=-5$

b) $-2-(-3) \Rightarrow -2+(+3)=1$ d) $2-(-3) \Rightarrow 2+(+3)=5$

Multiplying and dividing integers: If the signs are the same, the product will be positive. If the signs are different, the product will be negative.

a) $2 \cdot 3=6$ or $3 \cdot 2=6$ b) $-2 \cdot (-3)=6$ or $(+2) \cdot (+3)=6$

c) $2 \div 3 = \frac{2}{3}$ or $3 \div 2 = \frac{3}{2}$ d) $(-2) \div (-3) = \frac{2}{3}$ or $(-3) \div (-2) = \frac{3}{2}$

e) $(-2) \cdot 3 = -6$ or $3 \cdot (-2) = -6$ f) $(-2) \div 3 = -\frac{2}{3}$ or $3 \div (-2) = -\frac{3}{2}$

g) $9 \cdot (-7) = -63$ or $-7 \cdot 9 = -63$ h) $-63 \div 9 = -7$ or $9 \div (-63) = -\frac{1}{7}$

Follow the same rules for fractions and decimals.

Remember to apply the correct order of operations when you are working with more than one operation.

Simplify the following expressions using integer operations WITHOUT USING A CALCULATOR.

1. $5+2$ 2. $5+(-2)$ 3. $-5+2$ 4. $-5+(-2)$

5. $5-2$ 6. $5-(-2)$ 7. $-5-2$ 8. $-5-(-2)$

9. $5 \cdot 2$ 10. $-5 \cdot (-2)$ 11. $-5 \cdot 2$ 12. $2 \cdot (-5)$

13. $5 \div 2$ 14. $-5 \div (-2)$ 15. $5 \div (-2)$ 16. $-5 \div 2$

17. $17 + 14$ 18. $37 + (-16)$ 19. $-64 + 42$ 20. $-29 + (-18)$

21. $55 - 46$ 22. $37 - (-13)$ 23. $-42 - 56$ 24. $-37 - (-15)$

25. $16 \cdot 32$ 26. $-42 \cdot (-12)$ 27. $-14 \cdot 4$ 28. $53 \cdot (-10)$

29. $42 \div 6$ 30. $-72 \div (-12)$ 31. $34 \div (-2)$ 32. $-60 \div 15$

SKILL BUILDERS

Simplify the following expressions without a calculator. Rational numbers (fractions or decimals) follow the same rules as integers.

33. $(16 + (-12))3$

34. $(-63 \div 7) + (-3)$

35. $\frac{1}{2} + (-\frac{1}{4})$

36. $\frac{3}{5} - \frac{2}{3}$

37. $(-3 \div 1\frac{1}{2})2$

38. $(5 - (-2))(-3 + (-2))$

39. $\frac{1}{2}(-5 + (-7)) - (-3 + 2)$

40. $-(0.5 + 0.2) - (6 + (-0.3))$

41. $-2(-57 + 71)$

42. $33 \div (-3) + 11$

43. $-\frac{3}{4} + 1\frac{3}{8}$

44. $\frac{4}{5} - \frac{6}{8}$

45. $-2(-\frac{3}{2} \cdot \frac{2}{3})$

46. $(-4 + 3)(2 \cdot 3)$

47. $-\frac{3}{4}(3 - 2) - (\frac{1}{2} + (-3))$

48. $(0.8 + (-5.2)) - 0.3(-0.5 + 4)$

Answers

1. 7
2. 3
3. -3
4. -7
5. 3
6. 7
7. -7
8. -3
9. 10
10. 10
11. -10
12. -10
13. $\frac{5}{2}$ or $2\frac{1}{2}$ or 2.5
14. $\frac{5}{2}$ or $2\frac{1}{2}$ or 2.5
15. $-\frac{5}{2}$ or $-2\frac{1}{2}$ or -2.5
16. $-\frac{5}{2}$ or $-2\frac{1}{2}$ or -2.5
17. 31
18. 21
19. -22
20. -47
21. 9
22. 50
23. -98
24. -22
25. 512
26. 504
27. -56
28. -530
29. 7
30. 6
31. -17
32. -4
33. 12
34. -12
35. $\frac{1}{4}$
36. $-\frac{1}{15}$
37. -4
38. -35
39. -5
40. -6.4
41. -28
42. 0
43. $\frac{5}{8}$
44. $\frac{2}{40} = \frac{1}{20}$
45. 2
46. -6
47. $1\frac{3}{4}$
48. -5.45

COMBINING LIKE TERMS #2

Like terms are algebraic expressions with the same variables and the same exponents for each variable. Like terms may be combined by performing addition and/or subtraction of the coefficients of the terms. Combining like terms using algebra tiles is shown on page 42 of the textbook. Review problem SQ-67 now.

Example 1

$(3x^2 - 4x + 3) + (-x^2 - 3x - 7)$ means combine $3x^2 - 4x + 3$ with $-x^2 - 3x - 7$.

1. To combine horizontally, reorder the six terms so that you can add the ones that are the same: $3x^2 - x^2 = 2x^2$ and $-4x - 3x = -7x$ and $3 - 7 = -4$. The sum is $2x^2 - 7x - 4$.

2. Combining vertically:
$$\begin{array}{r} 3x^2 - 4x + 3 \\ -x^2 - 3x - 7 \\ \hline 2x^2 - 7x - 4 \end{array}$$ is the sum.

Example 2

Combine $(x^2 + 3x - 2) - (2x^2 + 3x - 1)$.

First apply the negative sign to each term in the second set of parentheses by distributing (that is, multiplying) the -1 to all three terms.

$-(2x^2 + 3x - 1) \Rightarrow (-1)(2x^2) + (-1)(3x) + (-1)(-1) \Rightarrow -2x^2 - 3x + 1$

Next, combine the terms. A complete presentation of the problem and its solution is:

$(x^2 + 3x - 2) - (2x^2 + 3x - 1) \Rightarrow x^2 + 3x - 2 - 2x^2 - 3x + 1$
$\Rightarrow -x^2 + 0x - 1 \Rightarrow -x^2 - 1.$

Combine like terms for each expression below.

1. $(x^2 + 3x + 4) + (x^2 + 4x + 3)$
2. $(2x^2 + x + 3) + (5x^2 + 2x + 7)$
3. $(x^2 + 2x + 3) + (x^2 + 4x)$
4. $(x + 7) + (3x^2 + 2x + 9)$
5. $(2x^2 - x + 3) + (x^2 + 3x - 4)$
6. $(-x^2 + 2x - 3) + (2x^2 - 3x + 1)$
7. $(-4x^2 - 4x - 3) + (2x^2 - 5x + 6)$
8. $(3x^2 - 6x + 7) + (-3x^2 + 4x - 7)$
9. $(9x^2 + 3x - 7) - (5x^2 + 2x + 3)$
10. $(3x^2 + 4x + 2) - (x^2 + 2x + 1)$
11. $(3x^2 + x + 2) - (-4x^2 + 3x - 1)$
12. $(4x^2 - 2x + 7) - (-5x^2 + 4x - 8)$
13. $(-x^2 - 3x - 6) - (7x^2 - 4x + 7)$
14. $(-3x^2 - x + 6) - (-2x^2 - x - 7)$
15. $(4x^2 + x) - (6x^2 - x + 2)$
16. $(-3x + 9) - (5x^2 - 6x - 1)$
17. $(3y^2 + x - 4) + (-x^2 + x - 3)$
18. $(5y^2 + 3x^2 + x - y) - (-2y^2 + y)$
19. $(x^3 + y^2 - y) - (y^2 + x)$
20. $(-3x^3 + 2x^2 + x) + (-x^2 + y)$

Answers

1. $2x^2 + 7x + 7$
2. $7x^2 + 3x + 10$
3. $2x^2 + 6x + 3$
4. $3x^2 + 3x + 16$
5. $3x^2 + 2x - 1$
6. $x^2 - x - 2$
7. $-2x^2 - 9x + 3$
8. $-2x$
9. $4x^2 + x - 10$
10. $2x^2 + 2x + 1$
11. $7x^2 - 2x + 3$
12. $9x^2 - 6x + 15$
13. $-8x^2 + x - 13$
14. $-x^2 + 13$
15. $-2x^2 + 2x - 2$
16. $-5x^2 + 3x + 10$
17. $3y^2 - x^2 + 2x - 7$
18. $7y^2 + 3x^2 + x - 2y$
19. $x^3 - y - x$
20. $-3x^3 + x^2 + x + y$

Algebra 1

ORDER OF OPERATIONS #3

The **order of operations** establishes the necessary rules so that expressions are evaluated in a consistent way by everyone. The rules, in order, are:
- When grouping symbols such as parentheses are present, do the operations within them first.
- Next, perform all operations with exponents.
- Then do multiplication and division in order from left to right.
- Finally, do addition and subtraction in order from left to right.

Example

Simplify the numerical expression at right: $\quad 12 \div 2^2 - 4 + 3(1 + 2)^3$

Start by simplifying the parentheses:	$3(1 + 2)^3 = 3(3)^3$	so $12 \div 2^2 - 4 + 3(3)^3$
Then perform the exponent operation:	$2^2 = 4$ and $3^3 = 27$	so $12 \div 4 - 4 + 3(27)$
Next, multiply and divide left to right:	$12 \div 4 = 3$ and $3(27) = 81$	so $3 - 4 + 81$
Finally, add and subtract left to right:	$3 - 4 = -1$	so $-1 + 81 = 80$

Simplify the following numerical expressions.

1. $29 + 16 \div 8 \cdot 25$
2. $36 + 16 - 50 \div 25$
3. $2(3 - 1) \div 8$
4. $\frac{1}{2}(6 - 2)^2 - 4 \cdot 3$
5. $3[2(1 + 5) + 8 - 3^2]$
6. $(8 + 12) \div 4 - 6$
7. $-6^2 + 4 \cdot 8$
8. $18 \cdot 3 \div 3^3$
9. $10 + 5^2 - 25$
10. $20 - (3^3 \div 9) \cdot 2$
11. $100 - (2^3 - 6) \div 2$
12. $22 + (3 \cdot 2)^2 \div 2$
13. $85 - (4 \cdot 2)^2 - 3$
14. $12 + 3\left(\frac{8-2}{12-9}\right) - 2\left(\frac{9-1}{19-15}\right)$
15. $15 + 4\left(\frac{11-2}{9-6}\right) - 2\left(\frac{12-4}{18-10}\right)$

Answers

1. 79
2. 50
3. $\frac{1}{2}$
4. -4
5. 33
6. -1
7. -4
8. 2
9. 10
10. 14
11. 99
12. 40
13. 18
14. 14
15. 25

DISTRIBUTIVE PROPERTY #4

The **Distributive Property** is used to regroup a numerical expression or a polynomial with two or more terms by multiplying each value or term of the polynomial. The resulting sum is an equivalent numerical or algebraic expression. See page 74 in the textbook for more information. In general, the Distributive Property is expressed as:

$$a(b + c) = ab + ac \text{ and } (b + c)a = ba + ca$$

Example 1

$2(x + 4) = (2 \cdot x) + (2 \cdot 4) = 2x + 8$

Example 2

$(x + 2y + 1)2 = (2 \cdot x) + (2 \cdot 2y) + 2(1) = 2x + 4y + 2$

Simplify each expression below by applying the Distributive Property. Follow the correct order of operations in problems 18 through 22.

1. $3(1 + 5)$
2. $4(3 + 2)$
3. $2(x + 6)$
4. $5(x + 4)$
5. $3(x - 4)$
6. $6(x - 6)$
7. $(3 + x)4$
8. $(2 + x)2$
9. $-x(3 - 1)$
10. $-4(x - 1)$
11. $x(y - z)$
12. $a(b - c)$
13. $3(x + y + 3)$
14. $5(y + 2x + 3)$
15. $2(-x + y - 3)$
16. $-4(3x - y + 2)$
17. $x(x + 3x)$
18. $4(x + 2^2 + x^2)$
19. $(2x^2 - 5x - 7)3$
20. $(a + b - c)d$
21. $5a\left(\frac{12-3}{3} + 2\left(\frac{1}{2} + \frac{1}{2}\right) - b^2\right)$
22. $b\left(2^2 + \frac{1}{3}(6 + 3) - ab\right)$

Answers

1. $(3 \cdot 1) + (3 \cdot 5)$ or $3(6) = 18$
2. $(4 \cdot 3) + (4 \cdot 2)$ or $4(5) = 20$
3. $2x + 12$
4. $5x + 20$
5. $3x - 12$
6. $6x - 36$
7. $12 + 4x$
8. $4 + 2x$
9. $-3x + x = -2x$
10. $-4x + 4$
11. $xy - xz$
12. $ab - ac$
13. $3x + 3y + 9$
14. $5y + 10x + 15$
15. $-2x + 2y - 6$
16. $-12x + 4y - 8$
17. $x^2 + 3x^2 = 4x^2$
18. $4x^2 + 4x + 16$
19. $6x^2 - 15x - 21$
20. $ad + bd - cd$
21. $25a - 5ab^2$
22. $7b - ab^2$

Algebra 1

SUBSTITUTION AND EVALUATION #5

Substitution is replacing one symbol with another (a number, a variable, or an expression). One application of the substitution property is replacing a variable name with a number in any expression or equation. In general, if $a = b$, then a may replace b and b may replace a. A **variable** is a letter used to represent one or more numbers (or other algebraic expression). The numbers are the values of the variable. A variable expression has numbers and variables and operations performed on it.

Examples

Evaluate each variable expression for $x = 2$.

a) $5x \Rightarrow 5(2) \Rightarrow 10$

b) $x + 10 \Rightarrow (2) + 10 \Rightarrow 12$

c) $\frac{18}{x} \Rightarrow \frac{18}{(2)} \Rightarrow 9$

d) $\frac{x}{2} \Rightarrow \frac{2}{2} \Rightarrow 1$

e) $3x - 5 \Rightarrow 3(2) - 5 \Rightarrow 6 - 5 \Rightarrow 1$

f) $5x + 3x \Rightarrow 5(2) + 3(2) \Rightarrow 10 + 6 \Rightarrow 16$

Evaluate each of the variable expressions below for the values $x = -3$ and $y = 2$. Be sure to follow the order of operations as you simplify each expression.

1. $x + 3$
2. $x - 2$
3. $x + y + 4$
4. $y - 2 + x$
5. $x^2 - 7$
6. $-x^2 + 4$
7. $x^2 + 2x - 1$
8. $-2x^2 + 3x$
9. $x + 2 + 3y$
10. $y^2 + 2x - 1$
11. $x^2 + y^2 + 2^2$
12. $3^2 + y^2 - x^2$

Evaluate the expressions below using the values of the variables in each problem. These problems ask you to evaluate each expression twice, once with each of the values.

13. $2x^2 - 3x + 4$ for $x = -2$ and $x = 5$
14. $-4x^2 + 8$ for $x = -2$ and $x = 5$
15. $3x^2 - 2x + 8$ for $x = -3$ and $x = 3$
16. $-x^2 + 3$ for $x = -3$ and $x = 3$

Evaluate the variable expressions for $x = -4$ and $y = 5$.

17. $x(x + 3x)$
18. $2(x + 4x)$
19. $2(x + y) + 4\left(\frac{y+3}{x}\right)$
20. $4\left(y^2 + 2\left(\frac{x+9}{5}\right)\right)$
21. $3y(x + x^2 - y)$
22. $(x + y)(3x + 4y)$

Answers
1. 0
2. -5
3. 3
4. -3
5. 2
6. -5
7. 2
8. -27
9. 5
10. -3
11. 17
12. 4
13. a) 18 b) 39
14. a) -8 b) -92
15. a) 41 b) 29
16. a) -6 b) -6
17. 64
18. -40
19. -6
20. 108
21. 105
22. 8

SKILL BUILDERS

TABLES, EQUATIONS, AND GRAPHS #6

An input/output table provides the opportunity to find the rule that determines the output value for each input value. If you already know the rule, the table is one way to find points to graph the equation, which in this course will usually be written in y-form as $y = mx + b$. Review the information in the Tool Kit entry on page 96 in the textbook, then use the following examples and problems to practice these skills.

Example 1

Use the input/output table below to find the pattern (rule) that pairs each x-value with its y-value. Write the rule below x in the table, then write the equation in y-form.

x (input)	-1	3	2	-3	1	0	-4	4	-2	x
y (output)		5			1	-1		7	-5	

Use a guess and check approach to test various patterns. Since (3, 5) is in the table, try $y = x + 2$ and test another input value, $x = 1$, to see if the same rule works. Unfortunately, this is not true. Next try $2x$ and add or subtract values. For (4, 7), $2(4) - 1 = 7$. Now try (-2, -5): $2(-2) - 1 = -5$. Test (3, 5): $2(3) - 1 = 5$. It appears that the equation for this table is $y = 2x - 1$.

Example 2

Find the missing values for $y = 2x + 1$ and graph the equation. Each output value is found by substituting the input value for x, multiplying it by 2, then adding 1.

x (input)	-3	-2	-1	0	1	2	3	4	5	x
y (output)	-5				3		7			2x+1

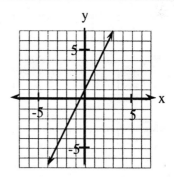

- x-values are referred to as inputs. The set of all input values is the domain.

- y-values are referred to as outputs. The set of all output values is the range.

Use the pairs of input/output values in the table to graph the equation. A portion of the graph is shown at right.

Algebra 1

For each input/output table below, find the missing values, write the rule for x, then write the equation in y-form.

1.
input x	-3	-2	-1	0	1	2	3	x
output y			-1	1			7	

2.
input x	-3	-2	-1	0	1	2	3	x
output y	0			2			5	

3.
input x	-3	-2	-1	0	1	2	3	x
output y		-4		-2		0		

4.
input x	-3	-2	-1	0	1	2	3	x
output y	-10			-1			8	

5.
input x	2	7		-3		-4	3	x
output y	10		8	-10	22			

6.
input x	0	5		-6		3	7	x
output y	3		1	-9	-1	9		-5

7.
input x	4	3	-2	0	1	-5	-1	x
output y		-11		-5			-3	

8.
input x	6		0	7		-2	-1	x
output y		-6	-3		2		-4	1

9.
input x	$-\frac{1}{2}$	0	0.3	0.5	0.75	$\frac{5}{4}$	3.2	x
output y	0		0.8			$\frac{7}{4}$		

10.
input x	$-\frac{3}{4}$	$-\frac{1}{2}$	$-\frac{1}{4}$	0	$\frac{1}{4}$	$\frac{1}{2}$	$\frac{3}{4}$	1	x
output y		$-\frac{1}{4}$		0			$\frac{3}{8}$		

11.
input x	-3	-2	-1	0	1	2	3	x
output y		5		3	1			

12.
input x	-3	-2	-1	0	1	2	3	4	x
output y		2		-2		-6			

13.
input x	5	-2		0		4		x
output y		5	-21	-3	13		9	

14.
input x	5	6	-3		7	4		2	x
output y	2		10	16			1	5	

15.
input x	-3	-2	-1	0	1	2	3	x
output y		4		0			9	

16.
input x	-3	-2	-1	0	1	2	3	x
output y	10		2			5		

Make an input/output table and use it to draw a graph for each of the following equations. Use inputs (domain values) of $-3 \leq x \leq 3$.

17. $y = x + 5$

18. $y = -x + 4$

19. $y = 2x + 3$

20. $y = \frac{1}{2}x - 2$

21. $y = -\frac{2}{3}x + 3$

22. $y = 2$

23. $y = x^2 + 3$

24. $y = -x^2 - 4$

Answers

1. y = 2x + 1

input x	-3	-2	-1	0	1	2	3	x
output y	-5	-3	-1	1	3	5	7	2x+1

2. y = x + 3

input x	-3	-2	-1	0	1	2	3	x
output y	0	1	2	3	4	5	6	x+3

3. y = x − 2

input x	-3	-2	-1	0	1	2	3	x
output y	-5	-4	-3	-2	-1	0	1	x−2

4. y = 3x − 1

input x	-3	-2	-1	0	1	2	3	x
output y	-10	-7	-4	-1	2	5	8	3x−1

5. y = 4x + 2

input x	2	7	$\frac{3}{2}$	-3	5	-4	3	x
output y	10	30	8	-10	22	-14	14	4x+2

6. y = 2x + 3

input x	0	5	-1	-6	-2	3	7	-4	x
output y	3	13	1	-9	-1	9	17	-5	2x+3

7. y = −2x − 5

input x	4	3	-2	0	1	-5	-1	x
output y	-13	-11	-1	-5	-7	5	-3	−2x−5

8. y = x − 3

input x	6	-3	0	7	5	-2	-1	4	x
output y	3	-6	-3	4	2	-5	-4	1	x−3

9. y = x + 0.5

input x	$-\frac{1}{2}$	0	0.3	0.5	0.75	$\frac{5}{4}$	3.2	x
output y	0	0.5	0.8	1	1.25	$\frac{7}{4}$	3.7	x+.5

10. y = $\frac{1}{2}$x

input x	$-\frac{3}{4}$	$-\frac{1}{2}$	$-\frac{1}{4}$	0	$\frac{1}{4}$	$\frac{1}{2}$	$\frac{3}{4}$	1	x
output y	$-\frac{3}{8}$	$-\frac{1}{4}$	$-\frac{1}{8}$	0	$\frac{1}{8}$	$\frac{1}{4}$	$\frac{3}{8}$	$\frac{1}{2}$	$\frac{1}{2}$x

11.

input x	-3	-2	-1	0	1	2	3	x
output y	6	5	4	3	2	1	0	-x+3

12.

input x	-3	-2	-1	0	1	2	3	4	x
output y	4	2	0	-2	-4	-6	-8	-10	−2x−2

13.

input x	5	-2	4.5	0	-4	4	-3	x
output y	-23	5	-21	-3	13	-19	9	−4x−3

14.

input x	5	6	-3	-9	7	4	6	2	x
output y	2	1	10	16	0	3	1	5	−x+7

15.

input x	-3	-2	-1	0	1	2	3	x
output y	9	4	1	0	1	4	9	x^2

16.

input x	-3	-2	-1	0	1	2	3	x
output y	10	5	2	1	2	5	10	x^2+1

17.

input x	-3	-2	-1	0	1	2	3
output y	2	3	4	5	6	7	8

18.

input x	-3	-2	-1	0	1	2	3
output y	7	6	5	4	3	2	1

19.

input x	-3	-2	-1	0	1	2	3
output y	-3	-1	1	3	5	7	9

20.

input x	-3	-2	-1	0	1	2	3
output y	-3.5	-3	-2.5	-2	-1.5	-1	-0.5

21.

input x	-3	-2	-1	0	1	2	3
output y	5	$4\frac{1}{3}$	$3\frac{2}{3}$	3	$2\frac{1}{3}$	$1\frac{2}{3}$	1

22.

input x	-3	-2	-1	0	1	2	3
output y	2	2	2	2	2	2	2

23.

input x	-3	-2	-1	0	1	2	3
output y	12	7	4	3	4	7	12

24.

input x	-3	-2	-1	0	1	2	3
output y	-13	-8	-5	-4	-5	-8	-13

Algebra 1

17.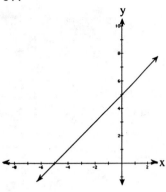

18.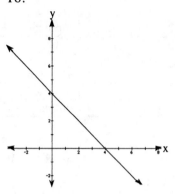

19.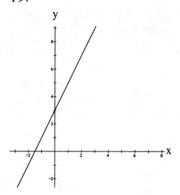

20.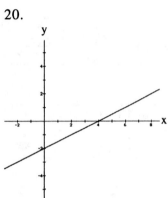

21.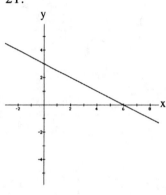

22.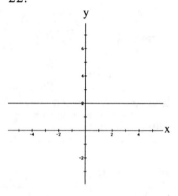

23.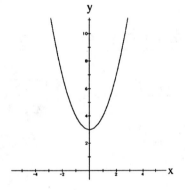

24.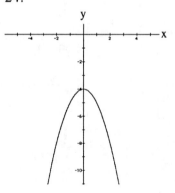

SOLVING LINEAR EQUATIONS #7

Solving equations involves "undoing" what has been done to create the equation. In this sense, solving an equation can be described as "working backward," generally known as using inverse (or opposite) operations. For example, to undo $x + 2 = 5$, that is, <u>adding</u> 2 to x, <u>subtract</u> 2 from <u>both</u> sides of the equation. The result is $x = 3$, which makes $x + 2 = 5$ true. For $2x = 17$, x is <u>multiplied</u> by 2, so <u>divide</u> both sides by 2 and the result is $x = 8.5$. For equations like those in the examples and exercises below, apply the idea of inverse (opposite) operations several times. Always follow the correct order of operations.

Example 1

Solve for x: $2(2x - 1) - 6 = -x + 2$

First distribute to remove parentheses, then combine like terms.

$4x - 2 - 6 = -x + 2$
$4x - 8 = -x + 2$

Next, move variables and constants by addition of opposites to get the variable term on one side of the equation.

$$\begin{array}{r} 4x - 8 = -x + 2 \\ +x +x \\ \hline 5x - 8 = 2 \end{array}$$

$$\begin{array}{r} 5x - 8 = 2 \\ +8 +8 \\ \hline 5x = 10 \end{array}$$

Now, divide by 5 to get the value of x.

$\frac{5x}{5} = \frac{10}{5} \quad \Rightarrow \quad x = 2$

Finally, check that your answer is correct.

$2(2(2) - 1) - 6 = -(2) + 2$
$2(4 - 1) - 6 = 0$
$2(3) - 6 = 0$
$6 - 6 = 0$ checks

Example 2

Solve for y: $2x + 3y - 9 = 0$

This equation has two variables, but the instruction says to isolate y. First move the terms without y to the other side of the equation by adding their opposites to both sides of the equation.

$$\begin{array}{r} 2x + 3y - 9 = 0 \\ +9 +9 \\ \hline 2x + 3y = +9 \end{array}$$

$$\begin{array}{r} 2x + 3y = 9 \\ -2x -2x \\ \hline 3y = -2x + 9 \end{array}$$

Divide by 3 to isolate y. Be careful to divide every term on the right by 3.

$\frac{3y}{3} = \frac{-2x+9}{3} \quad \Rightarrow \quad y = -\frac{2}{3}x + 3$

Algebra 1

Solve each equation below.

1. $5x + 2 = -x + 14$
2. $3x - 2 = x + 10$
3. $6x + 4x - 2 = 15$
4. $6x - 3x + 2 = -10$
5. $\frac{2}{3}y - 6 = 12$
6. $\frac{3}{4}x + 2 = -7$
7. $2(x + 2) = 3(x - 5)$
8. $3(m - 2) = -2(m - 7)$
9. $3(2x + 2) + 2(x - 7) = x + 3$
10. $2(x + 3) + 5(x - 2) = -x + 10$
11. $4 - 6(w + 2) = 10$
12. $6 - 2(x - 3) = 12$
13. $3(2z - 7) = 5z + 17 + z$
14. $-3(2z - 7) = -3z + 21 - 3z$

Solve for the named variable.

15. $2x + b = c$ (for x)
16. $3x - d = m$ (for x)
17. $3x + 2y = 6$ (for y)
18. $-3x + 5y = -10$ (for y)
19. $y = mx + b$ (for b)
20. $y = mx + b$ (for x)

Answers

1. 2
2. 6
3. $\frac{17}{10}$
4. -4
5. 27
6. -12
7. 19
8. 4
9. $\frac{11}{7}$
10. $\frac{7}{4}$
11. -3
12. 0
13. no solution
14. all numbers
15. $x = \frac{c-b}{2}$
16. $x = \frac{m+d}{3}$
17. $y = -\frac{3}{2}x + 3$
18. $y = \frac{3}{5}x - 2$
19. $b = y - mx$
20. $x = \frac{y-b}{m}$

WRITING EQUATIONS #8

You have used Guess and Check tables to solve problems. The patterns and organization of a Guess and Check table will also help to write equations. You will eventually be able to solve a problem by setting up the table headings, writing the equation, and solving it with few or no guesses and checks.

Example 1

The perimeter of a triangle is 51 centimeters. The longest side is twice the length of the shortest side. The third side is three centimeters longer than the shortest side. How long is each side? Write an equation that represents the problem.

First set up a table (with headings) for this problem and, if necessary, fill in numbers to see the pattern.

guess short side	long side	third side	perimeter	check 51?
10	2(10) = 20	10 + 3 = 13	10 + 20 + 13 = 43	too low
15	2(15) = 20	15 + 3 = 18	15 + 30 + 18 = 63	too high
13	2(13) = 26	13 + 3 = 16	13 + 26 + 16 = 55	too high
12	2(12) = 24	12 + 3 = 15	12 + 24 + 15 = 51	correct

The lengths of the sides are 12 cm, 24 cm, and 15 cm.

Since we could guess any number for the short side, use x to represent it and continue the pattern.

guess short side	long side	third side	perimeter	check 51?
x	$2x$	$x + 3$	$x + 2x + x + 3$	51

A possible equation is $x + 2x + x + 3 = 51$ (simplified, $4x + 3 = 51$). Solving the equation gives the same solution as the guess and check table.

Example 2

Darren sold 75 tickets worth $462.50 for the school play. He charged $7.50 for adults and $3.50 for students. How many of each kind of ticket did he sell? First set up a table with columns and headings for number of tickets and their value.

guess number of adult tickets sold	value of adult tickets sold	number of student tickets sold	value of student tickets sold	total value	check 462.50?
40	40(7.50) = 300	75 − 40 = 35	3.50(35) = 122.50	300 + 122.50 = 422.50	too low

By guessing different (and in this case, larger) numbers of adult tickets, the answer can be found. The <u>pattern</u> from the table can be generalized as an equation.

number of adult tickets sold	value of adult tickets sold	number of student tickets sold	value of student tickets sold	total value	check 462.50?
x	$7.50x$	$75 - x$	$3.50(75 - x)$	$7.50x + 3.50(75 - x)$	462.50

Algebra 1

A possible equation is $7.50x + 3.50(75 - x) = 462.50$. Solving the equation gives the solution--50 adult tickets and 25 student tickets--without guessing and checking.

$7.50x + 3.50(75 - x) = 462.50 \Rightarrow 7.50x + 262.50 - 3.50x = 462.50$
$\Rightarrow 4x = 200 \Rightarrow x = 50$ adult tickets $\Rightarrow 75 - x = 75 - 50 = 25$ student tickets

Find the solution and write a possible equation. You may make a Guess and Check table, solve it, then write the equation or use one or two guesses to establish a pattern, write an equation, and solve it to find the solution.

1. A box of fruit has four more apples than oranges. Together there are 52 pieces of fruit. How many of each type of fruit are there?

2. Thu and Cleo are sharing the driving on a 520 mile trip. If Thu drives 60 miles more than Cleo, how far did each of them drive?

3. Aimee cut a string that was originally 126 centimeters long into two pieces so that one piece is twice as long as the other. How long is each piece?

4. A full bucket of water weighs eight kilograms. If the water weighs five times as much as the bucket empty, how much does the water weigh?

5. The perimeter of a rectangle is 100 feet. If the length is five feet more than twice the width, find the length and width.

6. The perimeter of a rectangular city is 94 miles. If the length is one mile less than three times the width, find the length and width of the city.

7. Find three consecutive numbers whose sum is 138.

8. Find three consecutive even numbers whose sum is 468.

9. The perimeter of a triangle is 57. The first side is twice the length of the second side. The third side is seven more than the second side. What is the length of each side?

10. The perimeter of a triangle is 86 inches. The largest side is four inches less than twice the smallest side. The third side is 10 inches longer than the smallest side. What is the length of each side?

11. Thirty more student tickets than adult tickets were sold for the game. Student tickets cost $2, adult tickets cost $5, and $1460 was collected. How many of each kind of ticket were sold?

12. Fifty more "couples" tickets than "singles" tickets were sold for the dance. "Singles" tickets cost $10 and "couples" tickets cost $15. If $4000 was collected, how many of each kind of ticket was sold?

13. Helen has twice as many dimes as nickels and five more quarters than nickels. The value of her coins is $4.75. How many dimes does she have?

14. Ly has three more dimes than nickels and twice as many quarters as dimes. The value of his coins is $9.60. How many of each kind of coin does he have?

15. Enrique put his money in the credit union for one year. His money earned 8% simple interest and at the end of the year his account was worth $1350. How much was originally invested?

16. Juli's bank pays 7.5% simple interest. At the end of the year, her college fund was worth $10,965. How much was it worth at the start of the year?

17. Elisa sold 110 tickets for the football game. Adult tickets cost $2.50 and student tickets cost $1.10. If she collected $212, how many of each kind of ticket did she sell?

18. The first performance of the school play sold out all 2000 tickets. The ticket sales receipts totaled $8500. If adults paid $5 and students paid $3 for their tickets, how many of each kind of ticket was sold?

19. Leon and Jason leave Los Angeles going in opposite directions. Leon travels five miles per hour faster than Jason. In four hours they are 524 miles apart. How fast is each person traveling?

20. Keri and Yuki leave New York City going in opposite directions. Keri travels three miles per hour slower than Yuki. In six hours they are 522 miles apart. How fast is each person traveling?

Answers (equations may vary)

1. 24 oranges, 28 apples; $x + (x + 4) = 52$

2. Cleo 230 miles, Thu 290 miles; $x + (x + 60) = 520$

3. 42, 84; $x + 2x = 126$

4. $6\frac{2}{3}$ kg.; $x + 5x = 8$

5. 15, 35; $2x + 2(2x + 5) = 100$

6. 12, 35; $2x + 2(3x - 1) = 94$

7. 45, 46, 47; $x + (x + 1) + (x + 2) = 138$

8. 154, 156, 158; $x + (x + 2) + (x + 4) = 468$

9. 10, 20, 27; $x + 2x + (2x + 7) = 57$

10. 20, 36, 30; $x + (2x - 4) + (x + 10) = 86$

11. 200 adult, 230 students; $5x + 2(x + 30) = 1460$

12. 130 single, 180 couple; $10x + 15(x + 50) = 4000$

13. 7 nickels, 14 dimes, 12 quarters; $0.05x + 0.10(2x) + 0.25(x + 5) = 4.75$

14. 12 nickels, 15 dimes, 30 quarters; $0.05x + 0.10(x + 3) + 0.25(2x + 6) = 9.60$

15. $1250; $x + 0.08x = 1350$

16. 10,200; $x + 0.075x = 10,965$

17. 65 adult, 45 student; $2.50x + 1.10(110 - x) = 212$

18. 1250 adults, 750 students; $5.00x + 3.00(2000 - x) = 8500$

19. Jason 63, Leon 68; $4x + 4(x + 5) = 524$

20. Yuki 45, Keri 42; $6x + 6(x - 3) = 522$

Algebra 1

SOLVING PROPORTIONS #9

A **proportion** is an equation stating that two ratios (fractions) are equal. To solve a proportion, begin by eliminating fractions. This means using the inverse operation of division, namely, multiplication. Multiply both sides of the proportion by one or both of the denominators. Then solve the resulting equation in the usual way.

Example 1

$\frac{x}{3} = \frac{5}{8}$

Undo the division by 3 by multiplying both sides by 3.

$(3)\frac{x}{3} = \frac{5}{8}(3)$

$x = \frac{15}{8} = 1\frac{7}{8}$

Example 2

$\frac{x}{x+1} = \frac{3}{5}$

Multiply by 5 and (x+1) on both sides of the equation.

$5(x+1)\frac{x}{x+1} = \frac{3}{5}(5)(x+1)$

Note that $\frac{(x+1)}{(x+1)} = 1$ and $\frac{5}{5} = 1$, so $5x = 3(x+1)$

$5x = 3x + 3 \Rightarrow 2x = 3 \Rightarrow x = \frac{3}{2} = 1\frac{1}{2}$

Solve for x or y.

1. $\frac{2}{5} = \frac{y}{15}$
2. $\frac{x}{36} = \frac{4}{9}$
3. $\frac{2}{3} = \frac{x}{5}$
4. $\frac{5}{8} = \frac{x}{100}$
5. $\frac{3x}{10} = \frac{24}{9}$
6. $\frac{3y}{5} = \frac{24}{10}$
7. $\frac{x+2}{3} = \frac{5}{7}$
8. $\frac{x-1}{4} = \frac{7}{8}$
9. $\frac{4x}{5} = \frac{x-2}{7}$
10. $\frac{3x}{4} = \frac{x+1}{6}$
11. $\frac{9-x}{6} = \frac{24}{2}$
12. $\frac{7-y}{5} = \frac{3}{4}$
13. $\frac{1}{x} = \frac{5}{x+1}$
14. $\frac{3}{y} = \frac{6}{y-2}$
15. $\frac{4}{x} = \frac{x}{9}$
16. $\frac{25}{y} = \frac{y}{4}$

Answers

1. 6
2. 16
3. $\frac{10}{3} = 3\frac{1}{3}$
4. $62\frac{1}{2}$
5. $\frac{80}{9}$
6. 4
7. $\frac{1}{7}$
8. $4\frac{1}{2}$
9. $\frac{-10}{23}$
10. $\frac{2}{7}$
11. -63
12. $\frac{13}{4}$
13. $\frac{1}{4}$
14. -2
15. ±6
16. ±10

SKILL BUILDERS

RATIO APPLICATIONS   #10

Ratios and proportions are used to solve problems involving similar figures, percents, and relationships that vary directly.

Example 1

△ABC is similar to △DEF. Use ratios to find x.

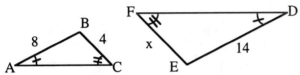

Since the triangles are similar, the ratios of the corresponding sides are equal.

$$\frac{8}{14} = \frac{4}{x} \Rightarrow 8x = 56 \Rightarrow x = 7$$

Example 2

a) What percent of 60 is 45?
b) Forty percent of what number is 45?

In percent problems use the following proportion: $\frac{part}{whole} = \frac{percent}{100}$.

a) $\frac{45}{60} = \frac{x}{100}$

$60x = 4500$

$x = 75 \ (75\%)$

b) $\frac{40}{100} = \frac{45}{x}$

$40x = 4500$

$x = 112$

Example 3

Amy usually swims 20 laps in 30 minutes. How long will it take to swim 50 laps at the same rate?

Since two units are being compared, set up a ratio using the unit words consistently. In this case, "laps" is on top (the numerator) and "minutes" is on the bottom (the denominator) in both ratios. Then solve as shown in Skill Builder #9.

$$\frac{laps}{minutes}: \quad \frac{20}{30} = \frac{50}{x} \quad \Rightarrow \quad 20x = 1500 \quad \Rightarrow \quad x = 75 \text{ minutes}$$

Each pair of figures is similar. Solve for the variable.

1.

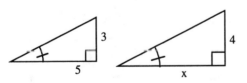

2.

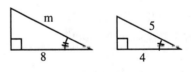

3.

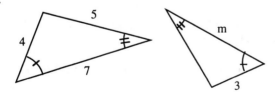

4.

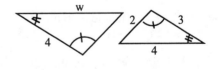

Algebra 1

5.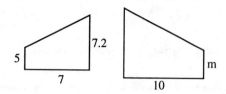

6.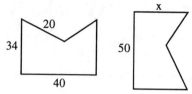

7.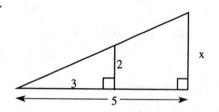

8.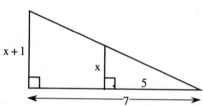

Write and solve a proportion to find the missing part.

9. 15 is 25% of what?

10. 12 is 30% of what?

11. 45% of 200 is what?

12. 32% of 150 is what?

13. 18 is what percent of 24?

14. What percent of 300 is 250?

15. What is 32% of $12.50?

16. What is 7.5% of $325.75?

Use ratios to solve each problem.

17. A rectangle has length 10 feet and width six feet. It is enlarged to a similar rectangle with length 18 feet. What is the new width?

18. If 200 vitamins cost $4.75, what should 500 vitamins cost?

19. The tax on a $400 painting is $34. What should the tax be on a $700 painting?

20. If a basketball player made 72 of 85 shots, how many shots could she expect to make in 200 shots?

21. A cookie recipe uses $\frac{1}{2}$ teaspoon of vanilla with $\frac{3}{4}$ cup of flour. How much vanilla should be used with five cups of flour?

22. My brother grew $1\frac{3}{4}$ inches in $2\frac{1}{2}$ months. At that rate, how much would he grow in one year?

23. The length of a rectangle is four centimeters more than the width. If the ratio of the length to width is seven to five, find the dimensions of the rectangle.

24. A class has three fewer girls than boys. If the ratio of girls to boys is four to five, how many students are in the class?

Answers

1. $\frac{20}{3} = 6\frac{2}{3}$
2. 10
3. $5\frac{1}{4}$
4. $\frac{16}{3} = 5\frac{1}{3}$
5. $\frac{50}{7} = 7\frac{1}{7}$
6. 42.5
7. $\frac{10}{3} = 3\frac{1}{3}$
8. $2\frac{1}{2}$
9. 60
10. 40
11. 90
12. 48
13. 75%
14. $83\frac{1}{3}$%
15. $4
16. $24.43
17. 10.8 ft.
18. $11.88
19. $59.50
20. About 169 shots
21. $3\frac{1}{3}$ teaspoons
22. $8\frac{2}{5}$ inches
23. 10 cm x 14 cm
24. 27 students

USING SUBSTITUTION TO FIND THE POINT OF INTERSECTION OF TWO LINES #11

To find where two lines intersect we could graph them, but there is a faster, more accurate algebraic method called the **substitution method**. This method may also be used to solve systems of equations in word problems.

Example 1

Start with two linear equations in y-form. $\qquad y = -2x + 5 \text{ and } y = x - 1$

Substitute the equal parts. $\qquad -2x + 5 = x - 1$

Solve for x. $\qquad 6 = 3x \Rightarrow x = 2$

The x-coordinate of the point of intersection is $x = 2$. To find the y-coordinate, substitute the value of x into either original equation. Solve for y, then write the solution as an ordered pair. Check that the point works in both equations.

$y = -2(2) + 5 = 1$ and $y = 2 - 1 = 1$, so $(2, 1)$ is where the lines intersect.

Check: $1 = -2(2) + 5$ √ and $1 = 2 - 1$ √.

Example 2

The sales of Gizmo Sports Drink at the local supermarket are currently 6,500 bottles per month. Since New Age Refreshers were introduced, sales of Gizmo have been declining by 55 bottles per month. New Age currently sells 2,200 bottles per month and its sales are increasing by 250 bottles per month. If these rates of change remain the same, in about how many months will the sales for both companies be the same? How many bottles will each company be selling at that time?

Let x = months from now and y = total monthly sales.

For Gizmo: $y = 6500 - 55x$; for New Age: $y = 2200 + 250x$.

Substituting equal parts: $6500 - 55x = 2200 + 250x \Rightarrow 3300 = 305x \Rightarrow 10.82 \approx x$.

Use either equation to find y: $y = 2200 + 250(10.82) \approx 4905$ and $y = 6500 - 55(10.82) \approx 4905$.

The solution is $(10.82, 4905)$. This means that in about 11 months, both drink companies will be selling 4,905 bottles of the sports drinks.

Find the point of intersection (x, y) for each pair of lines by using the substitution method.

1. $y = x + 2$
   $y = 2x - 1$

2. $y = 3x + 5$
   $y = 4x + 8$

3. $y = 11 - 2x$
   $y = x + 2$

4. $y = 3 - 2x$
   $y = 1 + 2x$

5. $y = 3x - 4$
   $y = \frac{1}{2}x + 7$

6. $y = -\frac{2}{3}x + 4$
   $y = \frac{1}{3}x - 2$

7. $y = 4.5 - x$
   $y = -2x + 6$

8. $y = 4x$
   $y = x + 1$

For each problem, define your variables, write a system of equations, and solve them by using substitution.

9. Janelle has $20 and is saving $6 per week. April has $150 and is spending $4 per week. When will they both have the same amount of money?

10. Sam and Hector are gaining weight for football season. Sam weighs 205 pounds and is gaining two pounds per week. Hector weighs 195 pounds but is gaining three pounds per week. In how many weeks will they both weigh the same amount?

11. PhotosFast charges a fee of $2.50 plus $0.05 for each picture developed. PhotosQuick charges a fee of $3.70 plus $0.03 for each picture developed. For how many pictures will the total cost be the same at each shop?

12. Playland Park charges $7 admission plus 75¢ per ride. Funland Park charges $12.50 admission plus 50¢ per ride. For what number of rides is the total cost the same at both parks?

Change one or both equations to y-form and solve by the substitution method.

13. $y = 2x - 3$
    $x + y = 15$

14. $y = 3x + 11$
    $x + y = 3$

15. $x + y = 5$
    $2y - x = -2$

16. $x + 2y = 10$
    $3x - 2y = -2$

17. $x + y = 3$
    $2x - y = -9$

18. $y = 2x - 3$
    $x - y = -4$

19. $x + 2y = 4$
    $x + 2y = 6$

20. $3x = y - 2$
    $6x + 4 = 2y$

Answers

1. (3, 5)
2. (3, 4)
3. (3, 5)
4. $\left(\frac{1}{2}, 2\right)$

5. (4.4, 9.2)
6. (6, 0)
7. (1.5, 3)
8. $\left(\frac{1}{3}, \frac{4}{3}\right)$

9. 13 weeks, $98
10. 10 weeks, 225 pounds
11. 60 pictures, $5.50
12. 22 rides, $23.50

13. (6, 9)
14. (-2, 5)
15. (4, 1)
16. (2, 4)

17. (-2, 5)
18. (7, 11)
19. none
20. infinite

Algebra 1

MULTIPLYING POLYNOMIALS #12

We can use generic rectangles as area models to find the products of polynomials. A generic rectangle helps us organize the problem. It does not have to be drawn accurately or to scale.

Example 1

Multiply $(2x + 5)(x + 3)$

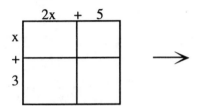 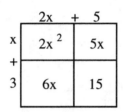

$(2x + 5)(x + 3) = 2x^2 + 11x + 15$
area as a product   area as a sum

Example 2

Multiply $(x + 9)(x^2 - 3x + 5)$

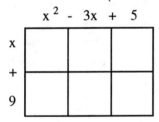 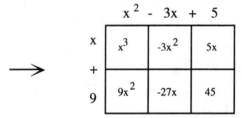

Therefore $(x + 9)(x^2 - 3x + 5) = x^3 + 9x^2 - 3x^2 - 27x + 5x + 45 = x^3 + 6x^2 - 22x + 45$

Another approach to multiplying binomials is to use the mnemonic "F.O.I.L." F.O.I.L. is an acronym for First, Outside, Inside, Last in reference to the positions of the terms in the two binomials.

Example 3

Multiply $(3x - 2)(4x + 5)$ using the F.O.I.L. method.

    F. multiply the FIRST terms of each binomial $(3x)(4x) = 12x^2$
    O. multiply the OUTSIDE terms $(3x)(5) = 15x$
    I. multiply the INSIDE terms $(-2)(4x) = -8x$
    L. multiply the LAST terms of each binomial $(-2)(5) = -10$

Finally, we combine like terms: $12x^2 + 15x - 8x - 10 = 12x^2 + 7x - 10$.

Multiply, then simplify each expression.

1. $x(2x - 3)$
2. $y(3y - 4)$
3. $2y(y^2 + 3y - 2)$
4. $3x(2x^2 - x + 3)$
5. $(x + 2)(x + 7)$
6. $(y - 3)(y - 9)$
7. $(y - 2)(y + 7)$
8. $(x + 8)(x - 7)$
9. $(2x + 1)(3x - 5)$
10. $(3m - 2)(2m + 1)$
11. $(2m + 1)(2m - 1)$
12. $(3y - 4)(3y + 4)$
13. $(3x + 7)^2$
14. $(2x - 5)^2$
15. $(3x + 2)(x^2 - 5x + 2)$
16. $(y - 2)(3y^2 + 2y - 2)$
17. $3(x + 2)(2x - 1)$
18. $-2(x - 2)(3x + 1)$
19. $x(2x - 3)(x + 4)$
20. $2y(2y - 1)(3y + 2)$

Answers

1. $2x^2 - 3x$
2. $3y^2 - 4y$
3. $2y^3 + 6y^2 - 4y$
4. $6x^3 - 3x^2 + 9x$
5. $x^2 + 9x + 14$
6. $y^2 - 12y + 27$
7. $y^2 + 5y - 14$
8. $x^2 + x - 56$
9. $6x^2 - 7x - 5$
10. $6m^2 - m - 2$
11. $4m^2 - 1$
12. $9y^2 - 16$
13. $9x^2 + 42x + 49$
14. $4x^2 - 20x + 25$
15. $3x^3 - 13x^2 - 4x + 4$
16. $3y^3 - 4y^2 - 6y + 4$
17. $6x^2 + 9x - 6$
18. $-6x^2 + 10x + 4$
19. $2x^3 + 5x^2 - 12x$
20. $12y^3 + 2y^2 - 4y$

WRITING AND GRAPHING LINEAR EQUATIONS ON A FLAT SURFACE #13

SLOPE is a number that indicates the steepness (or flatness) of a line, as well as its direction (up or down) left to right.

SLOPE is determined by the ratio: $\dfrac{\text{vertical change}}{\text{horizontal change}}$ between <u>any</u> two points on a line.

For lines that go **up** (from left to right), the sign of the slope is **positive.** For lines that go **down** (left to right), the sign of the slope is **negative**.

Any linear equation written as **y = mx + b**, where m and b are any real numbers, is said to be in **SLOPE-INTERCEPT FORM**. m is the **SLOPE** of the line. b is the **Y-INTERCEPT**, that is, the point (0, b) where the line intersects (crosses) the y-axis.

If two lines have the same slope, then they are parallel. Likewise, **PARALLEL LINES** have the same slope.

Two lines are **PERPENDICULAR** if the slope of one line is the negative reciprocal of the slope of the other line, that is, m and $-\dfrac{1}{m}$. Note that $m \cdot \left(\dfrac{-1}{m}\right) = -1$.

Examples: 3 and $-\dfrac{1}{3}$, $-\dfrac{2}{3}$ and $\dfrac{3}{2}$, $\dfrac{5}{4}$ and $-\dfrac{4}{5}$

Two distinct lines that are not parallel intersect in a single point. See "Solving Linear Systems" to review how to find the point of intersection.

Example 1

Write the slope of the line containing the points (-1, 3) and (4, 5).

First graph the two points and draw the line through them.

Look for and draw a slope triangle using the two given points.

Write the ratio $\dfrac{\text{vertical change in } y}{\text{horizontal change in } x}$ using the legs of the right triangle: $\dfrac{2}{5}$.

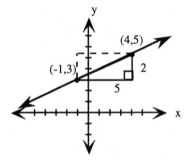

Assign a positive or negative value to the slope (this one is positive) depending on whether the line goes up (+) or down (−) from left to right.

If the points are inconvenient to graph, use a "Generic Slope Triangle", visualizing where the points lie with respect to each other.

Example 2

Graph the linear equation $y = \frac{4}{7}x + 2$

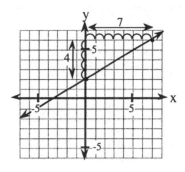

Using $y = mx + b$, the slope in $y = \frac{4}{7}x + 2$ is $\frac{4}{7}$ and the y-intercept is the point (0, 2). To graph, begin at the y-intercept (0, 2). Remember that slope is $\frac{\text{vertical change}}{\text{horizontal change}}$ so go up 4 units (since 4 is positive) from (0, 2) and then move right 7 units. This gives a second point on the graph. To create the graph, draw a straight line through the two points.

Example 3

A line has a slope of $\frac{3}{4}$ and passes through (3, 2). What is the equation of the line?

Using $y = mx + b$, write $y = \frac{3}{4}x + b$. Since (3, 2) represents a point (x, y) on the line, substitute 3 for x and 2 for y, $2 = \frac{3}{4}(3) + b$, and solve for b. $2 = \frac{9}{4} + b \Rightarrow 2 - \frac{9}{4} = b \Rightarrow -\frac{1}{4} = b$.

The equation is $y = \frac{3}{4}x - \frac{1}{4}$.

Example 4

Decide whether the two lines at right are parallel, perpendicular, or neither (i.e., intersecting).

$5x - 4y = -6$ and $-4x + 5y = 3$.

First find the slope of each equation. Then compare the slopes.

$5x - 4y = -6$ $-4y = -5x - 6$ $y = \frac{-5x - 6}{-4}$ $y = \frac{5}{4}x + \frac{3}{2}$ The slope of this line is $\frac{5}{4}$.	$-4x + 5y = 3$ $5y = 4x + 3$ $y = \frac{4x + 3}{5}$ $y = \frac{4}{5}x + \frac{3}{5}$ The slope of this line is $\frac{4}{5}$.	These two slopes are not equal, so they are not parallel. The product of the two slopes is 1, not -1, so they are not perpendicular. These two lines are neither parallel nor perpendicular, but do intersect.

Example 5

Find two equations of the line through the given point, one parallel and one perpendicular to the given line: $y = -\frac{5}{2}x + 5$ and (-4, 5).

For the parallel line, use $y = mx + b$ with the same slope to write $y = -\frac{5}{2}x + b$. Substitute the point (-4, 5) for x and y and solve for b. $5 = -\frac{5}{2}(-4) + b \Rightarrow 5 = \frac{20}{2} + b \Rightarrow -5 = b$ Therefore the parallel line through (-4, 5) is $y = -\frac{5}{2}x - 5$.	For the perpendicular line, use $y = mx + b$ where m is the negative reciprocal of the slope of the original equation to write $y = \frac{2}{5}x + b$. Substitute the point (-4, 5) and solve for b. $5 = \frac{2}{5}(-4) + b \Rightarrow \frac{33}{5} = b$ Therefore the perpendicular line through (-4, 5) is $y = \frac{2}{5}x + \frac{33}{5}$.

Write the slope of the line containing each pair of points.

1. (3, 4) and (5, 7)
2. (5, 2) and (9, 4)
3. (1, -3) and (-4, 7)
4. (-2, 1) and (2, -2)
5. (-2, 3) and (4, 3)
6. (8, 5) and (3, 5)

Use a Generic Slope Triangle to write the slope of the line containing each pair of points:

7. (51, 40) and (33, 72)
8. (20, 49) and (54, 90)
9. (10, -13) and (-61, 20)

Identify the y-intercept in each equation.

10. $y = \frac{1}{2}x - 2$
11. $y = -\frac{3}{5}x - \frac{5}{3}$
12. $3x + 2y = 12$
13. $x - y = -13$
14. $2x - 4y = 12$
15. $4y - 2x = 12$

Write the equation of the line with:

16. slope $= \frac{1}{2}$ and passing through (4, 3).
17. slope $= \frac{2}{3}$ and passing through (-3, -2).
18. slope $= -\frac{1}{3}$ and passing through (4, -1).
19. slope $= -4$ and passing through (-3, 5).

Determine the slope of each line using the highlighted points.

20.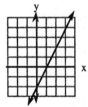
21.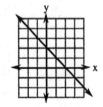
22.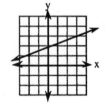

Using the slope and y-intercept, determine the equation of the line.

23.
24.
25.
26.

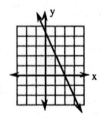

Graph the following linear equations on graph paper.

27. $y = \frac{1}{2}x + 3$
28. $y = -\frac{3}{5}x - 1$
29. $y = 4x$
30. $y = -6x + \frac{1}{2}$
31. $3x + 2y = 12$

State whether each pair of lines is parallel, perpendicular, or intersecting.

32. $y = 2x - 2$ and $y = 2x + 4$
33. $y = \frac{1}{2}x + 3$ and $y = -2x - 4$
34. $x - y = 2$ and $x + y = 3$
35. $y - x = -1$ and $y + x = 3$
36. $x + 3y = 6$ and $y = -\frac{1}{3}x - 3$
37. $3x + 2y = 6$ and $2x + 3y = 6$
38. $4x = 5y - 3$ and $4y = 5x + 3$
39. $3x - 4y = 12$ and $4y = 3x + 7$

Find an equation of the line through the given point and parallel to the given line.

40. $y = 2x - 2$ and $(-3, 5)$
41. $y = \frac{1}{2}x + 3$ and $(-4, 2)$
42. $x - y = 2$ and $(-2, 3)$
43. $y - x = -1$ and $(-2, 1)$
44. $x + 3y = 6$ and $(-1, 1)$
45. $3x + 2y = 6$ and $(2, -1)$
46. $4x = 5y - 3$ and $(1, -1)$
47. $3x - 4y = 12$ and $(4, -2)$

Find an equation of the line through the given point and perpendicular to the given line.

48. $y = 2x - 2$ and $(-3, 5)$
49. $y = \frac{1}{2}x + 3$ and $(-4, 2)$
50. $x - y = 2$ and $(-2, 3)$
51. $y - x = -1$ and $(-2, 1)$
52. $x + 3y = 6$ and $(-1, 1)$
53. $3x + 2y = 6$ and $(2, -1)$
54. $4x = 5y - 3$ and $(1, -1)$
55. $3x - 4y = 12$ and $(4, -2)$

Write an equation of the line parallel to each line below through the given point.

56.
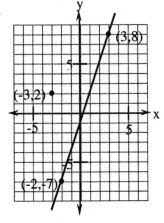

57.
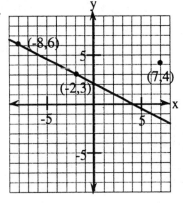

Algebra 1

Answers

1. $\dfrac{3}{2}$
2. $\dfrac{1}{2}$
3. -2
4. $-\dfrac{3}{4}$
5. 0
6. 0
7. $-\dfrac{16}{9}$
8. $\dfrac{41}{34}$
9. $\dfrac{-33}{71}$
10. $(0, -2)$
11. $\left(0, -\dfrac{5}{3}\right)$
12. $(0, 6)$
13. $(0, 13)$
14. $(0, -3)$
15. $(0, 3)$
16. $y = \dfrac{1}{2}x + 1$
17. $y = \dfrac{2}{3}x$
18. $y = -\dfrac{1}{3}x + \dfrac{1}{3}$
19. $y = -4x - 7$
20. $-\dfrac{1}{2}$
21. $\dfrac{3}{4}$
22. -2
23. $y = 2x - 2$
24. $y = -x + 2$
25. $y = \dfrac{1}{3}x + 2$
26. $y = -2x + 4$
27. line with slope $\dfrac{1}{2}$ and y-intercept $(0, 3)$
28. line with slope $-\dfrac{3}{5}$ and y-intercept $(0, -1)$
29. line with slope 4 and y-intercept $(0, 0)$
30. line with slope -6 and y-intercept $\left(0, \dfrac{1}{2}\right)$
31. line with slope $-\dfrac{3}{2}$ and y-intercept $(0, 6)$
32. parallel
33. perpendicular
34. perpendicular
35. perpendicular
36. parallel
37. intersecting
38. intersecting
39. parallel
40. $y = 2x + 11$
41. $y = \dfrac{1}{2}x + 4$
42. $y = x + 5$
43. $y = x + 3$
44. $y = -\dfrac{1}{3}x + \dfrac{2}{3}$
45. $y = -\dfrac{3}{2}x + 2$
46. $y = \dfrac{4}{5}x - \dfrac{9}{5}$
47. $y = \dfrac{3}{4}x - 5$
48. $y = -\dfrac{1}{2}x + \dfrac{7}{2}$
49. $y = -2x - 6$
50. $y = -x + 1$
51. $y = -x - 1$
52. $y = 3x + 4$
53. $y = \dfrac{2}{3}x - \dfrac{7}{3}$
54. $y = -\dfrac{5}{4}x + \dfrac{1}{4}$
55. $y = -\dfrac{4}{3}x + \dfrac{10}{3}$
56. $y = 3x + 11$
57. $y = -\dfrac{1}{2}x + \dfrac{15}{2}$

FACTORING POLYNOMIALS #14

Often we want to un-multiply or **factor** a polynomial P(x). This process involves finding a constant and/or another polynomial that evenly divides the given polynomial. In formal mathematical terms, this means $P(x) = q(x) \cdot r(x)$, where q and r are also polynomials. For elementary algebra there are three general types of factoring.

1) **Common term** (finding the largest common factor):

   $6x + 18 = 6(x + 3)$ where 6 is a common factor of both terms.

   $2x^3 - 8x^2 - 10x = 2x(x^2 - 4x - 5)$ where 2x is the common factor.

   $2x^2(x - 1) + 7(x - 1) = (x - 1)(2x^2 + 7)$ where x - 1 is the common factor.

2) **Special products**

   $a^2 - b^2 = (a + b)(a - b)$ $x^2 - 25 = (x + 5)(x - 5)$

    $9x^2 - 4y^2 = (3x + 2y)(3x - 2y)$

   $x^2 + 2xy + y^2 = (x + y)^2$ $x^2 + 8x + 16 = (x + 4)^2$

   $x^2 - 2xy + y^2 = (x - y)^2$ $x^2 - 8x + 16 = (x - 4)^2$

3a) **Trinomials** in the form $x^2 + bx + c$ where the coefficient of x^2 is 1.

   Consider $x^2 + (d + e)x + d \cdot e = (x + d)(x + e)$, where the coefficient of x is the <u>sum</u> of two numbers d and e AND the constant is the <u>product</u> of the same two numbers, d and e. A quick way to determine all of the possible pairs of integers d and e is to factor the constant in the original trinomial. For example, 12 is $1 \cdot 12$, $2 \cdot 6$, and $3 \cdot 4$. The signs of the two numbers are determined by the combination you need to get the sum. The "sum and product" approach to factoring trinomials is the same as solving a "Diamond Problem" in CPM's Algebra 1 course (see below).

   $x^2 + 8x + 15 = (x + 3)(x + 5); \quad 3 + 5 = 8, \quad 3 \cdot 5 = 15$

   $x^2 - 2x - 15 = (x - 5)(x + 3); \quad -5 + 3 = -2, \quad -5 \cdot 3 = -15$

   $x^2 - 7x + 12 = (x - 3)(x - 4); \quad -3 + (-4) = -7, \quad (-3)(-4) = 12$

   The sum and product approach can be shown visually using rectangles for an area model. The figure at far left below shows the "Diamond Problem" format for finding a sum and product. Here is how to use this method to factor $x^2 + 6x + 8$.

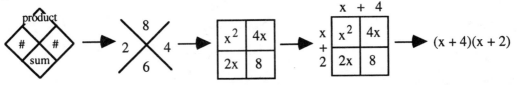

>> Explanation and examples continue on the next page. >>

3b) **Trinomials** in the form $ax^2 + bx + c$ where $a \neq 1$.

Note that the upper value in the diamond is no longer the constant. Rather, it is the product of a and c, that is, the coefficient of x^2 and the constant.

$2x^2 + 7x + 3$

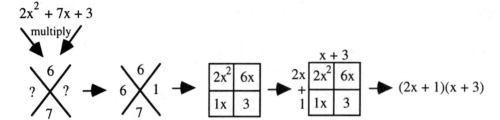

Below is the process to factor $5x^2 - 13x + 6$.

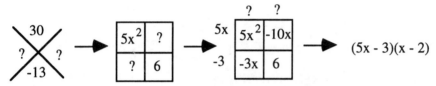

Polynomials with four or more terms are generally factored by grouping the terms and using one or more of the three procedures shown above. Note that polynomials are usually factored completely. In the second example in part (1) above, the trinomial also needs to be factored. Thus, the complete factorization of $2x^3 - 8x^2 - 10x = 2x(x^2 - 4x - 5) = 2x(x - 5)(x + 1)$.

Factor each polynomial completely.

1. $x^2 - x - 42$
2. $4x^2 - 18$
3. $2x^2 + 9x + 9$
4. $2x^2 + 3xy + y^2$
5. $6x^2 - x - 15$
6. $4x^2 - 25$
7. $x^2 - 28x + 196$
8. $7x^2 - 847$
9. $x^2 + 18x + 81$
10. $x^2 + 4x - 21$
11. $3x^2 + 21x$
12. $3x^2 - 20x - 32$
13. $9x^2 - 16$
14. $4x^2 + 20x + 25$
15. $x^2 - 5x + 6$
16. $5x^3 + 15x^2 - 20x$
17. $4x^2 + 18$
18. $x^2 - 12x + 36$
19. $x^2 - 3x - 54$
20. $6x^2 - 21$
21. $2x^2 + 15x + 18$
22. $16x^2 - 1$
23. $x^2 - 14x + 49$
24. $x^2 + 8x + 15$
25. $3x^3 - 12x^2 - 45x$
26. $3x^2 + 24$
27. $x^2 + 16x + 64$

Factor completely.

28. $75x^3 - 27x$
29. $3x^3 - 12x^2 - 36x$
30. $4x^3 - 44x^2 + 112x$
31. $5y^2 - 125$
32. $3x^2y^2 - xy^2 - 4y^2$
33. $x^3 + 10x^2 - 24x$
34. $3x^3 - 6x^2 - 45x$
35. $3x^2 - 27$
36. $x^4 - 16$

Factor each of the following completely. Use the modified diamond approach.

37. $2x^2 + 5x - 7$
38. $3x^2 - 13x + 4$
39. $2x^2 + 9x + 10$
40. $4x^2 - 13x + 3$
41. $4x^2 + 12x + 5$
42. $6x^3 + 31x^2 + 5x$
43. $64x^2 + 16x + 1$
44. $7x^2 - 33x - 10$
45. $5x^2 + 12x - 9$

Answers

1. $(x + 6)(x - 7)$
2. $2(2x^2 - 9)$
3. $(2x + 3)(x + 3)$
4. $(2x + y)(x + y)$
5. $(2x + 3)(3x - 5)$
6. $(2x - 5)(2x + 5)$
7. $(x - 14)^2$
8. $7(x - 11)(x + 11)$
9. $(x + 9)^2$
10. $(x + 7)(x - 3)$
11. $3x(x + 7)$
12. $(x - 8)(3x + 4)$
13. $(3x - 4)(3x + 4)$
14. $(2x + 5)^2$
15. $(x - 3)(x - 2)$
16. $5x(x + 4)(x - 1)$
17. $2(2x^2 + 9)$
18. $(x - 6)^2$
19. $(x - 9)(x + 6)$
20. $3(2x^2 - 7)$
21. $(2x + 3)(x + 6)$
22. $(4x + 1)(4x - 1)$
23. $(x - 7)^2$
24. $(x + 3)(x + 5)$
25. $3x(x^2 - 4x - 15)$
26. $3(x^2 + 8)$
27. $(x + 8)^2$
28. $3x(5x - 3)(5x + 3)$
29. $3x(x - 6)(x + 2)$
30. $4x(x - 7)(x - 4)$
31. $5(y + 5)(y - 5)$
32. $y^2(3x - 4)(x + 1)$
33. $x(x + 12)(x - 2)$
34. $3x(x - 5)(x + 3)$
35. $3(x - 3)(x + 3)$
36. $(x - 2)(x + 2)(x^2 + 4)$
37. $(2x + 7)(x - 1)$
38. $(3x - 1)(x - 4)$
39. $(x + 2)(2x + 5)$
40. $(4x - 1)(x - 3)$
41. $(2x + 5)(2x + 1)$
42. $x(6x + 1)(x + 5)$
43. $(8x + 1)^2$
44. $(7x + 2)(x - 5)$
45. $(5x - 3)(x + 3)$

Algebra 1

ZERO PRODUCT PROPERTY AND QUADRATICS #15

If $a \cdot b = 0$, then either $a = 0$ or $b = 0$.

Note that this property states that <u>at least</u> one of the factors MUST be zero. It is also possible that all of the factors are zero. This simple statement gives us a powerful result which is most often used with equations involving the products of binomials. For example, solve $(x + 5)(x - 2) = 0$.

By the Zero Product Property, since $(x + 5)(x - 2) = 0$, either $x + 5 = 0$ or $x - 2 = 0$. Thus, $x = -5$ or $x = 2$.

The Zero Product Property can be used to find where a quadratic crosses the x-axis. These points are the x-intercepts. In the example above, they would be (-5, 0) and (2, 0).

Here are two more examples. Solve each quadratic equation and check each solution.

Example 1

$(x + 4)(x - 7) = 0$

By the Zero Product Property,
either $x + 4 = 0$ or $x - 7 = 0$
Solving, $x = -4$ or $x = 7$.

Checking,

$(-4 + 4)(-4 - 7) \stackrel{?}{=} 0$
$(0)(-11) = 0 \checkmark$

$(7 + 4)(7 - 7) \stackrel{?}{=} 0$
$(11)(0) = 0 \checkmark$

Example 2

$x^2 + 3x - 10 = 0$

First factor $x^2 + 3x - 10 = 0$
into $(x + 5)(x - 2) = 0$
then $x + 5 = 0$ or $x - 2 = 0$,
so $x = -5$ or $x = 2$

Checking,

$(-5 + 5)(-5 - 2) \stackrel{?}{=} 0$
$(0)(-7) = 0 \checkmark$

$(2 + 5)(2 - 2) \stackrel{?}{=} 0$
$(7)(0) = 0 \checkmark$

Solve each of the following quadratic equations.

1. $(x + 7)(x + 1) = 0$
2. $(x + 2)(x + 3) = 0$
3. $x(x - 2) = 0$
4. $x(x - 7) = 0$
5. $(3x - 3)(4x + 2) = 0$
6. $(2x + 5)(4x - 3) = 0$
7. $x^2 + 4x + 3 = 0$
8. $x^2 + 6x + 5 = 0$
9. $x^2 - 6x + 8 = 0$
10. $x^2 - 8x + 15 = 0$
11. $x^2 + x = 6$
12. $x^2 - x = 6$
13. $x^2 - 10x = -16$
14. $x^2 - 11x = -28$

Without graphing, find where each parabola crosses the x-axis.

15. $y = x^2 - 2x - 3$
16. $y = x^2 + 2x - 8$
17. $y = x^2 - x - 30$
18. $y = x^2 + 4x - 5$
19. $x^2 + 4x = 5 + y$
20. $x^2 - 3x = 10 + y$

Answers

1. $x = -7$ and $x = -1$
2. $x = -2$ and $x = -3$
3. $x = 0$ and $x = 2$
4. $x = 0$ and $x = 7$
5. $x = 1$ and $x = -\frac{1}{2}$
6. $x = \frac{-5}{2}$ and $x = \frac{3}{4}$
7. $x = -1$ and $x = -3$
8. $x = -1$ and $x = -5$
9. $x = 4$ and $x = 2$
10. $x = 5$ and $x = 3$
11. $x = -3$ and $x = 2$
12. $x = 3$ and $x = -2$
13. $x = 2$ and $x = 8$
14. $x = 4$ and $x = 7$
15. $(-1, 0)$ and $(3, 0)$
16. $(-4, 0)$ and $(2, 0)$
17. $(6, 0)$ and $(-5, 0)$
18. $(-5, 0)$ and $(1, 0)$
19. $(1, 0)$ and $(-5, 0)$
20. $(5, 0)$ and $(-2, 0)$

PYTHAGOREAN THEOREM #16

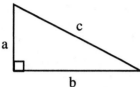

Any triangle that has a right angle is called a **right triangle**. The two sides that form the right angle, a and b, are called **legs**, and the side opposite (that is, across the triangle from) the right angle, c, is called the **hypotenuse**.

For any right triangle, the sum of the squares of the legs of the triangle is equal to the square of the hypotenuse, that is, $a^2 + b^2 = c^2$. This relationship is known as the **Pythagorean Theorem**. In words, the theorem states that:

$$(\text{leg})^2 + (\text{leg})^2 = (\text{hypotenuse})^2.$$

Example

Draw a diagram, then use the Pythagorean Theorem to write an equation or use area pictures (as shown on page 22, problem RC-1) on each side of the triangle to solve each problem.

a) Solve for the missing side.

$c^2 + 13^2 = 17^2$
$c^2 + 169 = 289$
$c^2 = 120$
$c = \sqrt{120}$
$c = 2\sqrt{30}$
$c \approx 10.95$

b) Find x to the nearest tenth:

$(5x)^2 + x^2 = 20^2$
$25x^2 + x^2 = 400$
$26x^2 = 400$
$x^2 \approx 15.4$
$x \approx \sqrt{15.4}$
$x \approx 3.9$

c) One end of a ten foot ladder is four feet from the base of a wall. How high on the wall does the top of the ladder touch?

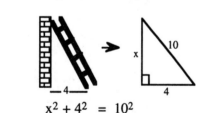

$x^2 + 4^2 = 10^2$
$x^2 + 16 = 100$
$x^2 = 84$
$x \approx 9.2$

The ladder touches the wall about 9.2 feet above the ground.

d) Could 3, 6 and 8 represent the lengths of the sides of a right triangle? Explain.

$3^2 + 6^2 \stackrel{?}{=} 8^2$

$9 + 36 \stackrel{?}{=} 64$

$45 \neq 64$

Since the Pythagorean Theorem relationship is not true for these lengths, they cannot be the side lengths of a right triangle.

Write an equation and solve for each unknown side. Round to the nearest hundredth.

1.
2.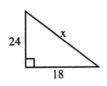
3.
4.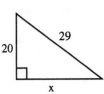

5.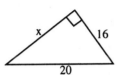
6.
7.
8.

9.
10.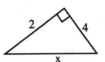
11.
12.

13.
14.

Be careful! Remember to square the whole side. For example, $(2x)^2 = (2x)(2x) = 4x^2$.

15.
16.
17.

18.
19.
20.

For each of the following problems draw and label a diagram. Then write an equation using the Pythagorean Theorem and solve for the unknown. Round answers to the nearest hundredth.

21. In a right triangle, the length of the hypotenuse is four inches. The length of one leg is two inches. Find the length of the other leg.

22. The length of the hypotenuse of a right triangle is six cm. The length of one leg is four cm. Find the length of the other leg.

23. Find the diagonal length of a television screen 30 inches wide by 20 inches long.

24. Find the length of a path that runs diagonally across a 53 yard by 100 yard field.

25. A mover must put a circular mirror two meters in diameter through a one meter by 1.8 meter doorway. Find the length of the diagonal of the doorway. Will the mirror fit?

Algebra 1

26. A surveyor walked eight miles north, then three miles west. How far was she from her starting point?

27. A four meter ladder is one meter from the base of a building. How high up the building will the ladder reach?

28. A 12-meter loading ramp rises to the edge of a warehouse doorway. The bottom of the ramp is nine meters from the base of the warehouse wall. How high above the base of the wall is the doorway?

29. What is the longest line you can draw on a paper that is 15 cm by 25 cm?

30. How long an umbrella will fit in the bottom of a suitcase that is 2.5 feet by 3 feet?

31. How long a guy wire is needed to support a 10 meter tall tower if it is fastened five meters from the foot of the tower?

32. Find the diagonal distance from one corner of a 30 foot square classroom floor to the other corner of the floor.

33. Harry drove 10 miles west, then five miles north, then three miles west. How far was he from his starting point?

34. Linda can turn off her car alarm from 20 yards away. Will she be able to do it from the far corner of a 15 yard by 12 yard parking lot?

35. The hypotenuse of a right triangle is twice as long as one of its legs. The other leg is nine inches long. Find the length of the hypotenuse.

36. One leg of a right triangle is three times as long as the other. The hypotenuse is 100 cm. Find the length of the shorter leg.

Answers

1. $x = 10$
2. $x = 30$
3. $x = 12$
4. $x = 21$
5. $x = 12$
6. $x = 40$
7. $x = 16$
8. $x = 20$
9. $x \approx 6.71$
10. $x \approx 4.47$
11. $x \approx 7.07$
12. $x \approx 9.9$
13. $x \approx 8.66$
14. $x \approx 5.66$
15. $x = 2$
16. $x = 2$
17. $x \approx 22.36$
18. $x \approx 3.16$
19. $x \approx 4.44$
20. $x \approx 4.33$
21. 3.46 inches
22. 4.47 cm
23. 36.06 inches
24. 113.18 yards
25. The diagonal is 2.06 meters, so yes.
26. 8.54 miles
27. 3.87 meters
28. 7.94 meters
29. 29.15 cm
30. 3.91 feet
31. 11.18 meters
32. 42.43 feet
33. 13.93 miles
34. The corner is 19.21 yards away so yes!
35. 10.39 inches
36. 31.62 cm

SOLVING EQUATIONS CONTAINING ALGEBRAIC FRACTIONS #17

Fractions that appear in algebraic equations can usually be eliminated in one step by multiplying each term on both sides of the equation by the common denominator for all of the fractions. If you cannot determine the common denominator, use the product of all the denominators. Multiply, simplify each term as usual, then solve the remaining equation. For more information, read the Tool Kit information on page 313 (problem BP-46) in the textbook. In this course we call this method for eliminating fractions in equations "fraction busting."

Example 1

Solve for x: $\frac{x}{9} + \frac{2x}{5} = 3$

$45\left(\frac{x}{9} + \frac{2x}{5}\right) = 45(3)$

$45\left(\frac{x}{9}\right) + 45\left(\frac{2x}{5}\right) = 135$

$5x + 18x = 135$

$23x = 135$

$x = \frac{135}{23}$

Example 2

Solve for x: $\frac{5}{2x} + \frac{1}{6} = 8$

$6x\left(\frac{5}{2x} + \frac{1}{6}\right) = 6x(8)$

$6x\left(\frac{5}{2x}\right) + 6x\left(\frac{1}{6}\right) = 48x$

$15 + x = 48x$

$15 = 47x$

$x = \frac{15}{47}$

Solve the following equations using the fraction busters method.

1. $\frac{x}{6} + \frac{2x}{3} = 5$
2. $\frac{x}{3} + \frac{x}{2} = 1$
3. $\frac{16}{x} + \frac{16}{40} = 1$
4. $\frac{5}{x} + \frac{5}{3x} = 1$

5. $\frac{x}{2} - \frac{x}{5} = 9$
6. $\frac{x}{3} - \frac{x}{5} = \frac{2}{3}$
7. $\frac{x}{2} - 4 = \frac{x}{3}$
8. $\frac{x}{8} = \frac{x}{12} + \frac{1}{3}$

9. $5 - \frac{7x}{6} = \frac{3}{2}$
10. $\frac{2x}{3} - x = 4$
11. $\frac{x}{8} = \frac{x}{5} - \frac{1}{3}$
12. $\frac{2x}{3} - \frac{3x}{5} = 2$

13. $\frac{4}{x} + \frac{2}{x} = 1$
14. $\frac{3}{x} + 2 = 4$
15. $\frac{5}{x} + 6 = \frac{17}{x}$
16. $\frac{2}{x} - \frac{4}{3x} = \frac{2}{9}$

17. $\frac{x+2}{3} + \frac{x-1}{6} = 5$
18. $\frac{x}{4} + \frac{x+5}{3} = 4$
19. $\frac{x-1}{2x} + \frac{x+3}{4x} = \frac{5}{8}$
20. $\frac{2-x}{x} - \frac{x+3}{3x} = \frac{-1}{3}$

Answers

1. x = 6
2. x = $\frac{6}{5}$
3. x = $26\frac{2}{3}$
4. x = $6\frac{2}{3}$

5. x = 30
6. x = 5
7. x = 24
8. x = 8

9. x = 3
10. x = -12
11. x = $\frac{40}{9}$
12. x = 30

13. x = 6
14. x = 1.5
15. x = 2
16. x = 3

17. x = 9
18. x = 4
19. x = -2
20. x = 1

Algebra 1

LAWS OF EXPONENTS #18

BASE, EXPONENT, AND VALUE

In the expression 2^5, 2 is the **base**, 5 is the **exponent**, and the **value** is 32.

2^5 means $2 \cdot 2 \cdot 2 \cdot 2 \cdot 2 = 32$ $\qquad\qquad$ x^3 means $x \cdot x \cdot x$

LAWS OF EXPONENTS

Here are the basic patterns with examples:

1) $x^a \cdot x^b = x^{a+b}$ $\qquad$ examples: $x^3 \cdot x^4 = x^{3+4} = x^7$; $\qquad$ $2^7 \cdot 2^4 = 2^{11}$

2) $\dfrac{x^a}{x^b} = x^{a-b}$ $\qquad$ examples: $x^{10} \div x^4 = x^{10-4} = x^6$; $\qquad$ $\dfrac{2^4}{2^7} = 2^{-3}$ or $\dfrac{1}{2^3}$

3) $(x^a)^b = x^{ab}$ $\qquad$ examples: $(x^4)^3 = x^{4 \cdot 3} = x^{12}$; $\qquad$ $(2x^3)^4 = 2^4 \cdot x^{12} = 16x^{12}$

4) $x^{-a} = \dfrac{1}{x^a}$ and $\dfrac{1}{x^{-b}} = x^b$ $\qquad$ examples: $3x^{-3}y^2 = \dfrac{3y^2}{x^3}$; $\qquad$ $\dfrac{2x^5}{y^{-2}} = 2x^5y^2$

5) $x^0 = 1$ $\qquad$ examples: $5^0 = 1$; $\qquad$ $(2x)^0 = 1$

Example 1

Simplify: $(2xy^3)(5x^2y^4)$

Multiply the coefficients: $2 \cdot 5 \cdot xy^3 \cdot x^2y^4 = 10xy^3 \cdot x^2y^4$

Add the exponents of x, then y: $10x^{1+2}y^{3+4} = 10x^3y^7$

Example 2

Simplify: $\qquad \dfrac{14x^2y^{12}}{7x^5y^7}$

Divide the coefficients: $\dfrac{(14 \div 7)x^2y^{12}}{x^5y^7} = \dfrac{2x^2y^{12}}{x^5y^7}$

Subtract the exponents: $2x^{2-5}y^{12-7} = 2x^{-3}y^5$ OR $\dfrac{2y^5}{x^3}$

Example 3

Simplify: $\qquad (3x^2y^4)^3$

Cube each factor: $\qquad 3^3 \cdot (x^2)^3 \cdot (y^4)^3 = 27(x^2)^3(y^4)^3$

Multiply the exponents: $27x^6y^{12}$

Example 4
Simplify: $\quad 3x^{-4}y^2z^{-3} \Rightarrow \dfrac{3y^2}{x^4z^3}$

Simplify each expression:

1. $y^5 \cdot y^7$
2. $b^4 \cdot b^3 \cdot b^2$
3. $8^6 \cdot 8^2$
4. $(y^5)^2$
5. $(3a)^4$
6. $\dfrac{m^8}{m^3}$
7. $\dfrac{12x^9}{4x^4}$
8. $(x^3y^2)^3$
9. $\dfrac{(y^4)^2}{(y^3)^2}$
10. $\dfrac{15x^2y^7}{3x^4y^5}$
11. $(4c^4)(ac^3)(3a^5c)$
12. $(7x^3y^5)^2$
13. $(4xy^2)(2y)^3$
14. $\left(\dfrac{4}{x^2}\right)^3$
15. $\dfrac{(2a^7)(3a^2)}{6a^3}$
16. $\left(\dfrac{5m^3n}{m^5}\right)^3$
17. $(3a^2x^3)^2(2ax^4)^3$
18. $\left(\dfrac{x^3y}{y^4}\right)^4$
19. $\left(\dfrac{6y^2x^8}{12x^3y^7}\right)^2$
20. $\dfrac{(2x^5y^3)^3(4xy^4)^2}{8x^7y^{12}}$

Write the following expressions without negative exponents.

21. x^{-2}
22. $y^{-3}y^2$
23. $\dfrac{x^5}{x^{-2}}$
24. $(y^{-2})^5$

Note: More practice with negative exponents is available in Skill Builder #21.

Answers

1. y^{12}
2. b^9
3. 8^8
4. y^{10}
5. $81a^4$
6. m^5
7. $3x^5$
8. x^9y^6
9. y^2
10. $\dfrac{5y^2}{x^2}$
11. $12a^6c^8$
12. $49x^6y^{10}$
13. $32xy^5$
14. $\dfrac{64}{x^6}$
15. a^6
16. $\dfrac{125n^3}{m^6}$
17. $72a^7x^{18}$
18. $\dfrac{x^{12}}{y^{12}}$
19. $\dfrac{x^{10}}{4y^{10}}$
20. $16x^{10}y^5$
21. $\dfrac{1}{x^2}$
22. $\dfrac{1}{y}$
23. x^7
24. $\dfrac{1}{y^{10}}$

SIMPLIFYING RADICALS #19

Sometimes it is convenient to leave square roots in radical form instead of using a calculator to find approximations (decimal values). Look for perfect squares (i.e., 4, 9, 16, 25, 36, 49, ...) as **factors** of the number that is inside the radical sign (**radicand**) and take the square root of any perfect square factor. Multiply the root of the perfect square times the reduced radical. When there is an existing value that multiplies the radical, multiply any root(s) times that value.

For example:

$\sqrt{9} = 3$ $\qquad\qquad 5\sqrt{9} = 5 \cdot 3 = 15$

$\sqrt{18} = \sqrt{9 \cdot 2} = \sqrt{9} \cdot \sqrt{2} = 3\sqrt{2} \qquad 3\sqrt{98} = 3\sqrt{49 \cdot 2} = 3 \cdot 7\sqrt{2} = 21\sqrt{2}$

$\sqrt{80} = \sqrt{16 \cdot 5} = \sqrt{16} \cdot \sqrt{5} = 4\sqrt{5} \qquad \sqrt{45} + 4\sqrt{20} = \sqrt{9 \cdot 5} + 4\sqrt{4 \cdot 5} = 3\sqrt{5} + 4 \cdot 2\sqrt{5} = 11\sqrt{5}$

When there are no more perfect square factors inside the radical sign, the product of the whole number (or fraction) and the remaining radical is said to be in **simple radical form**.

Simple radical form does not allow radicals in the denominator of a fraction. If there is a radical in the denominator, **rationalize the denominator** by multiplying the numerator and denominator of the fraction by the radical in the original denominator. Then simplify the remaining fraction. Examples:

$$\frac{2}{\sqrt{2}} = \frac{2}{\sqrt{2}} \cdot \frac{\sqrt{2}}{\sqrt{2}} = \frac{2\sqrt{2}}{2} = \sqrt{2} \qquad\qquad \frac{4\sqrt{5}}{\sqrt{6}} = \frac{4\sqrt{5}}{\sqrt{6}} \cdot \frac{\sqrt{6}}{\sqrt{6}} = \frac{4\sqrt{30}}{6} = \frac{2\sqrt{30}}{3}$$

In the first example, $\sqrt{2} \cdot \sqrt{2} = \sqrt{4} = 2$ and $\frac{2}{2} = 1$. In the second example, $\sqrt{6} \cdot \sqrt{6} = \sqrt{36} = 6$ and $\frac{4}{6} = \frac{2}{3}$.

The rules for radicals used in the above examples are shown below. Assume that the variables represent non-negative numbers.

(1) $\sqrt{x} \cdot \sqrt{y} = \sqrt{xy}$ $\qquad$ (2) $\sqrt{x \cdot y} = \sqrt{x} \cdot \sqrt{y}$ $\qquad$ (3) $\frac{\sqrt{x}}{\sqrt{y}} = \sqrt{\frac{x}{y}}$

(4) $\sqrt{x^2} = (\sqrt{x})^2 = x$ $\qquad$ (5) $a\sqrt{x} + b\sqrt{x} = (a+b)\sqrt{x}$

Write each expression in simple radical (square root) form.

1. $\sqrt{32}$
2. $\sqrt{28}$
3. $\sqrt{54}$
4. $\sqrt{68}$
5. $2\sqrt{24}$
6. $5\sqrt{90}$
7. $6\sqrt{132}$
8. $5\sqrt{200}$
9. $2\sqrt{6} \cdot 3\sqrt{2}$
10. $3\sqrt{12} \cdot 2\sqrt{3}$
11. $\dfrac{\sqrt{12}}{\sqrt{3}}$
12. $\dfrac{\sqrt{20}}{\sqrt{5}}$
13. $\dfrac{8\sqrt{12}}{2\sqrt{3}}$
14. $\dfrac{14\sqrt{8}}{7\sqrt{2}}$
15. $\dfrac{2}{\sqrt{3}}$
16. $\dfrac{4}{\sqrt{5}}$
17. $\dfrac{6}{\sqrt{3}}$
18. $\dfrac{2\sqrt{3}}{\sqrt{6}}$
19. $2\sqrt{3} + 3\sqrt{12}$
20. $4\sqrt{12} - 2\sqrt{3}$
21. $6\sqrt{3} + 2\sqrt{27}$
22. $2\sqrt{45} - 2\sqrt{5}$
23. $2\sqrt{8} - \sqrt{18}$
24. $3\sqrt{48} - 4\sqrt{27}$

Answers

1. $4\sqrt{2}$
2. $2\sqrt{7}$
3. $3\sqrt{6}$
4. $2\sqrt{17}$
5. $4\sqrt{6}$
6. $15\sqrt{10}$
7. $12\sqrt{33}$
8. $50\sqrt{2}$
9. $12\sqrt{3}$
10. 36
11. 2
12. 2
13. 8
14. 4
15. $\dfrac{2\sqrt{3}}{3}$
16. $\dfrac{4\sqrt{5}}{5}$
17. $2\sqrt{3}$
18. $\sqrt{2}$
19. $8\sqrt{3}$
20. $6\sqrt{3}$
21. $12\sqrt{3}$
22. $4\sqrt{5}$
23. $\sqrt{2}$
24. 0

Algebra 1

THE QUADRATIC FORMULA #20

You have used factoring and the Zero Product Property to solve quadratic equations. You can solve any quadratic equation by using the **quadratic formula.**

If $ax^2 + bx + c = 0$, then $x = \dfrac{-b \pm \sqrt{b^2 - 4ac}}{2a}$.

For example, suppose $3x^2 + 7x - 6 = 0$. Here $a = 3$, $b = 7$, and $c = -6$. Substituting these values into the formula results in:

$$x = \dfrac{-(7) \pm \sqrt{7^2 - 4(3)(-6)}}{2(3)} \Rightarrow x = \dfrac{-7 \pm \sqrt{121}}{6} \Rightarrow x = \dfrac{-7 \pm 11}{6}$$

Remember that non-negative numbers have both a positive and negative square root. The sign $\pm$ represents this fact for the square root in the formula and allows us to write the equation once (representing two possible solutions) until later in the solution process.

Split the numerator into the two values: $x = \dfrac{-7 + 11}{6}$ or $x = \dfrac{-7 - 11}{6}$

Thus the solution for the quadratic equation is: $x = \dfrac{2}{3}$ or -3.

Example 1

Solve: $x^2 + 7x + 5 = 0$

First make sure the equation is in standard form with zero on one side of the equation. This equation is already in standard form.

Second, list the numerical values of the coefficients a, b, and c. Since $ax^2 + bx + c = 0$, then $a = 1$, $b = 7$, and $c = 5$ for the equation $x^2 + 7x + 5 = 0$.

Write out the quadratic formula (see above). Substitute the numerical values of the coefficients a, b, and c in the quadratic formula, $x = \dfrac{-7 \pm \sqrt{7^2 - 4(1)(5)}}{2(1)}$.

Simplify to get the exact solutions.

$x = \dfrac{-7 \pm \sqrt{49 - 20}}{2} \Rightarrow x = \dfrac{-7 \pm \sqrt{29}}{2}$,

so $x = \dfrac{-7 + \sqrt{29}}{2}$ or $\dfrac{-7 - \sqrt{29}}{2}$

Use a calculator to get approximate solutions.

$x \approx \dfrac{-7 + 5.39}{2} \approx \dfrac{-1.61}{2} \approx -0.81$

$x \approx \dfrac{-7 - 5.39}{2} \approx \dfrac{-12.39}{2} \approx -6.20$

Example 2

Solve: $6x^2 + 1 = 8x$

First make sure the equation is in standard form with zero on one side of the equation.

$\begin{array}{r} 6x^2 + 1 = 8x \\ -8x \quad -8x \end{array} \Rightarrow 6x^2 - 8x + 1 = 0$

Second, list the numerical values of the coefficients a, b, and c: $a = 6$, $b = -8$, and $c = 1$ for this equation.

Write out the quadratic formula, then substitute the values in the formula.

$x = \dfrac{-(-8) \pm \sqrt{(-8)^2 - 4(6)(1)}}{2(6)}$

Simplify to get the exact solutions.

$x = \dfrac{8 \pm \sqrt{64 - 24}}{12} \Rightarrow x = \dfrac{8 \pm \sqrt{40}}{12} \Rightarrow \dfrac{8 \pm 2\sqrt{10}}{12}$

so $x = \dfrac{4 + \sqrt{10}}{6}$ or $\dfrac{4 - \sqrt{10}}{6}$

Use a calculator with the original answer to get approximate solutions.

$x \approx \dfrac{8 + 6.32}{12} \approx \dfrac{14.32}{12} \approx 1.19$

$x \approx \dfrac{8 - 6.32}{12} \approx \dfrac{1.68}{12} \approx 0.14$

SKILL BUILDERS

Use the quadratic formula to solve the following equations.

1. $x^2 + 8x + 6 = 0$
2. $x^2 + 6x + 4 = 0$
3. $x^2 - 2x - 30 = 0$
4. $x^2 - 5x - 2 = 0$
5. $7 = 13x - x^2$
6. $15x - x^2 = 5$
7. $x^2 = -14x - 12$
8. $6x = x^2 + 3$
9. $3x^2 + 10x + 5 = 0$
10. $2x^2 + 8x + 5 = 0$
11. $5x^2 + 5x - 7 = 0$
12. $6x^2 - 2x - 3 = 0$
13. $2x^2 + 9x = -1$
14. $-6x + 6x^2 = 8$
15. $3x - 12 = -4x^2$
16. $10x^2 + 2x = 7$
17. $2x^2 - 11 = 0$
18. $3x^2 - 6 = 0$
19. $3x^2 + 0.75x - 1.5 = 0$
20. $0.1x^2 + 5x + 2.6 = 0$

Answers

1. $x \approx -0.84$ and -7.16
2. $x \approx -0.76$ and -5.24
3. $x \approx 6.57$ and -4.57
4. $x \approx 5.37$ and -0.37
5. $x \approx 12.44$ and 0.56
6. $x \approx 14.66$ and 0.34
7. $x \approx -0.92$ and -13.08
8. $x \approx 5.45$ and 0.55
9. $x \approx -0.61$ and -2.72
10. $x \approx -0.78$ and -3.22
11. $x \approx 0.78$ and -1.78
12. $x \approx 0.89$ and -0.56
13. $x \approx -0.11$ and -4.39
14. $x \approx 1.76$ and -0.76
15. $x \approx 1.40$ and -2.15
16. $x \approx 0.74$ and -0.94
17. $x \approx -2.35$ and 2.35
18. $x \approx -1.41$ and 1.41
19. $x \approx 0.59$ and -0.84
20. $x \approx -0.53$ and -49.47

SIMPLIFYING RATIONAL EXPRESSIONS

Rational expressions are fractions that have algebraic expressions in their numerators and/or denominators. To simplify rational expressions find **factors** in the numerator and denominator that are the same and then write them as fractions equal to 1. For example,

$$\frac{6}{6} = 1 \qquad \frac{x^2}{x^2} = 1 \qquad \frac{(x+2)}{(x+2)} = 1 \qquad \frac{(3x-2)}{(3x-2)} = 1$$

Notice that the last two examples involved binomial sums and differences. **Only** when sums or differences are **exactly** the same does the fraction equal 1. Rational expressions such as the examples below **cannot** be simplified:

$$\frac{(6+5)}{6} \qquad \frac{x^3 + y}{x^3} \qquad \frac{x}{x+2} \qquad \frac{3x-2}{2}$$

Most problems that involve rational expressions will require that you **factor** the numerator and denominator. For example:

$$\frac{12}{54} = \frac{2 \cdot 2 \cdot 3}{2 \cdot 3 \cdot 3 \cdot 3} = \frac{2}{9} \qquad \text{Notice that } \frac{2}{2} \text{ and } \frac{3}{3} \text{ each equal 1.}$$

$$\frac{6x^3y^2}{15x^2y^4} = \frac{2 \cdot 3 \cdot x^2 \cdot x \cdot y^2}{5 \cdot 3 \cdot x^2 \cdot y^2 \cdot y^2} = \frac{2x}{5y^2} \qquad \text{Notice that } \frac{3}{3}, \frac{x^2}{x^2}, \text{ and } \frac{y^2}{y^2} = 1.$$

$$\frac{x^2 - x - 6}{x^2 - 5x + 6} = \frac{(x+2)(x-3)}{(x-2)(x-3)} = \frac{x+2}{x-2} \qquad \text{where } \frac{x-3}{x-3} = 1.$$

All three examples demonstrate that **all parts** of the numerator and denominator--whether constants, monomials, binomials, or factorable trinomials--must be written as products **before** you can look for factors that equal 1.

One special situation is shown in the following examples:

$$\frac{-2}{2} = -1 \qquad \frac{-x}{x} = -1 \qquad \frac{-x-2}{x+2} = \frac{-(x+2)}{x+2} = -1 \qquad \frac{5-x}{x-5} = \frac{-(x-5)}{x-5} = -1$$

Note that in all cases we assume the denominator does not equal zero.

Example 1

Simplify: $\dfrac{(a^3b^{-2})^2}{a^4}$

Rewrite the numerator and denominator without negative exponents and parentheses.

$$\dfrac{(a^3b^{-2})^2}{a^4} \Rightarrow \dfrac{a^6b^{-4}}{a^4} \Rightarrow \dfrac{a^6}{a^4b^4}$$

Then look for the same pairs of factors that equal one (1) when divided. Writing out all of the factors can be helpful.

$$\dfrac{a \cdot a \cdot a \cdot a \cdot a \cdot a}{a \cdot a \cdot a \cdot a \cdot b \cdot b \cdot b \cdot b} = 1 \cdot 1 \cdot 1 \cdot 1 \cdot \dfrac{a \cdot a}{b \cdot b \cdot b \cdot b}$$

Write the simplified expression with exponents.

$\dfrac{(a^3b^{-2})^2}{a^4} = \dfrac{a^2}{b^4}$, $b \neq 0$. Note that $\dfrac{a}{a} = 1$.

Example 2

Simplify: $\dfrac{2x^2 - 13x - 7}{x^2 - 4x - 21}$

To simplify some rational expressions, the numerator and/or denominator may need to be factored before you may simplify the expression.

$$\dfrac{2x^2 - 13x - 7}{x^2 - 4x - 21} \Rightarrow \dfrac{(2x+1)(x-7)}{(x-7)(x+3)}$$

Then look for the same pairs of factors that equal one (1) when divided.

$$\dfrac{(2x+1)(x-7)}{(x+3)(x-7)} \Rightarrow \dfrac{2x+1}{x+3} \cdot 1 \Rightarrow \dfrac{2x+1}{x+3} \text{ for } x \neq -3 \text{ or } 7.$$

Note that $\dfrac{(x-7)}{(x-7)} = 1$.

Simplify the following expressions. Assume that the denominator is not equal to zero.

1. $\dfrac{12x^2y^4}{3x^2y^3}$
2. $\dfrac{10a^6b^8}{40a^2b^2}$
3. $\dfrac{(x^5y^3)^3}{x^{12}y}$
4. $\dfrac{(a^5)^2}{a^{13}b^6}$

5. $\dfrac{(5x^3)^2 y^3}{10xy^9}$
6. $\dfrac{3(a^3)^5 b}{(3a^4)^3 b^{10}}$
7. $\dfrac{4ab^{-5}}{a^8 b}$
8. $\dfrac{2x^{-3}y^8}{4x^{-2}}$

9. $\dfrac{(x^8 y^{-3})^{-2}}{x^2}$
10. $\dfrac{2x^3 y^{-1}}{6(4x)^{-2} y^7}$
11. $\dfrac{(2x-1)(x+3)}{(x-5)(x+3)}$
12. $\dfrac{(5x-1)(x+2)}{(x+7)(5x-1)}$

13. $\dfrac{3x+1}{3x^2+10x+3}$
14. $\dfrac{x^2-x-20}{x-5}$
15. $\dfrac{3x-6}{x^2+4x-12}$
16. $\dfrac{2x^2-x-3}{10x-15}$

17. $\dfrac{3x^2+x-10}{x^2+6x+8}$
18. $\dfrac{x^2-64}{x^2+16x+64}$
19. $\dfrac{4x^2-x}{4x^3+11x^2-3x}$
20. $\dfrac{2x^3+2x^2-12x}{8x^2-8x-16}$

Answers

1. $4y$
2. $\dfrac{a^4 b^6}{4}$
3. $\dfrac{x^{15}y^9}{x^{12}y} = x^3 y^8$

4. $\dfrac{a^{10}}{a^{13}b^6} = \dfrac{1}{a^3 b^6}$
5. $\dfrac{25x^6 y^3}{10xy^9} = \dfrac{5x^5}{2y^6}$
6. $\dfrac{3a^{15}b}{27a^{12}b^{10}} = \dfrac{a^3}{9b^9}$

7. $\dfrac{4a}{a^8 b^6} = \dfrac{4}{a^7 b^6}$
8. $\dfrac{2x^2 y^8}{4x^3} = \dfrac{y^8}{2x}$
9. $\dfrac{x^{-16}y^6}{x^2} = \dfrac{y^6}{x^{18}}$

10. $\dfrac{32x^5}{6y^8} = \dfrac{16x^5}{3y^8}$
11. $\dfrac{2x-1}{x-5}$
12. $\dfrac{x+2}{x+7}$

13. $\dfrac{3x+1}{(x+3)(3x+1)} = \dfrac{1}{x+3}$
14. $\dfrac{(x-5)(x+4)}{x-5} = x+4$
15. $\dfrac{3(x-2)}{(x-2)(x+6)} = \dfrac{3}{x+6}$

16. $\dfrac{(2x-3)(x+1)}{5(2x-3)} = \dfrac{x+1}{5}$
17. $\dfrac{(3x-5)(x+2)}{(x+4)(x+2)} = \dfrac{3x-5}{x+4}$
18. $\dfrac{(x+8)(x-8)}{(x+8)(x+8)} = \dfrac{x-8}{x+8}$

19. $\dfrac{x(4x-1)}{x(4x-1)(x+3)} = \dfrac{1}{x+3}$
20. $\dfrac{2x(x+3)(x-2)}{8(x-2)(x+1)} = \dfrac{x(x+3)}{4(x+1)} = \dfrac{x^2+3x}{4x+4}$

Algebra 1

MULTIPLICATION AND DIVISION OF RATIONAL EXPRESSIONS   #22

To multiply or divide rational expressions, follow the same procedures used with numerical fractions. However, it is often necessary to factor the polynomials in order to simplify the rational expression.

Example 1

Multiply $\dfrac{x^2+6x}{(x+6)^2} \cdot \dfrac{x^2+7x+6}{x^2-1}$ and simplify the result.

After factoring, the expression becomes: $\dfrac{x(x+6)}{(x+6)(x+6)} \cdot \dfrac{(x+6)(x+1)}{(x+1)(x-1)}$

After multiplying, reorder the factors: $\dfrac{(x+6)}{(x+6)} \cdot \dfrac{(x+6)}{(x+6)} \cdot \dfrac{x}{(x-1)} \cdot \dfrac{(x+1)}{(x+1)}$

Since $\dfrac{(x+6)}{(x+6)} = 1$ and $\dfrac{(x+1)}{(x+1)} = 1$, simplify: $1 \cdot 1 \cdot \dfrac{x}{x-1} \cdot 1 \;\Rightarrow\; \dfrac{x}{x-1}$.

Note: $x \neq -6, -1,$ or 1.

Example 2

Divide $\dfrac{x^2-4x-5}{x^2-4x+4} \div \dfrac{x^2-2x-15}{x^2+4x-12}$ and simplify the result.

First, change to a multiplication expression by inverting (flipping) the second fraction: $\dfrac{x^2-4x-5}{x^2-4x+4} \cdot \dfrac{x^2+4x-12}{x^2-2x-15}$

After factoring, the expression is: $\dfrac{(x-5)(x+1)}{(x-2)(x-2)} \cdot \dfrac{(x+6)(x-2)}{(x-5)(x+3)}$

Reorder the factors (if you need to): $\dfrac{(x-5)}{(x-5)} \cdot \dfrac{(x-2)}{(x-2)} \cdot \dfrac{(x+1)}{(x-2)} \cdot \dfrac{(x+6)}{(x+3)}$

Since $\dfrac{(x-5)}{(x-5)} = 1$ and $\dfrac{(x-2)}{(x-2)} = 1$, simplify: $\dfrac{(x+1)(x+6)}{(x-2)(x+3)}$

Thus, $\dfrac{x^2-4x-5}{x^2-4x+4} \div \dfrac{x^2-2x-15}{x^2+4x-12} = \dfrac{(x+1)(x+6)}{(x-2)(x+3)}$ or $\dfrac{x^2+7x+6}{x^2+x-6}$. Note: $x \neq -3, 2,$ or 5.

Multiply or divide each expression below and simplify the result. Assume the denominator is not equal to zero.

1. $\dfrac{3x+6}{5x} \cdot \dfrac{x+4}{x^2+2x}$

2. $\dfrac{8a}{a^2-16} \cdot \dfrac{a+4}{4}$

3. $\dfrac{x^2-1}{3} \cdot \dfrac{2}{x^2-x}$

4. $\dfrac{x^2-x-12}{x^2} \cdot \dfrac{x}{x-4}$

5. $\dfrac{x^2-16}{(x-4)^2} \cdot \dfrac{x^2-3x-18}{x^2-2x-24}$

6. $\dfrac{x^2+6x+8}{x^2-4x+3} \cdot \dfrac{x^2-5x+4}{5x+10}$

SKILL BUILDERS

7. $\dfrac{x^2-x-6}{x^2-x-20} \cdot \dfrac{x^2+6x+8}{x^2-x-6}$

8. $\dfrac{x^2-x-30}{x^2+13x+40} \cdot \dfrac{x^2+11x+24}{x^2-9x+18}$

9. $\dfrac{3x+12}{x^2} \div \dfrac{x+4}{x}$

10. $\dfrac{2a+6}{a^3} \div \dfrac{a+3}{a}$

11. $\dfrac{15-5x}{x^2-x-6} \div \dfrac{5x}{x^2+6x+8}$

12. $\dfrac{17x+119}{x^2+5x-14} \div \dfrac{9x-1}{x^2-3x+2}$

13. $\dfrac{x^2+8x}{9x} \div \dfrac{x^2-64}{3x^2}$

14. $\dfrac{x^2-1}{x^2-6x-7} \div \dfrac{x^3+x^2-2x}{x-7}$

15. $\dfrac{2x^2-5x-3}{3x^2-10x+3} \div \dfrac{4x^2+4x+1}{9x^2-1}$

16. $\dfrac{x^2+3x-10}{x^2+3x} \div \dfrac{x^2-4x+4}{4x+12}$

17. $\dfrac{x^2-x-6}{x^2+3x-10} \cdot \dfrac{x^2+2x-15}{x^2-6x+9} \cdot \dfrac{x^2+4x-21}{x^2+9x+14}$

18. $\dfrac{3x^2+21x}{x^2-49} \cdot \dfrac{x^2-x}{6x^3-9x^2} \cdot \dfrac{4x^2-9}{3x-3}$

19. $\dfrac{4x^3+7x-2x}{2x^2-162} \div \dfrac{4x^2+15x-4}{12x-60} \cdot \dfrac{x^2+9x}{x^2-3x-10}$

20. $\dfrac{10x^2-11x+3}{x^2-6x-40} \cdot \dfrac{x^2+11x+28}{2x^2-x} \div \dfrac{x+7}{2x^2-20x}$

Answers

1. $\dfrac{3(x+4)}{5x^2} = \dfrac{3x+12}{5x^2}$

2. $\dfrac{2a}{a-4}$

3. $\dfrac{2(x+1)}{3x} = \dfrac{2x+2}{3x}$

4. $\dfrac{x+3}{x}$

5. $\dfrac{(x+3)}{(x-4)}$

6. $\dfrac{(x+4)(x-4)}{5(x-3)} = \dfrac{x^2-16}{5x-15}$

7. $\dfrac{(x+2)}{(x-5)}$

8. $\dfrac{(x+3)}{(x-3)}$

9. $\dfrac{3}{x}$

10. $\dfrac{2}{a^2}$

11. $\dfrac{-(x+4)}{x} = \dfrac{-x-4}{x}$

12. $\dfrac{17(x-1)}{9x-1} = \dfrac{17x-17}{9x-1}$

13. $\dfrac{x^2}{3(x-8)} = \dfrac{x^2}{3x-24}$

14. $\dfrac{1}{x(x+2)}$

15. $\dfrac{(3x+1)}{(2x+1)}$

16. $\dfrac{4(x+5)}{x(x-2)} = \dfrac{4x+20}{x^2-2x}$

17. $\dfrac{(x-3)}{(x-2)}$

18. $\dfrac{2x+3}{3(x-7)} = \dfrac{2x+3}{3x-21}$

19. $\dfrac{6x^2}{(x-9)(x+4)} = \dfrac{6x^2}{x^2-5x-36}$

20. $\dfrac{2(5x-3)}{1} = 10x-6$

Algebra 1

ABSOLUTE VALUE EQUATIONS #23

Absolute value means the distance from a reference point. In the simplest case, the absolute value of a number is its distance from zero on the number line. Since absolute value is a distance, the result of finding an absolute value is zero or a positive number. All distances are positive.

Example 1

Solve $|2x + 3| = 7$.

Because the result of $(2x + 3)$ can be 7 or -7, we can write and solve two different equations. (Remember that the absolute value of 7 and -7 will be 7.)

$$2x + 3 = 7 \text{ or } 2x + 3 = -7$$
$$2x = 4 \text{ or } 2x = -10$$
$$x = 2 \text{ or } x = -5$$

Example 2

Solve $2|2x + 13| = 10$.

First the equation must have the absolute value isolated on one side of the equation.

$$2|2x + 13| = 10 \implies |2x + 13| = 5$$

Because the result of $2x + 13$ can be 5 or -5, we can write and solve two different equations.

$$2x + 13 = 5 \text{ or } 2x + 13 = -5$$
$$2x = -8 \text{ or } 2x = -18$$
$$x = -4 \text{ or } x = -9$$

Note that while some x-values of the solution are negative, the goal is to find values that make the original absolute value statement true. For $x = -5$ in example 1, $|2(-5) + 3| = 7 \implies |-10 + 3| = 7 \implies |-7| = 7$, which is true. Verify that the two negative values of x in example 2 make the original absolute value equation true.

Solve for x.

1. $|x + 2| = 4$
2. $|3x| = 27$
3. $|x - 5| = 2$
4. $|x - 8| = 2$
5. $\left|\frac{x}{5}\right| = 2$
6. $|-3x| = 4$
7. $|3x + 4| = 10$
8. $|12x - 6| = 6$
9. $|x| + 3 = 20$
10. $|x| - 8 = -2$
11. $2|x| - 5 = 3$
12. $4|x| - 5 = 7$
13. $|x + 2| - 3 = 7$
14. $|x + 5| + 4 = 12$
15. $|2x - 3| + 2 = 11$
16. $-3|x| + 5 = -4$
17. $-3|x + 6| + 12 = 0$
18. $15 - |x + 1| = 3$
19. $14 + 2|3x + 5| = 26$
20. $4|x - 10| - 23 = 37$

Answers

1. x = 2, -6
2. x = 9, -9
3. x = 7, 3
4. x = 10, 6
5. x = 10, -10
6. $x = -\frac{4}{3}, \frac{4}{3}$
7. $x = 2, -\frac{14}{3}$
8. x = 1, 0
9. x = 17, -17
10. x = 6, -6
11. x = 4, -4
12. x = 3, -3
13. x = 8, -12
14. x = 3, -13
15. x = 6, -3
16. x = 3, -3
17. x = -2, -10
18. x = 11, -13
19. $x = \frac{1}{3}, -\frac{11}{3}$
20. x = 25, -5

USING ELIMINATION (ADDITION) TO FIND THE POINT OF INTERSECTION OF TWO LINES #24

The **elimination** method can be used to solve a system of linear equations. By adding or subtracting the two linear equations in a way that eliminates one of the variables, a single variable equation is left.

Example 1

Solve: $\begin{array}{l} x + 2y = 16 \\ x + y = 2 \end{array}$

First decide whether to add or subtract the equations. Remember that the addition or subtraction should <u>eliminate</u> one variable. In the system above, the x in each equation is positive, so we need to subtract, that is, change all the signs of the terms in the second equation.

$\begin{array}{l} x + 2y = 16 \\ -(x + y = 2) \end{array} \Rightarrow \begin{array}{l} x + 2y = 16 \\ -x - y = -2 \end{array} \Rightarrow y = 14$

Substitute the solution for y into either of the original equations to solve for the other variable, x.

$x + 2(14) = 16 \Rightarrow x = -12$

Check your solution (-12, 14) in the second equation. You could also use the first equation to check your solution.

$-12 + 14 = 2 \Rightarrow 2 = 2 \checkmark$

Example 2

Solve: $\begin{array}{l} 2x + 3y = 10 \\ 3x - 4y = -2 \end{array}$

Sometimes the equations need to be adjusted by multiplication before they can be added or subtracted to eliminate a variable. Multiply one or both equations to set them up for elimination.

Multiply the first equation by 3:
$3(2x + 3y) = 10(3) \Rightarrow 6x + 9y = 30$

Multiply the second equation by -2:
$-2(3x - 4y) = -2 \cdot (-2) \Rightarrow -6x + 8y = 4$

Decide whether to add or subtract the equations to eliminate one variable. Since the x-terms are additive opposites, add these equations.

$\begin{array}{l} 6x + 9y = 30 \\ -6x + 8y = 4 \\ \hline 17y = 34 \end{array}$ so $y = 2$.

Substitute the solution for y into either of the original equations to solve for the other variable.

$2x + 3(2) = 10 \Rightarrow 2x = 4 \Rightarrow x = 2 \checkmark$

Check the solution (2, 2) in the second equation.
$3(2) - 4(2) = -2 \Rightarrow 6 - 8 = -2 \Rightarrow -2 = -2 \checkmark$

Solve each system of linear equations using the Elimination Method.

1. $x + y = -4$
   $-x + 2y = 13$

2. $3x - y = 1$
   $-2x + y = 2$

3. $2x + 5y = 1$
   $2x - y = 19$

4. $x + 3y = 1$
   $2x + 3y = -4$

5. $x - 5y = 1$
   $x - 4y = 2$

6. $3x - 2y = -2$
   $5x - 2y = 10$

7. $x + y = 10$
   $15x + 28y = 176$

8. $x + 2y = 21$
   $9x + 24y = 243$

9. $4x + 3y = 7$
   $2x - 9y = 35$

10. $2x + 3y = 0$
    $6x - 5y = -28$

11. $7x - 3y = 37$
    $2x - y = 12$

12. $5x - 4y = 10$
    $3x - 2y = 6$

13. $x - 7y = 4$
    $3x + y = -10$

14. $y = -4x + 3$
    $3x + 5y = -19$

15. $2x - 3y = 50$
    $7x + 8y = -10$

16. $5x + 6y = 16$
    $3x = 4y + 2$

17. $3x + 2y = 14$
    $3y = -2x + 1$

18. $2x + 3y = 10$
    $5x - 4y = 2$

19. $5x + 2y = 9$
    $2x + 3y = -3$

20. $10x + 3y = 15$
    $3x - 2y = -10$

Answers

1. (-7, 3)
2. (3, 8)
3. (8, -3)
4. (-5, 2)
5. (6, 1)
6. (6, 10)
7. (8, 2)
8. (3, 9)
9. (4, -3)
10. (-3, 2)
11. (1, -10)
12. (2, 0)
13. (-3, -1)
14. (2, -5)
15. (10, -10)
16. (2, 1)
17. (8, -5)
18. (2, 2)
19. (3, -3)
20. (0, 5)

Algebra 1

SOLVING INEQUALITIES   #25

When an equation has a solution, depending on the type of equation, the solution can be represented as a point on a line or a point, line, or curve in the coordinate plane. Dividing points, lines, and curves are used to solve inequalities.

If the inequality has one variable, the solution can be represented on a line. To solve any type of inequality, first solve it as you would if it were an equation. Use the solution(s) as dividing point(s) of the line. Then test a value from each region on the number line. If the test value makes the inequality true, then that region is part of the solution. If it is false then the value and thus that region is not part of the solution. In addition, if the inequality is ≥ or ≤ then the dividing point is part of the solution and is indicated by a solid dot. If the inequality is > or <, then the dividing point is not part of the solution and is indicated by an open dot.

Example 1

Solve $-2x - 3 \geq x + 6$

Solve the equation

$$-2x - 3 = x + 6$$
$$-2x = x + 9$$
$$-3x = 9$$
$$x = -3$$

Draw a number line and put a solid dot at $x = -3$, which is the dividing point.

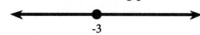

Test a value from each region. Here we test -4 and 0. Be sure to use the <u>original</u> inequality.

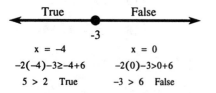

The region(s) that are true represent the solution. The solution is -3 and all numbers in the left region, written: $x \leq -3$.

Example 2

Solve $x^2 - 2x + 2 < 5$

Solve the equation

$$x^2 - 2x + 2 = 5$$
$$x^2 - 2x - 3 = 0$$
$$(x - 3)(x + 1) = 0$$
$$x = 3 \quad \text{or} \quad x = -1$$

Draw a number line and put open dots at $x = 3$ and $x = -1$, the dividing points.

Test a value from each region in the original inequality. Here we test -3, 0, and 4.

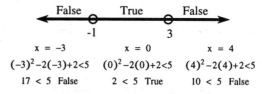

The region(s) that are true represent the solution. The solution is the set of all numbers greater than -1 but less than 3, written: $-1 < x < 3$.

If the inequality has two variables, then the solution is represented by a graph in the xy-coordinate plane. The graph of the inequality written as an equation (a line or curve) divides the coordinate plane into regions which are tested in the same manner described above using an ordered pair for a point on a side of the dividing line or curve. If the inequality is > or <, then the boundary line or curve is dashed. If the inequality is ≥ or ≤, then the boundary line or curve is solid.

SKILL BUILDERS

Example 3

Graph and shade the solution to this system of inequalities $\begin{cases} y \leq \frac{2}{5}x \\ y > 5 - x \end{cases}$

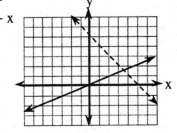

Graph each equation. For $y = \frac{2}{5}x$ the slope of the solid line is $\frac{2}{5}$ and y-intercept is 0. For $y = 5 - x$ the slope of the dashed line is -1 and the y-intercept is 5.

Test a point from each region in <u>both</u> of the original inequalities.

(0, 2)	(2, 0)	(4, 5)	(5, 1)
False in both	True in first, False in second	False in first, True in second	True in both

The region that makes <u>both</u> statements (inequalities) true is the solution.

The solution is the region below the solid line $y = \frac{2}{5}x$ and above the dashed line $y = 5 - x$.

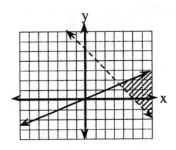

Solve and graph each inequality.

1. $x + 12 \geq 2x - 5$
2. $-16 + 4x > 10 - x$
3. $7x - 2x - x \geq 24 + 3x$
4. $3(x - 4) - 9x \geq 2x - 4$
5. $|x - 1| < 5$
6. $|x + 10| > 5$
7. $|12x| \geq 24$
8. $\left|\frac{x}{3}\right| < 8$
9. $x^2 + 3x - 10 \leq 0$
10. $x^2 - 7x + 6 > 0$
11. $x^2 + 2x - 8 \leq 7$
12. $x^2 - 5x - 16 > -2$
13. $y < 2x + 1$
14. $y \leq -\frac{2}{3}x + 3$
15. $y \geq \frac{1}{4}x - 2$
16. $2x - 3y \leq 5$
17. $y \geq -2$
18. $-3x - 4y > 4$
19. $y \leq \frac{1}{2}x + 2$ and $y > -\frac{2}{3}x - 1$.
20. $y \leq -\frac{3}{5}x + 4$ and $y \leq \frac{1}{3}x + 3$
21. $y < 3$ and $y \leq -\frac{1}{2}x + 2$
22. $x \leq 3$ and $y < \frac{3}{4}x - 4$
23. $y \leq x^2 + 4x + 3$
24. $y > x^2 - x - 2$

Answers

1.
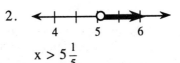

   16 17 18

   $x \leq 17$

2.
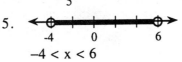

   4   5   6

   $x > 5\frac{1}{5}$

3.
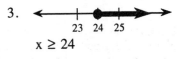

   23 24 25

   $x \geq 24$

4.

   -2 -1 0

   $x \leq -1$

5.
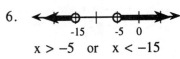

   -4   0   6

   $-4 < x < 6$

6.

   -15   -5 0

   $x > -5$ or $x < -15$

Algebra 1

7.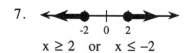
   $x \geq 2$ or $x \leq -2$

8.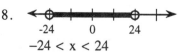
   $-24 < x < 24$

9.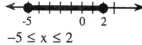
   $-5 \leq x \leq 2$

10.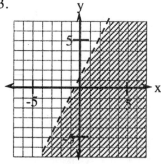
    $x < 1$ or $x > 6$

11.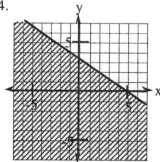
    $-5 \leq x \leq 3$

12.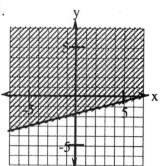
    $x < -2$ or $x > 7$

13.

14.

15.

16.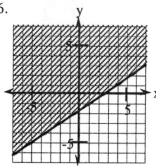

17.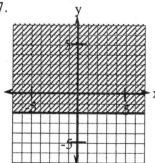

18.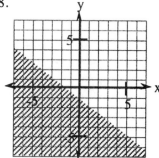

19.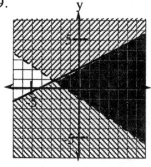

20.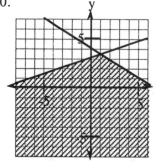

21.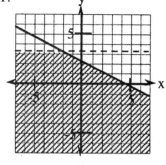

22.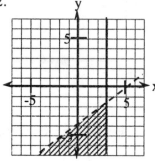

23.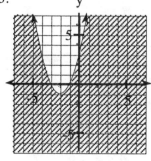

24.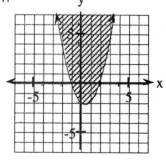

SKILL BUILDERS

ADDITION AND SUBTRACTION OF RATIONAL EXPRESSIONS #26

Addition and subtraction of rational expressions is done the same way as addition and subtraction of numerical fractions. Change to a common denominator (if necessary), combine the numerators, and then simplify.

Example 1

The Least Common Multiple (lowest common denominator) of $(x + 3)(x + 2)$ and $(x + 2)$ is $(x + 3)(x + 2)$.

$$\frac{4}{(x+2)(x+3)} + \frac{2x}{x+2}$$

The denominator of the first fraction already is the Least Common Multiple. To get a common denominator in the <u>second</u> fraction, multiply the fraction by $\frac{x+3}{x+3}$, a form of one (1).

$$= \frac{4}{(x+2)(x+3)} + \frac{2x}{x+2} \cdot \frac{(x+3)}{(x+3)}$$

Multiply the numerator and denominator of the second term:

$$= \frac{4}{(x+2)(x+3)} + \frac{2x(x+3)}{(x+2)(x+3)}$$

Distribute in the second numerator.

$$= \frac{4}{(x+2)(x+3)} + \frac{2x^2+6x}{(x+2)(x+3)}$$

Add, factor, and simplify. Note: $x \neq -2$ or -3.

$$= \frac{2x^2+6x+4}{(x+2)(x+3)} = \frac{2(x+1)(x+2)}{(x+2)(x+3)} = \frac{2(x+1)}{(x+3)}$$

Example 2

Subtract $\frac{3}{x-1} - \frac{2}{x-2}$ and simplify the result.

Find the lowest common denominator of $(x - 1)$ and $(x - 2)$. It is $(x - 1)(x - 2)$.

In order to change each denominator into the lowest common denominator, we need to multiply each fraction by factors that are equal to one.

$$\frac{(x-2)}{(x-2)} \cdot \frac{3}{x-1} - \frac{2}{(x-2)} \cdot \frac{(x-1)}{(x-1)}$$

Multiply the denominators.

$$\frac{3(x-2)}{(x-2)(x-1)} - \frac{2(x-1)}{(x-2)(x-1)}$$

Multiply and distribute the numerators.

$$\frac{3x-6}{(x-2)(x-1)} - \frac{2x-2}{(x-2)(x-1)}$$

When adding fractions, the denominator does not change. The numerators need to be added or subtracted and like terms combined.

$$\frac{3x-6-(2x-2)}{(x-2)(x-1)} \Rightarrow \frac{3x-6-2x+2}{(x-2)(x-1)} \Rightarrow \frac{x-4}{(x-2)(x-1)}$$

Check that both the numerator and denominator are completely factored. If the answer can be simplified, simplify it. This answer is already simplified. Note: $x \neq 1$ or 2.

$$\frac{x-4}{(x-2)(x-1)} = \frac{x-4}{x^2-3x+2}$$

Algebra 1

Add or subtract the expressions and simplify the result.

1. $\dfrac{x}{(x+2)(x+3)} + \dfrac{2}{(x+2)(x+3)}$
2. $\dfrac{x}{x^2+6x+8} + \dfrac{4}{x^2+6x+8}$
3. $\dfrac{b^2}{b^2+2b-3} + \dfrac{-9}{b^2+2b-3}$
4. $\dfrac{2a}{a^2+2a+1} + \dfrac{2}{a^2+2a+1}$
5. $\dfrac{x+10}{x+2} + \dfrac{x-6}{x+2}$
6. $\dfrac{a+2b}{a+b} + \dfrac{2a+b}{a+b}$
7. $\dfrac{3x-4}{3x+3} - \dfrac{2x-5}{3x+3}$
8. $\dfrac{3x}{4x-12} - \dfrac{9}{4x-12}$
9. $\dfrac{6a}{5a^2+a} - \dfrac{a-1}{5a^2+a}$
10. $\dfrac{x^2+3x-5}{10} - \dfrac{x^2-2x+10}{10}$
11. $\dfrac{6}{x(x+3)} + \dfrac{2}{x+3}$
12. $\dfrac{5}{x-7} + \dfrac{3}{4(x-7)}$
13. $\dfrac{5x+6}{x^2} - \dfrac{5}{x}$
14. $\dfrac{2}{x+4} - \dfrac{x-4}{x^2-16}$
15. $\dfrac{10a}{a^2+6a} - \dfrac{3}{3a+18}$
16. $\dfrac{3x}{2x^2-8x} + \dfrac{2}{(x-4)}$
17. $\dfrac{5x+9}{x^2-2x-3} + \dfrac{6}{x^2-7x+12}$
18. $\dfrac{x+4}{x^2-3x-28} - \dfrac{x-5}{x^2+2x-35}$
19. $\dfrac{3x+1}{x^2-16} - \dfrac{3x+5}{x^2+8x+16}$
20. $\dfrac{7x-1}{x^2-2x-3} - \dfrac{6x}{x^2-x-2}$

Answers

1. $\dfrac{1}{x+3}$
2. $\dfrac{1}{x+2}$
3. $\dfrac{b-3}{b-1}$
4. $\dfrac{2}{a+1}$
5. 2
6. 3
7. $\dfrac{1}{3}$
8. $\dfrac{3}{4}$
9. $\dfrac{1}{a}$
10. $\dfrac{x-3}{2}$
11. $\dfrac{2}{x}$
12. $\dfrac{23}{4(x-7)} = \dfrac{23}{4x-28}$
13. $\dfrac{6}{x^2}$
14. $\dfrac{1}{x+4}$
15. $\dfrac{9}{(a+6)}$
16. $\dfrac{7}{2(x-4)} = \dfrac{7}{2x-8}$
17. $\dfrac{5(x+2)}{(x-4)(x+1)} = \dfrac{5x+10}{x^2-3x-4}$
18. $\dfrac{14}{(x+7)(x-7)} = \dfrac{14}{x^2-49}$
19. $\dfrac{4(5x+6)}{(x-4)(x+4)^2}$
20. $\dfrac{x+2}{(x-3)(x-2)} = \dfrac{x+2}{x^2-5x+6}$

SKILL BUILDERS

SOLVING MIXED EQUATIONS AND INEQUALITIES #27

Solve these various types of equations.

1. $2(x - 3) + 2 = -4$
2. $6 - 12x = 108$
3. $3x - 11 = 0$
4. $0 = 2x - 5$
5. $y = 2x - 3$
   $x + y = 15$
6. $ax - b = 0$ (solve for x)
7. $0 = (2x - 5)(x + 3)$
8. $2(2x - 1) = -x + 5$
9. $x^2 + 5^2 = 13^2$
10. $2x + 1 = 7x - 15$
11. $\frac{5 - 2x}{3} = \frac{x}{5}$
12. $2x - 3y + 9 = 0$ (solve for y)
13. $x^2 + 5x + 6 = 0$
14. $x^2 = y$
    $100 = y$
15. $x - y = 7$
    $y = 2x - 1$
16. $x^2 - 4x = 0$
17. $x^2 - 6 = -2$
18. $\frac{x}{2} + \frac{x}{3} = 2$
19. $x^2 + 7x + 9 = 3$
20. $y = x + 3$
    $x + 2y = 3$
21. $3x^2 + 7x + 2 = 0$
22. $\frac{x}{x + 1} = \frac{5}{7}$
23. $x^2 + 2x - 4 = 0$
24. $\frac{1}{x} + \frac{1}{3x} = 2$
25. $3x + y = 5$
    $x - y = 11$
26. $y = -\frac{3}{4}x + 4$
    $\frac{1}{4}x - y = 8$
27. $3x^2 = 8x$
28. $|x| = 4$
29. $\frac{2}{3}x + 1 = \frac{1}{2}x - 3$
30. $x^2 - 4x = 5$
31. $3x + 5y = 15$ (solve for y)
32. $(3x)^2 + x^2 = 15^2$
33. $y = 11$
    $y = 2x^2 + 3x - 9$
34. $(x + 2)(x + 3)(x - 4) = 0$
35. $|x + 6| = 8$
36. $2(x + 3) = y + 2$
    $y + 2 = 8x$
37. $2x + 3y = 13$
    $x - 2y = -11$
38. $2x^2 = -x + 7$
39. $1 - \frac{5}{6x} = \frac{x}{6}$
40. $\frac{x - 1}{5} = \frac{3}{x + 1}$
41. $\sqrt{2x + 1} = 5$
42. $2|2x - 1| + 3 = 7$
43. $\sqrt{3x - 1} + 1 = 7$
44. $(x + 3)^2 = 49$
45. $\frac{4x - 1}{x - 1} = x + 1$

Algebra 1

Solve these various types of inequalities.

46. $4x - 2 \leq 6$
47. $4 - 3(x + 2) \geq 19$
48. $\frac{x}{2} > \frac{3}{7}$
49. $3(x + 2) \geq -9$
50. $-\frac{2}{3}x < 6$
51. $y < 2x - 3$
52. $|x| > 4$
53. $x^2 - 6x + 8 \leq 0$
54. $|x + 3| > 5$
55. $2x^2 - 4x \geq 0$
56. $y \leq -\frac{2}{3}x + 2$
57. $y \leq -x + 2$
    $y \leq 3x - 6$
58. $|2x - 1| \leq 9$
59. $5 - 3(x - 1) \geq -x + 2$
60. $y \leq 4x + 16$
    $y > -\frac{4}{3}x - 4$

Answers

1. 0
2. -8.5
3. $\frac{11}{3}$
4. $\frac{5}{2}$
5. (6, 9)
6. $x = \frac{b}{a}$
7. $\frac{5}{2}$, -3
8. $\frac{7}{5}$
9. ±12
10. $\frac{16}{5}$
11. $\frac{25}{13}$
12. $y = \frac{2}{3}x + 3$
13. -2, -3
14. (±10, 100)
15. (-6, -13)
16. 0, 4
17. ±2
18. $\frac{12}{5}$
19. -1, -6
20. (-1, 2)
21. $-\frac{1}{3}$, -2
22. $\frac{5}{2}$
23. $\frac{-2 \pm \sqrt{20}}{2}$
24. $\frac{2}{3}$
25. (4, -7)
26. (12, -5)
27. 0, $\frac{8}{3}$
28. ±4
29. -24
30. 5, -1
31. $y = -\frac{3}{5}x + 3$
32. ≈±4.74
33. (−4, 11) and $\left(\frac{5}{2}, 11\right)$
34. -2, -3, 4
35. 2, -14
36. (1, 6)
37. (-1, 5)
38. $\frac{1 \pm \sqrt{57}}{4}$
39. 1, 5
40. ±4
41. 12
42. $\frac{3}{2}$, $-\frac{1}{2}$
43. $\frac{37}{3}$
44. 4, -10
45. 0, 4
46. $x \leq 2$
47. $x \leq -7$
48. $x > \frac{6}{7}$
49. $x \geq -5$
50. $x > -9$
51. below
52. $x > 4$, $x < -4$
53. $2 \leq x \leq 4$
54. $x > 2$ or $x < -8$
55. $x \leq 0$ or $x \geq 2$
56. below
57. below
58. $-4 \leq x \leq 5$
59. $x \leq 3$
60. below

51.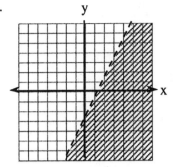

56.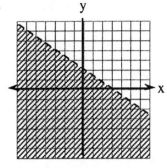

57.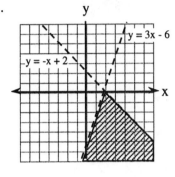

60.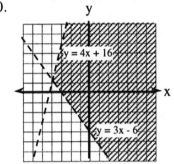

Algebra 1

Glossary

absolute value The absolute value of a number is the distance of the number from zero. Since the absolute value represents a distance, without regard to direction, it is always non-negative. (381)

additive identity The Additive Identity Property states that any term added to zero (0) remains unchanged; a + 0 = a. (393)

additive inverse The Additive Inverse Property states that when opposites are added, the result is always zero: a + (-a) = 0. (393)

additive property of equality The Additive Property of Equality states that equality is maintained if you add the same amount to both sides of an equation. If a = b, then a + c = b + c. (393)

algebra tiles The algebra tiles used in this course consist of large squares with dimensions x by x, rectangles with dimensions x by 1, and small squares with dimensions 1 by 1. The areas of these tiles are x^2, x and 1 respectively. We call the smallest squares unit squares. (42)

area For this course, area is the number of square units needed to fill up a region on a flat surface. The idea can be extended to cones, spheres, and more complex surfaces. (63)

area of a triangle To find the area of a triangle, multiply the length of the base (b) by the height (h) and divide by two. $A = \frac{1}{2}bh$.

associative property The Associative Property states that if a sum or product contains terms that are grouped, then the sum or product can be grouped differently with no effect on the result; a + (b + c) = (a + b) + c and a(bc) = (ab)c. (393)

average See mean.

base In the expression 2^5, 2 is called the base. Also, 5 is the exponent and 32 is the value. The term "base" may also refer to sides of a triangle, rectangle, parallelogram, trapezoid, prism, cylinder, pyramid, and cone. (340)

binomial The sum or difference of two monomials is called a binomial. (217, 269)

circle A circle is the set of all points that are the same distance r from a fixed point P. The fixed point is called the center of the circle and the distance from the center to the points on the circle is called the radius. (61)

circle (area of) $A = \pi r^2$, where r is the length of the radius of the circle. See area. (63)

circumference The circumference C of a circle is its perimeter, that is, the distance around the circle. $C = 2\pi r = \pi d$ (r = radius, d = 2r = diameter). (63)

coefficient (numerical) The numeral part of a term, such as 6 in 6x. (350)

common factor Factors which are the same for two or more terms. (137, 291)

common term factoring Factoring out a common (term) factor means identifying the common factor of the terms of a polynomial and then writing it outside the parentheses containing the sum of the factored terms. For example, $6x^2y - 9xy^2 = 3xy(2x - 3y)$. Factoring usually means using the Distributive Property: $ab + ac = a(b + c)$. (137, 391)

Commutative Property The Commutative Property states that if two terms are added or multiplied, the order is reversible. $a + b = b + a$ and $ab = ba$. (393)

complete graph A complete graph has the following components: (1) the x-axis and y-axis labeled, clearly showing the scale. (2) Equation of the graph written near the line or curve. (3) Line or curve extended as far as possible on the graph. (4) x- and y-intercepts labeled. (5) Coordinates of points stated in (x, y) form. (96)

completing the square A method for solving quadratic equations or writing quadratic functions in graphing form. (444)

congruent Two shapes (for example, triangles) are congruent if they have exactly the same size and shape.

conjecture An educated guess, based on data, patterns, and relationships. Scientists use the term hypothesis. (161)

constant A symbol representing a value that does not change. For example, in the equation $y = 2x + 5$, "5" is referred to as the constant. (202)

coordinate The number paired with a point on the number line or an ordered pair (x, y) that corresponds to a point in xy-coordinate system. (96)

coordinate system A system of graphing ordered pairs of numbers in relation to axes (horizontal and vertical) that intersect at right angles at their zero points (origin). (96)

corresponding parts Points, edges (sides), or angles in congruent or similar figures that are arranged in similar ways. For instance if $\triangle ABC$ is similar to $\triangle XYZ$, side $\overline{AB}$ corresponds to (matches) the side $\overline{XY}$. (158)

degree (of a polynomial) (1) The degree of a monomial is the sum of the exponents of its variables, such as $3x^2y^5$ has degree 7; (2) The degree of a polynomial is: (a) in one variable, the degree of the term with the highest exponent, such as $3x^5 - 4x^2 - x + 7$ has degree 5; (b) in more than one variable, the highest sum of the exponents among the terms, such as $2x^5y^3 - 4x^2x^4z^3 - xy^5 + 3y^2z - 12$ has degree 9.

dependent variable The output variable, y, of a relation or function is called the dependent variable because its values are determined by the value of x that is used in the relation or function. (378)

diagram A problem-solving technique based on drawing a picture or diagram representing the problem. Including all the known information on the diagram helps us to see what is important and necessary in solving the problem. (308)

diameter A line segment drawn through the center of the circle with both endpoints on the circle is called a diameter of the circle, usually denoted d. Note: $d = 2r$, where r is the radius of the circle. (61)

difference of squares A special polynomial that can be factored as the product of the sum and difference of two terms. The general pattern is $x^2 - y^2 = (x + y)(x - y)$. (287)

discriminant For quadratic equations in standard form $ax^2 + bx + c = 0$, the discriminant is $b^2 - 4ac$.

distributive property For any numbers or expressions a, b, and c, $a(b + c) = ab + ac$. (74, 391, 393)

dividing line (or boundary line) A line on a two dimensional graph that divides the graph into two regions. We use a dividing line or boundary line when graphing linear inequalities such as $y > 3x - 1$. (418)

dividing point The endpoint of a segment on a number line where an inequality is true. For strict inequalities, that is, < or >, the point is not part of the solution. (411)

domain The set of all input values for a relation or function. For variables, the set of numbers the variable may represent. For ordered pairs (x, y), all x-values. (111-12, 363, 378)

elimination method (systems of equations) A method for solving a system of equations by adding or subtracting the equations to eliminate one of the variables. (367, 370)

ellipsis The symbol "..." is called an ellipsis. It indicates that certain values in an established pattern have not been written, although they are part of the pattern. (110)

enlargement ratio The ratio of similarity comparing a figure to a similar larger one is often called the enlargement ratio. This number tells you by what factor the first figure is enlarged by to get the second. (165)

equation A mathematical sentence with an equal sign (=).

evaluate To evaluate an expression, substitute the value(s) given for the variable(s) and perform the operations according to the order of operations. (63)

exponent In the expression 2^5, 5 is called the exponent. Also, 2 is the base and 32 is the value. The exponent indicates how many times to use the number 2 as a multiplier, in this case, 5 times, $2 \cdot 2 \cdot 2 \cdot 2 \cdot 2 = 32$. (340)

exponents (laws of) There are several basic laws of exponents: (1) $x^a \cdot x^b = x^{a+b}$; (2) $\dfrac{x^a}{x^b} = x^{a-b}$; (3) $(x^a)^b = x^{ab}$. (342)

exponents (negative) For any number $x \neq 0$, $x^{-n} = \dfrac{1}{x^n}$ and $\dfrac{1}{x^{-n}} = x^n$. (343)

exponents (zero) For any number $x \neq 0$, $x^0 = 1$. (343)

expression An algebraic expression consists of one or more variables. It may also contain some constants. Each part of the expression separated by addition or subtraction signs is called a term. (42)

factor (1) In arithmetic: when two or more numbers are multiplied, each of the numbers is a factor of the product. (2) In algebra: where two or more algebraic expressions are multiplied together, each of the expressions is a factor of the product.

factored completely A polynomial is factored completely if none of the resulting factors can be factored further. (291)

Fibonacci Numbers The numbers in the sequence 1, 1, 2, 3, 5, 8, 13, ... are called the Fibonacci Numbers. (201)

Algebra 1

F.O.I.L. An approach for multiplying two binomials is to use the mnemonic "F.O.I.L." which stands for "First, Outer, Inner, Last." It describes the order in which to multiply the terms of two binomials to be sure to get all the products. (324)

fraction busters A method of simplifying equations involving fractions that uses the Multiplication Property of Equality to rearrange the equation so that no fractions remain. (313)

function A function is a relation in which for each input value there is one and only one output value. In terms of ordered pairs (x, y), no two ordered pairs have the same first member (x). (378)

function notation Functions are given names, most commonly "f", "g" or "h." The notation f(x) represents the output of a function, named f, when x is the input. It is pronounced "f of x". The notation g(2) is pronounced "g of 2" and represents the output of the function g when x = 2. (378)

generic rectangle In this course, used as an organizational device for multiplying polynomials. For example, the figure below shows how to use generic rectangles to multiply binomials. (215-17)

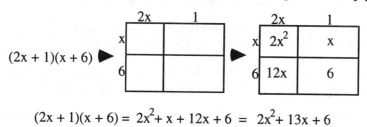

$$(2x + 1)(x + 6) = 2x^2 + x + 12x + 6 = 2x^2 + 13x + 6$$

greatest common factor (1) For integers, the greatest positive integer that is a common factor of two or more integers. (2) For two or more algebraic monomials, the product of the greatest common integer factor of the coefficients of the monomials and the variable(s) in each algebraic term with the greatest degree of that variable in every term. For example, the greatest common factor of $12x^3y^2$ and $8xy^4$ is $4xy^2$. (3) For a polynomial, the greatest common monomial factor of its terms. For example, the greatest common factor of $16x^4 + 8x^3 + 12x$ is $4x$. (279)

guess and check Guess and check is a problem solving strategy in which you begin by making a guess and then check whether or not your guess is correct. In the process of checking, you gain information about how close your guess might be and make adjustments to your guess. The second guess is then tested. This continues until the correct answer is discovered. Being organized is crucial to the success of this method, as well as writing a usable table. Guess and check also leads us to writing equations to represent word problems. (37, 127)

horizontal lines Horizontal lines are "flat" and run left to right in the same direction as the x-axis. All horizontal lines have equations of the form y = b, where b can be any number. Their slope is 0. The x-axis has the equation y = 0 because y = 0 everywhere on the x-axis. (174)

hypotenuse In a right triangle, the longest side of the triangle, opposite the right angle, is called the hypotenuse. (303)

hypothesis A conjecture is what mathematicians call an educated guess, based on data, patterns, and relationships. Scientists use the term hypothesis. (161)

identity element The identity element for addition is 0 because adding 0 leaves the number unchanged: a + 0 = 0. The identity element for multiplication is 1, because multiplying by 1 leaves a number unchanged: a(1) = a. (393)

independent variable The input variable, x, of a relation or function is called the independent variable. (378)

inequality symbols The symbol "≤" read from left to right means "less than or equal to." The symbol "≥" read from left to right means "greater than or equal to." The symbols < and > mean "less than: and "greater than" respectively. (110, 408)

input value In a relation or function where y = f(x), the values used for x (the domain) and substituted into the relationship are the input values. The input values are the numbers represented by the independent variables, the first numbers in ordered pairs (x, y). (93, 378)

integers The set of numbers { . . . . -3, -2, -1, 0, 1, 2, 3, . . . }.

inverse operations Addition and subtraction are inverse operations, as are multiplication and division. (139, 393)

irrational numbers The set of numbers that cannot be expressed in the form $\frac{a}{b}$ where a and b are integers and b ≠ 0. For example, π and √2 are irrational numbers.

justify To use facts, definitions, rules, and/or previously proven statements in an organized way to convince your audience that what you claim (or your answer) is valid (true). (161)

legs of a right triangle The two sides of a right triangle which are not the hypotenuse are called the legs of the triangle. Note that the legs meet to form the right angle of the triangle.

like terms Two or more terms that contain the same variable(s), with corresponding variables raised to the same power are called like terms. For example, 5x and 2x are like terms. Combine like terms by adding them: 5x + 2x = (5 + 2)x = 7x. (67)

linear equation Any equation equivalent to ax + by = c (standard form), where a, b, and c are real numbers and a and b are not both zero. Slope-intercept form, y = mx + b, is one equivalent form. (250, 292)

linear graphs Graphs that are straight lines are called linear graphs. (108)

line segment A portion of a line between two points. We name a line segment by its endpoints, A and B, and write $\overline{AB}$. (167)

mean The mean (average) of several numbers (or data points) is a statistical measurement that describes one way of defining the middle of the numbers. It is found by adding the numbers together and dividing by the number of data points in the set. (144)

monomial An expression with only one term. It can be a numeral, a variable, or the product of a number and a one or more variables. for example, 7, 3x, -4ab, or $3x^2y$. (269)

multiplicative identity The Multiplicative Identity Property states that any term multiplied by one (1) remains unchanged; a(1) = a. (393)

multiplicative inverse The Multiplicative Inverse Property states that when multiplying a term by its reciprocal, the result is always one: $a \cdot \frac{1}{a} = 1$ and $\frac{a}{b} \cdot \frac{b}{a} = 1$ for a (not equal) 0. (393)

multiplying binomials See generic rectangle. (217)

neutral field The number 0 can be represented by the same number of positive and negative tiles, known as a neutral field. (24)

numeral A symbol that names a number.

numerical coefficient See coefficient.

opposite The opposite of a number is its additive inverse. For example, -5 is the opposite of 5.

ordered pairs Points on the x-y coordinate grid written as (x, y). The first coordinate, x, represents the horizontal distance and direction from the origin; the second coordinate, y, represents the vertical distance and direction from the origin. (104)

order of operations We use the order of operations to simplify complex arithmetic and algebraic expressions, by performing certain operations in a specific order. The order is parentheses (or other grouping symbols), exponents (powers or roots), multiplication and division (left to right) and addition and subtraction (left to right). (77)

origin The point assigned to zero on the number line or the point where the x- and y-axes intersect in a coordinate system. (36, 96)

output values In a relation or function where y = f(x), the y-values (range) are the output values. (93, 378)

parabola (equation from its x-intercepts) A parabola with x-intercepts (b, 0) and (c, 0) can be written in the form: y = a(x - b)(x - c). The coefficient a determines the shape (wide or narrow) and the direction (open upward or downward) of the parabola. (439)

parabolic graphs Graphs of quadratic equations are parabolas and are called parabolic. (108

percent A notation for a ratio with the denominator 100.

perfect square trinomials Trinomials of the form $x^2 + 2ax + a^2$ are known as perfect square trinomials and factor as $(x + a)^2$. (289)

perimeter Perimeter is the distance around a figure on a flat surface. (10, 101)

perpendicular Two lines or segments on a flat surface meet (intersect) to form a 90° angle. (387)

pi (π) Pi is the name for the ratio of the circumference (c) of a circle, to its diameter (d), so $\pi = \frac{c}{d}$. (63)

polynomial The sum or difference of two or more monomials. (269)

power A number or variable raised to an exponent in the form x^n. See exponent. (3340)

probability Probability is a mathematical way to predict how likely it is that an event will occur. If all the outcomes of an event are equally likely to occur, then the probability (or likelihood) that a specified result occurs is expressed by the fraction:

$$P(event) = \frac{\text{number of outcomes in the specified event}}{\text{total number of possible outcomes}}$$ (15)

problem solving strategies This course deals with numerous problem solving strategies. Specifically, Making a Guess and Checking It, Using Manipulatives (such as algebra tiles), Making Systematic Lists, Graphing, Drawing a Diagram, Breaking a Large Problem into Smaller Subproblems, Working Backward, and Writing and Solving an Equation. (422)

proportion An equation stating that two fractions (or ratios) are equal. (169)

Pythagorean Theorem In a right triangle with legs of length a and b and hypotenuse of length c, $a^2 + b^2 = c^2$. This can also be written as $(\text{leg})^2 + (\text{leg})^2 = (\text{hypotenuse})^2$. (305)

quadratic A polynomial is quadratic if the largest exponent in the polynomial is two (that is, the polynomial has degree 2). (269)

quadratic equation (standard form) A quadratic equation is in standard form if it is written as $ax^2 + bx + c = 0$. (350)

quadratic formula If $ax^2 + bx + c = 0$ and $a \neq 0$, then $x = \dfrac{-b \pm \sqrt{b^2 - 4ac}}{2a}$. (353)

radical An expression in the form $\sqrt{a}$ (square root). Other roots, such as cube root, will be studied in other courses. (319, 326)

radical (simplified form) A number $r\sqrt{s}$ is in simple radical form if no square of an integer divides s and s is not a fraction; that is, there are no more perfect square factors (square numbers such as 4, 9, 16, etc.) under the radical sign and no radicals in the denominator. For example, $5\sqrt{12}$ is not in simple radical form since 4 (the square of 2) divides 12. But $5\sqrt{12} = 10\sqrt{3}$ is in simple radical form. (319, 326)

radicand The expression under the radical sign.

radius The distance from the center to the points on the circle is called the radius, usually denoted r. (61)

range The set of all second members of a function or relation, that is, all possible output values for a relation or function. For ordered pairs (x, y), all output (y) values. (378)

ratio A ratio is a comparison of two quantities by division. (158-59)

ratio of similarity The ratio of similarity between any two similar figures is the ratio of any pair of corresponding sides. In this course ratios will always be listed in the order that compares the "new" to the "original" figure. (158-59)

rational expressions A rational expression is a fraction in which the numerator and/or denominator contain polynomials. (348)

rational numbers Numbers that can be expressed in the form $\dfrac{a}{b}$ where a and b are integers, $b \neq 0$.

ratio of similar figures The ratio of similarity between any two similar figures is the ratio of any pair of corresponding sides. This means that once it is determined that two figures are similar, all of their pairs of corresponding sides have the same ratio. (165)

real numbers Irrational numbers together with rational numbers form the set of the real numbers. All real numbers are represented on the number line.

reciprocal The reciprocal of a non-zero number is its multiplicative inverse. For x, the reciprocal is $\dfrac{1}{x}$; for $\dfrac{a}{b}$, the reciprocal is $\dfrac{b}{a}$.

rectangular numbers The numbers in the pattern 2, 6, 12, 20, ... are known as the rectangular numbers. (40)

reduction ratio The ratio of similarity comparing a figure to a similar smaller one is often called the reduction ratio. This number tells you by what factor the first figure is reduced to get the second. (165)

reference point When solving equations involving absolute values, it is sometimes useful to start with a reference point. For an equation such as $|x - a| = b$, a is the reference point. After locating the point a on a number line, we find the numbers which are a distance of b from a. This usually gives us the two solutions to the equation. (384)

reflexive property The Reflexive Property states that a term is always equal to itself: $a = a$. (393)

relation An equation which relates inputs to outputs is called a relation. As such, a relation is a set of ordered pairs. The set of first values is the domain, the set of second values is the range. (363)

right angle A right angle is an angle that measures 90°. (171)

right triangle A right triangle is a triangle with one right angle. (171)

roots of an equation A solution of the equation. The x-intercepts of a parabola are also referred to as the roots of the quadratic equation. (284)

scientific notation A number is expressed in scientific notation when it is in the form $a \times 10^n$, where $1 \leq a < 10$ and n is an integer. (96)

similar figures Similar geometric figures are figures that have the same shape but are not necessarily the same size. In similar figures, the measures of corresponding angles are equal <u>and</u> the lengths of corresponding sides have the same ratio. (158)

simplest form (of a numerical expression) Replacing a numerical expression by the simplest form of its value.

simplest form (of a variable expression) A variable expression in simplest form has no like terms and no parentheses. (42, 348, 373)

slope The slope of a line is a ratio that describes how steep (or flat) the line is. Slope can be positive, negative, or even zero, but a straight line has only one slope. Slope is the ratio $\frac{\text{change in y value}}{\text{change in x value}}$ or $\frac{\text{vertical change}}{\text{horizontal change}}$. The symbol used to represent slope is the letter "m." Some texts refer to slope as the ratio of the "rise over the run."
A line has positive slope if it slopes upward from left to right on a graph; negative slope if it slopes downward from left to right. A vertical line has undefined or no slope. (243)

slope-intercept form Any non-vertical line can be described by an equation written in the form $y = mx + b$. The "m" represents the slope of the line (it is the coefficient of x). The "b" represents the y-value of the y-intercept (where $x = 0$). Ordered pairs (x, y) that make the equation true are coordinates of points on that line. (250)

slope triangle A slope triangle is a right triangle drawn on a graph of a line so that the hypotenuse of the triangle is part of the line. The vertical leg length is the change in the y-value; the horizontal leg length is the change in the x-value. The length of the legs of the triangle are the values used in the slope ratio (change in y and change in x). (252)

solution A replacement for the variable that makes an open sentence (equation) true.

solve Find the solution(s) of equations or inequalities. (134, 142, 205, 283, 387, 412)

square numbers The numbers in the pattern 1, 4, 9, 16, 25, ..., that is, the squares of the counting numbers 1, 2, 3, 4, 5, ..., are known as square numbers. (39-40)

standard form (of a linear equation) See linear equation.

standard notation A number written out completely, showing all digits and without use of exponents is written in standard notation. (39)

subproblem Breaking down problems into smaller, simpler parts is a technique for solving problems. The smaller, simpler problems are called subproblems. Solving the simpler, smaller problems first allows us to then put the results together to complete a larger problem. (54)

substitution Replacing one symbol by another (a number, a variable, or other algebraic expression) without changing the value of the expression. (63)

substitution method (systems of equations) A method of solving a system of equations by replacing one variable with an expression involving the remaining variable(s). One variable is expressed in terms of the other variable, such as y-form, that is, $y = mx + b$, then the expression $mx + b$ replaces that variable (y) in a second equation involving x and y. (206, 254)

substitution property If $a = b$, then either a or b can be replaced by the other. (392-93)

symmetric property The Symmetric Property states that if two terms are equal it does not matter which is stated first. If $a = b$ then $b = a$. (393)

systems of linear equations For this course, two equations in two variables that describe lines that may or may not intersect. The equations together are called a system of equations and the process of finding where, if at all, the lines intersect is called solving the system. (206)

term Each part of the expression separated by addition or subtraction signs is called a term. (42)

triangular numbers The numbers in the pattern 1, 3, 6, 10, 15, ... are known as triangular numbers. (41)

transitive property of equality The Transitive Property of Equality states that if $a = b$ and $b = c$, then $a = c$. (393)

trinomial A polynomial of three terms. (269)

two-point graphing method Using only two points to graph a linear equation. Often one of the points is the y-intercept, but it does not necessarily need to be. (203)

value In the expression 2^5, 32 is the value. Also, 5 is the exponent and 2 is the base. (340)

variable For this course a variable is a symbol used in a mathematical sentence to represent a number. (74)

vertex (of a parabola) The highest point or lowest point on a parabola (depending on its orientation) is called the vertex. (439)

vertical lines   Vertical lines run up and down in the same direction as the y-axis and parallel to it. All vertical lines have equations of the form x = a, where a can be any number. The y-axis has the equation x = 0 because x = 0 everywhere on the y-axis. Vertical lines have undefined slope. (174)

working backward   In many cases, to solve a problem we must "undo" something that has been "done." This is true when we solve equations. The notion of being able to "undo" something is part of the problem solving strategy of working backward and the mathematical operation is called an inverse operation. Working backwards uses inverse operations to find the original form of an expression or value(s) of a variable. Solving an equation is an example of this process. (139)

x-axis   The horizontal axis on a coordinate plane (graph) is called the x-axis. (96)

x-coordinate   The first component (coordinate) in an ordered pair.

x-intercepts   The point(s) where a graph crosses the x-axis is (are) called the x-intercepts. The x-intercept always has coordinates (x, 0). (96, 106, 239, 284)

y-axis   The vertical axis on a coordinate plane (graph) is called the y-axis. (96)

y-coordinate   The second component (coordinate) in an ordered pair.

y-form   An equation is written in y-form if the equation is solved for y, and so is written as "y = _____". For example, the y-form of a linear equation is y = mx + b. (202, 205)

y-intercepts   The point(s) where the graph crosses the y-axis is (are) called the y-intercepts. The y-intercept always has a coordinate (0, y). (96, 106, 239)

zero product property   The Zero Product Property states that when the product of two or more factors is zero, one of these factors must equal zero; that is, if a • b = 0 then either a = 0 or b = 0. Note that we can use the Zero Product Property to solve quadratic equations that are factorable. (283, 285)

Unit Prefixes

Unit 0 GS Getting Started: Working in Teams
Unit 1 SQ Difference of Squares: Organizing Data
Unit 2 KF Tiling the Kitchen Floor: Area and Subproblems
Unit 3 BC The Burning Candle: Patterns and Graphs
Unit 4 CP Choosing a Phone Plan: Writing and Solving Equations
Unit 5 EF Estimating Fish Populations: Numerical, Geometric, and Algebraic Ratios
Unit 6 WR World Records: Graphing and Systems of Linear Equations
Unit 7 BR The Big Race: Slopes and Rates of Change
Unit 8 AP The Amusement Park: Factoring Quadratics
Unit 9 BP The Birthday Party Piñata: Using Diagrams to Write Equations
Unit 10 YS Yearbook Sales: Exponents and Quadratics
Unit 11 CM The Cola Machine: Functions and Equality
Unit 12 GG The Grazing Goat: Problem Solving and Inequality
Unit 13 RS The Rocket Show: More About Quadratic Equations

Many of the problems listed here contain definitions or examples of the topic listed. It may be necessary, however, to read text preceding or following the problem, or additional problems to fully understand the topic. Also, some problems listed here are "good examples" of the topic and may not offer any explanation. It is very important, therefore, for you to be sure you correct and complete your homework and keep it organized. Your record of the problem may be your best index to understanding the mathematics of this course.

Absolute value **CM-71**, CM-72
    equations CM-83, CM-87, CM-97, **CM-99**
    graph **CM-76**, CM-111, CM-115
Addition
    of algebraic fractions GG-16, GG-19, GG-28, **GG-30**, GG-31
    of fractions KF-73, **GG-15**
    of integers SQ-9, SQ-10, **SQ-18**
    of like terms **SQ-67**
    of polynomials **AP-5**
    of square roots **BP-96**
    to solve pairs of equations **CM-14**, CM-15
Additive Inverse **CM-119**
Additive Property of Equality **CM-119**
Algebra walk **BC-1**
Algebra tiles **SQ-65**
    combining like terms with (grouping) **SQ-67**, KF-55
    multiplication with **KF-52**, KF-78
    factoring with **AP-18**
Algebraic Properties **CM-119**
Amusement Park, The AP-0, **AP-79**
Angles, right **EF-54**
Arc, circular **GG-65**

Area **GS-7**
    approximating **SQ-22**
    as a product **WR-71**, AP-3
    as a sum **WR-71**, AP-2, AP-3

    meaning **GS-7**
    of a circle KF-26, **KF-33**
    of a rectangle **GS-7**
    of a triangle **KF-14**
Associative Property **CM-119**
Average (mean) **SQ-20**, SQ-33, SQ-41, SQ-75
Axes
    horizontal and vertical **GS-13** (example), BC-1, BC-8
    scaling BC-28, BC-31, **BR-2**
Axis
    horizontal **GS-13**, SQ-3, BC-71
    vertical **GS-13**, SQ-3, BC-61

Bar graph (histogram) GS-13(f), **GS-16**
Base **YS-26**
Big Race, The BR-0, BR-1, **BR-91**
Birthday Party Piñata, The BP-0, **BP-95**
Burning Candle Problem, The **BC-72**

Binomial **WR-72**, AP-4
    using F.O.I.L. to multiply **BP-90**
Break up the problem (subproblems) KF-0, **KF-5**, KF-43

INDEX

Calculator
- $+/-$ **KF-10**
- error **BC-65**
- exponents KF-93, **BC-6**
- fraction key **KF-17**
- order of operations **SQ-13**, KF-93
- scientific notation **SQ-19**, SQ-56
- squares KF-90, **KF-103**
- square roots **KF-103**
- use with percents, fractions, and decimals **GS-9**

Center of a circle **KF-22**
Checking solutions CP-27, **CP-38**
Choosing a Phone Plan **CP-92**
Circle
- area of KF-26, **KF-33**
- circumference of KF-23, KF-24, **KF-33**
- center of **KF-22**
- definition **KF-22**
- diameter of **KF-22**
- radius of **KF-22**

Circular arc **GG-65**
Circumference of a circle KF-23, KF-24, **KF-33**
Coefficient **YS-75**
Combining like terms **SQ-67**
- with algebra tiles **KF-55**

Common denominator KF-73, GG-15, **GG-30**
Common factor
- greatest AP-17, **AP-31**
- and factoring **CP-51**

Commutative Property **CM-119**
Complete composite square **RS-58**
Complete graph **BC-9**
Completely factored **AP-70**
Completing the square **RS-67**, RS-68, RS-77
Compute **SQ-12**, SQ-17
Conjecture **EF-18**
Consecutive integers **SQ-46**
Constant **WR-29(d)**
Coordinates of a point **BC-9**, BC-12, BC-22
Corresponding sides **EF-9**
Cups and tiles (to represent and solve equations) CP-27, **CP-38**
Decimal, on calculator **GS-9**
Denominator,
- common KF-73, GG-15, **GG-30**
- lowest (least) common **KF-73**

Dependent variable **CM-60**
Derivation of the Quadratic Formula **RS-89**
Diagonal **CM-125**

Diagram
- in word problems **SQ-45**
- to represent percent **GS-9**
- to represent integer operations **SQ-18**, SQ-28, SQ-36

Diameter of a circle **KF-22**
Diamond problems **GS-3**, SQ-5
- and factoring **AP-19**, YS-1

Difference of two squares SQ-0, **SQ-85**
- factoring **AP-50**

Dimensions of a rectangle **KF-14**, KF-41
Distance-rate-time problems **GG-1**, GG-2, GG-3
Distributive property **KF-57**, KF-67, KF-68, **KF-78**, KF-81, KF-84, CM-119
- and factoring **CP-51**

Dividing line **GG-90**
Division
- by zero BC-65
- with integers SQ-29, **SQ-30**
- with powers YS-38, **YS-40**
- with square roots **BP-96**

Domain BC-66, BP-44, **CM-2**, CM-46, **CM-60**
Dot paper **EF-6**, EF-7, EF-8
Draw a picture or diagram (see Problem Solving)

Elimination method for solving pairs of equations **CM-14**, CM-15
Ellipsis (...) **BC-57**
Enlargement of a geometric figure **EF-7**,
Enlargement ratio EF-7, **EF-32**
Equality
- Additive Property of **CM-119**
- Multiplicative Property of **CM-119**
- Transitive Property of **CM-119**

Equations **BC-51**
- absolute value CM-83, CM-87, CM-97, **CM-99**
- graphing **BC-1**
- linear **AP-71**
- of lines, writing **BP-83**, BP-86
- pairs of **AP-71**
- parabola from x-intercepts, RS-20, **RS-42**
- quadratic **AP-71**
- solving with
  - cups and tiles CP-27, **CP-38**
  - mental math **CP-37**
  - undoing **CP-49**
- square roots in **CM-108**, CM-109, CM-110

summary of types of **AP-71**
system of **WR-39**
understanding **BR-49**
with fractions **BP-46**
with parentheses **CP-77**
writing from guess and check tables **CP-1**
Equivalent expressions **SQ-71**
Estimating Fish Populations EF-0, **EF-108**
Evaluating expressions **KF-50**
Exponent(s) SQ-55, **YS-26**, YS-28
    in multiplication and division **YS-25**, YS-53
    Laws of YS-38, **YS-40**
    negative BC-6, YS-30, YS-42, **YS-43**
    zero YS-30, YS-41, **YS-43**
Expressions
    binomials **WR-72**, AP-4
    compute (calculate) **SQ-17**
    equivalent **SQ-71**
    evaluate **KF-50**
    representing a diagram **SQ-67**
    simplifying **SQ-67**
    squaring **WR-82**
    trinomials **AP-4**

Factor(s)
    greatest common AP-17, **AP-31**
    of a number **AP-17**
    of a polynomial **AP-4**
Factoring
    and Diamond problems **AP-19**, YS-1
    as working backward **AP-11**, AP-12
    common factors **CP-51**
    completely **AP-70**
    difference of two squares **AP-50**
    perfect square trinomial **AP-58(g)**
    polynomials **AP-59**
    quadratics of the form $x^2 + bx + c$ **AP-10**
    quadratics of the form $ax^2 + bx + c$ **YS-1**
    to solve quadratic equations **AP-41**
    using algebra tiles to **AP-18**
    using diamond problems to **AP-19**, YS-1
    using the distributive property to **CP-51**
    writing the area of a rectangle as a product **AP-10**
Fibonacci Numbers **WR-26**
F.O.I.L. method for multiplying binomials **BP-90**

Fraction busters **BP-46**
Fractions
    addition and subtraction of KF-73, **GG-30**
    algebraic **GG-31**
    alternative representation of **GS-9**, SQ-6, SQ-34, SQ-42
    as probability **GS-20**
    in equations, **BP-46**
    on calculator **GS-9**
Function CM-46, CM-56, **CM-60**

Generic
    rectangle and its dimensions **KF-78**, KF-79, WR-71
    slope triangle **BR-63**
Graph interpretation GS-4, SQ-21
Graphing
    absolute value **CM-76**, CM-111, CM-115
    approximating values **BC-9**
    collected data for **SQ-1**, SQ-3
    equations **BC-1**
    horizontal lines **EF-66**
    human **BC-1**
    input-output **BC-1**
    ordered pairs (points) **BC-10**
    solving by BC-4, **WR-39**, AP-40, **AP-41**
    square roots **BP-15**, BP-44
    two point (y-form) method **WR-31**
    using a table **BC-1**, BC-9
    using slope and a point **BR-59**, BR-60
    vertical lines **EF-66**
Grazing Goat, The **GG-127**
Greatest Common Factor AP-17, **AP-31**
Grouping with algebra tiles **KF-55**
Guess and check table
    to solve a word problem **SQ-45**, CP-1
    to write an equation **CP-1**, CP-2

Histogram (example) **GS-13**, GS-16
Horizontal
    axis **GS-13**
    line, equation of **EF-66**
    lines, graphing **EF-66**
Hypotenuse **BP-1**, BP-2
Hypothesis **EF-18**

Identity
    Additive **CM-119**
    Multiplicative **CM-119**
Independent variable **CM-60**

Inequalities
    graphing **GG-90**
    linear **GG-55**, GG-58
    quadratic **GG-66**, GG-67
    solving GG-40, GG-41, GG-42, GG-44, GG-55, **GG-58**
    system of **GG-101**, GG102, GG-103, GG-104
Inequality (notation) symbols BC-56, **GG-39**
Input **BC-1**
Inspection, solve by (mental math) **CP-37**
Integers
    calculations with, SQ-9, SQ-10, **SQ-18**, SQ-28, SQ-29, SQ-30, SQ-31, **SQ-39**, SQ-40
    consecutive **SQ-46**
Intercept(s) **BC-22**
    x- BC-37, **BC-42**, **BR-13**, RS-42
    y- **BC-42**, WR-29, **BR-13**
Intersection, point of WR-20, **AP-8**
Inverse **CP-64**
    Additive **CM-119**
    Multiplicative **CM-119**
Isosceles triangle **CM-45**

Justify **EF-18**

Largest common factor AP-17, **AP-31**
Laws of exponents **YS-40**
Laws of simplifying square roots **BP-96**
Learning reflection **SQ-90**
Least common denominator **KF-73**
Least common multiple GG-28, GG-29, **GG-30**
Legs of a right triangle **BP-1**, BP-2
Length **EF-44**
Like terms
    addition of **SQ-67**
    combining **SQ-67**
Linear equations **AP-71**
    slope-intercept form BR-51, **BR-52**, BP-35
    two point method for graphing **WR-31**
    x-intercept **BC-37**, BR-13
    y-intercept **WR-29**, BR-13
Linear
    equations **AP-71**
        graph **BC-47**
    inequalities GG-55, **GG-58**
        graphing **GG-90**
Line segment
    length of **EF-44**

Line(s)
    dividing **GG-90**
    equations of, writing **BP-83**, BP-86
    graphing using a table **BC-1**
    graphing using slope and y-intercept **BR-59**, BR-60
    horizontal, equation of **EF-66**
    of best fit RS-1, **RS-2**, RS-11, RS-12, RS-13, RS-97
    parallel **BC-83**, BR-55
    point of intersection WR-20, **AP-8**
    slope of BR-22, **BR-25**
    vertical, equation of **EF-66**
Look for a pattern (see Problem Solving)

Make a list to solve a problem **GS-20**
Make a table **GS-22**, SQ-45
Making a model **BP-36**
Match-a-Graph **WR-1**, WR-2, WR-10, BR-36,
Mean (average) **CP-91**
Measure **GS-25**
Mental math (to solve equations) **CP-37**
Monomial **AP-4**
Money Matters **BR-12**
Multiplication
    and exponents YS-25, **YS-40**, YS-47
    of integers SQ-28, **SQ-39**
    of rational expressions CM-35, CM-36, **CM-39**
    of two binomials WR-70, **WR-72**, AP-2, AP-3
    products to sums **CP-51**
    with algebra tiles **KF-52**, KF-79, WR-70,
Multiplicative
    Identity **CM-119**
    Inverse **CM-119**
    Property of Equality **CM-119**
Multiplying binomials **WR-72**
    as an historical perspective **BP-90**

Neutral field **SQ-7**
Number line **SQ-43**
Number sequence **SQ-57**

Order of operations (arithmetic) SQ-13, **KF-93**
    on calculator **SQ-13**
Ordered pair (x, y) **BC-37**
Origin **BC-10**
Output **BC-1**

Pairs of equations **AP-71**

Parabola **BC-11**, BC-47
    equation from its intercepts RS-20, **RS-42**
Parallel lines BC-83, **BR-55**, BR-73, BR-83
Parentheses
    equations with **CP-77**
    with the distributive rule **KF-70**
Patterns (recognizing using a table) **SQ-57**, SQ-58, SQ-59
Percent on calculator **GS-9**
Perfect square trinomial **AP-58(g)**
Perimeter **GS-7**
    of a polygon **BC-24**
    of a rectangle **GS-7**
Perpendicular **CM-101**, CM-130
Pi (π) **KF-33**
Point(s)
    coordinates of **BC-10**, BC-12, BC-22
    on a graph **GS-4**
    reference **CM-84**
Point of intersection of two graphs WR-20, **AP-8**
Polynomials
    as sums and products **AP-4**
    factoring **AP-59**
    factors of **AP-4**
Powers of ten **SQ-55**
Predict (make a prediction) **GS-4**, GS-11, GS-19
Prime notation (A', B") **EF-9**, EF-44
Prime **AP-17**
Probability **GS-19**, GS-21
Problem solving strategies **GG-113**
    break up the problem (subproblems) KF-0, **KF-5**, KF-43
    draw a picture or diagram **GS-9**, SQ-18, BP-24
    guess and check **SQ-45**, CP-1
    look for a pattern **GS-3**, SQ-57, SQ-58, SQ-59
    make a list **GS-20**
    make a table **GS-22**, SQ-45
Product
    and area **WR-71**, AP-3
    as a sum **WR-70**, WR-72, AP-2, AP-3
    notation for **SQ-27**
    of two binomials WR-70, WR-71, **WR-72**
Progress Reports **GS-4**
Properties, Algebraic **CM-119**
Property (of real numbers)
    Additive of Equality **CM-119**
    Additive Inverse **CM-119**
    Associative **CM-119**
    Commutative **CM-119**
    Distributive **CM-119**
    Multiplicative of Equality **CM-119**
    Multiplicative Inverse **CM-119**
    Reflexive **CM-119**
    Substitution **CM-119**
    Symmetric **CM-119**
    Transitive of Equality **CM-119**
Proportion **EF-47**
Pythagorean Theorem BP-1, BP-2, BP-3, **BP-12**, BP-13

Quadratic equation
    solve by factoring (zero product property) AP-41, **AP-44**
    solve by graphing AP-40, **AP-41**, AP-42
    solve by using the quadratic formula **YS-86**
    standard form of **YS-75**
    when can it be solved by factoring YS-95, **YS-96**
Quadratic formula **YS-86**
    derivation of **RS-89**

Quadratic
    equations **AP-71**
    inequalities **GG-66**, GG-67
    polynomial **AP-4**
    factoring **AP-10**, YS-1
Quadrilateral **BC-15**

Radius (plural: radii) of a circle **KF-22**
Range **CM-2**, CM-46, **CM-60**
Ratio(s) **EF-1**
    as a percent **EF-53**, EF-70, EF-71
    enlargement EF-7, **EF-32**
    in equations **EF-47**
    in areas **EF-7**, EF-32
    of corresponding sides **EF-7**, EF-45
    of perimeters **EF-7**, EF-32
    of two sides within a triangle **EF-45**, EF-46
    reduction EF-20, **EF-32**
    slope as a ratio **BR-22**
Rational expressions YS-63, **YS-64**, CM-7, **CM-39**
    adding and subtracting **GG-30**
    multiplying and dividing, **CM-39**
Rectangle **GS-7**, KF-2
    area **GS-7**
    dimensions of **KF-14**, KF-41
    generic **KF-78**, KF-79, WR-71
    perimeter **GS-7**
Rectangular **SQ-58**
Rectangular numbers **SQ-58**, SQ-60

Reducing geometric figures **EF-20**
Reduction ratio EF-20, **EF-32**
Reference point **CM-84**
Relation(s) **CM-2**, CM-3, CM-56, **CM-60**
Reflexive Property **CM-119**
Right angle **EF-54**
Right triangle **EF-54**, BP-1
    hypotenuse of **BP-1**
    legs of **BP-1**, BP-2
Rocket Show, The RS-0, **RS-77**
Roll and Win **GS-13**
Roots **AP-42**, RS-75
Rule, from a table **BC-1**
    of a function **CM-60**

Sampling **EF-108**
Scale of a graph **BC-28**, BC-31
Scientific notation SQ-19, **KF-96**, YS-44
    on calculator **SQ-19**, SQ-56
Sector **KF-25**
Segment, line **EF-44**
Semi-circle **GG-20**
Sequence **SQ-57**
Sides, corresponding **EF-9**
Similar figures **EF-6**, EF-32
Similar triangles **EF-44**
Simplifying
    algebraic expressions (by collecting like terms) **SQ-67**
    fractions (using subproblems) **KF-73**
    rational expressions YS-63, **YS-64**, CM-7, CM-39
    square roots **BP-70**
Slide **EF-76**
Slope of a line BR-22, **BR-25**
    and zero **BR-53**
    as a ratio **BR-22**
    definition **BR-25**
    negative BR-16, **BR-25**
    positive BR-16, **BR-25**
    steepness **EF-118**, BR-10
Slope triangle BR-22, BR-24, **BR-25**, BR-63
Slope-intercept form of a linear equation BR-51, **BR-52**, BP-35
Solution to a system of equations **WR-39**
Solutions, checking CP-27, **CP-38**
Solving an equation
    by graphing BC-4, **WR-39**, AP-40, **AP-41**
    by inspection (mental math) **CP-37**

containing absolute value CM-83, CM-87, CM-97, **CM-99**
containing fractions **BP-46**
containing parentheses **CP-77**
mental math **CP-37**
undoing **CP-49**
using the zero product property **AP-44**, AP-60
with cups and tiles **CP-27**, CP-38
Solving an inequality
    linear GG-40, GG-41, GG-42, GG-44, GG-55, **GG-58**
    quadratic **GG-66**, GG-67
    with absolute value GG-77, **GG-79**
Solving a quadratic equation
    by graphing AP-40, **AP-41**, AP-42
    using the zero product property (by factoring) AP-41, **AP-44**
    using the quadratic formula **YS-86**
Solving a system (pair) of equations
    by elimination **CM-14**, **CM-27**
    by graphing **WR-39**
    by substitution **WR-41**, BR-70, **BR-71**, BR-81
Solving a system of inequalities **GG-101**, GG-102, GG-103, GG-104
Square numbers **SQ-57**, SQ-60
    on calculator KF-44, **KF-103**
Square roots
    addition, subtraction, multiplication and division of **BP-96**, BP-108
    approximation of **KF-103**, KF-105
    exact value of **BP-70**
    graph of $y = \sqrt{x}$ **BP-15**
    in equations **CM-108**, CM-109, CM-110
    notation **KF-103**
    of negative numbers **BC-68**
    on calculator **KF-103**
    simplifying **BP-70**, BP-96
Squaring expressions **WR-82**
Standard (notation) form
    of a number SQ-56, **KF-96** YS-44
    of a quadratic equation **YS-75**
Steepness number (slope) BR-10, **BR-22**
Study teams **GS-1**
Study team reflection **SQ-90**
Subproblems (break up the problem) KF-1, **KF-5**, KF-106

Substitution **KF-30**
    method of solving pairs of equations
        WR-41, BR-70, **BR-71**, BR-81
    Property **CM-119**
Subtraction
    of fractions **GG-30**
    of integers SQ-9, SQ-10, **SQ-18**
    of polynomials **SQ-70**
    square roots **BP-96**, BP-108
Sum
    and area **WR-71**, AP-2, AP-3,
    as a product (factoring quadratics)
        AP-2, AP-4, **AP-10**, AP-18
Summary **BC-79**, CP-111, EF-116,
    WR-85, BR-97, AP-85,
    BP-106, YS-103, CM-127,
    GG-128, RS-96
Summary posters **KF-114**
Symmetric Property **CM-119**
System (pair) of equations
    a solution of **WR-39**
    elimination method of solving
        CM-14, CM-15
    graphing method of solving
        WR-39
    no solution **BR-73**
    substitution method of solving **WR-41**,
        BR-70, **BR-71**
System of inequalities **GG-101**,
    GG-102, GG-103, GG-104
Systematic list **GS-20**

Tables
    and Graphs **BC-10**
    Guess and Check **SQ-45**
Terms, like **SQ-67**
Tiles SQ-8, **SQ-18**
Tiling the Kitchen Floor **KF-106**
Tool kit **SQ-18**
Transitive Property of Equality **CM-119**
Translations of $y = x^2$ **RS-33**
Triangles
    area **KF-14**
    isosceles **CM-45**
    right EF-54, **BP-1**
    similar **EF-44**
    slope triangle BR-22, BR-24,
        BR-25, BR-63
Triangular numbers **SQ-59**, SQ-60
Trinomial **AP-4**
Two-point method of graphing **WR-31**

Undoing an equation CP-49, **CP-64**
Unit summary problem **BC-79**, CP-111,
    EF-116, WR-85,
    BR-97, AP-85, BP-106,
    YS-103, CM-127, GG-128,
    RS-96

Value
    absolute **CM-71**, CM-72
    of an expression (compute; evaluate)
        SQ-17, KF-50,
        YS-26
Variables **KF-77**
    dependent **CM-60**
    independent **CM-60**
Vertex RS-23, RS-31, RS-32, **RS-42**
Vertical
    axis **GS-13**, SQ-3, BC-61
    line, equation of **EF-66**
    lines, graphing **EF-66**

Writing equations
    from guess and check tables
        CP-1, CP-2
    of lines **BP-83**, BP-86
Working backwards **CP-64**
    and factoring **AP-11**, AP-12
World Records **WR-3**, **WR-95**

x-axis **BC-9**
    slope of **BR-53**
x-intercept **BC-37**, BC-42, BC-43,
    BR-13
    of a parabola **AP-41**, RS-45

Yearbook Sales YS-0, **YS-104**
y-axis BC-9
y-intercept **BC-42**, BC-43, WR-29,
    BR-13
y-form of an equation WR-29, **WR-43**

Zero
    as an exponent YS-30, YS-41,
        YS-43
    as slope of a line **BR-53**
    in division **BC-65**
Zero product property AP-41, **AP-44**
    in solving quadratic equations
        AP-41, **AP-44**, AP-60